The control of fluid power

The control of fluid power

D. McCloy M.Sc., Ph.D., D.I.C., C.Eng., M.I.Mech.E.
Reader in Mechanical Engineering,
Queen's University,
Belfast, N. Ireland

H. R. Martin M.Sc., Ph.D., C.Eng., M.I.Mech.E.
Associate Professor in Fluid Control Systems,
University of Waterloo,
Ontario, Canada

Longman

LONGMAN GROUP LIMITED
London

Associated companies, branches and representatives throughout the world

First published 1973

ISBN 0 582 47003 x

Set in Times New Roman and Univers and printed in Great Britain by Lowe & Brydone (Printers) Ltd., Thetford, Norfolk

Contents

Preface

Fluid power has been used to advantage in such widely differing areas as transportation, machine tools, ship control, agricultural equipment, space technology, aircraft, mining and in other applications too numerous to mention.

In spite of industry's increasing awareness of the potential of fluid power, properly trained people are still at a premium. In the not too far distant past, education in the subject usually amounted to a few snippets included in Fluid Mechanics or Automatic Control courses and, in an attempt to improve matters, many fluid power manufacturers undertook to educate their own and their customers' staff. Fortunately the academic world is now giving some recognition to this important and rapidly growing industry. Universities and Technical colleges in the United Kingdom have started to introduce courses in this area and, in the United States, a fluid power option is offered in many universities and Technical Colleges.

The British Government has also appreciated that fluid power can play a major part in increasing productivity and efficiency, especially in the smaller firms. A network of Low Cost Automation Centres has been established and these provide training courses, advisory and consultancy services for industry, with particular reference to low cost applications of pneumatic and hydraulic systems.

In this book, we have collected and related several aspects of fluid power systems and components from widely spread sources. The book should be useful to both practising engineers and students of the subject who have a basic knowledge of mathematics and feedback control theory. It is concerned mainly with the control aspects of fluid power and deals only briefly with power generation. Hydraulic, pneumatic and fluidic components are discussed in detail, but in addition an attempt has been made to provide a comprehensive understanding of the fluid power system and where it fits into today's technology.

In our experience, units have been a continual source of confusion to the student, but fortunately the introduction of the International System of Units (SI) is well under way and this will alleviate this problem considerably. Several systems of units are used in the present book and an appendix is devoted to a discussion of units. This approach was thought to be necessary at this critical transition period when engineers are faced with working in several different systems of units.

Standardisation of fluid power symbols has also been attempted and the symbols recommended by the Comité Europeen des Transmissions Oleo-hydrauliques et Pneumatiques (C.E.T.O.P.) are used throughout.

The material for the text has been based largely on our research work over the past ten years, together with lecture courses presented by us at

The Department of Mechanical Engineering,
Queen's University of Belfast,
Belfast.

The School of Mechanical Engineering,
Georgia Institute of Technology,
Atlanta, Georgia.

The Department of Mechanical Engineering,
University of Waterloo,
Waterloo, Ontario.

We also gathered valuable experience of low cost applications as a result of our activities as Directors of the Northern Ireland Low Cost Automation Centre.

Acknowledgement is gratefully made to Professor B. Crossland and Professor S. A. Alpay for providing the opportunity to develop the subject; to Mr. J. Carrington, of Short Brothers and Harland Limited, for his invaluable comments; to Dr. E. K. Beatty, manager of the Northern Ireland Low Cost Automation Centre, for his loyal support; to the Institution of Mechanical Engineers, London, the International Federation of Automatic Control, the British Hydromechanics Research Association and the American Society of Mechanical Engineers for the use of information published in their proceedings; to the McGraw-Hill Publishing Company, *Hydraulic Pneumatic Power* and *Fluid Power International* magazines for permission to use information published by us over the past ten years.

Lastly we would like to thank our wives, Joanne and Pat, for gladly giving up many a sunny week-end.

D. McCloy
H. R. Martin

Chapter 1

Introduction to fluid power

1.1. History [1, 2, 3, 4]*

The first hydraulic system known to man was a double pump delivering up to 4·5 gal min^{-1} at 3 lbf in^{-2}, and feeding 7000 miles of self-sealing piping. The system is still in everyday use, but was not designed or manufactured by man and has never been duplicated by man. This unique and oldest fluid power system is the human heart and its network of blood vessels.

Fluid power technology began about 300 years ago with the discovery in 1650 of Pascal's law:

Pressure in a fluid at rest is transmitted equally in all directions.

Water has been used as a means of transmitting power from much earlier times. For example, Ctesibus, the son of a barber in Alexandria, has been accredited with inventing the first pump (in the second century before Christ). This device was a two-piston positive displacement unit used for pumping water. However, even earlier than this, the Egyptians used the principle of siphoning liquid. They also devised a piston type syringe in connection with the embalming art. This concept was undoubtedly the forerunner of the positive displacement pump.

Another famous 'pump of nature' is the rising of sap in trees every springtime, reaching sometimes heights of 300 ft. The pressure in a 50 ft vine, for example, is about fourteen times that produced by the human heart. Nature does not actually pump the sap up a tree, but very cleverly draws it up by means of surface tension.

Traditionally, hydraulics devices are thought of as using water, an example being the water turbine. These devices require vast quantities of fluid at a relatively low pressure which is usually generated by a natural head of water. It is only within the last few hundred years that pumps and compressors have been developed, capable of developing artificial pressures considerably greater than those that nature provides. The higher the pressure, the smaller the flow required to produce power, and systems become increasingly efficient.

* Figures in square brackets indicate bibliographical details. These details are to be found in Appendix II.

Around 1750, Bernoulli introduced his law concerning the conservation of energy in a flowing fluid and this well known concept forms the basis of our present day calculations on orifice sizes and pressure drop in pipe lengths.

By the 1850s the Industrial Revolution in the United Kingdom was well underway. Fluid power was being used extensively for powering presses, cranes, winches and extruding machines. Many cities had central industrial hydraulic distribution networks. The London Hydraulic Power Co. still exists in service, having at present pumps totalling 7000 horsepower, connected to 180 miles of pressure main. Examples of machines using this network are the revolving stage of the Coliseum, the bascules of the Tower Bridge and the front lift in the Institution of Mechanical Engineers building.

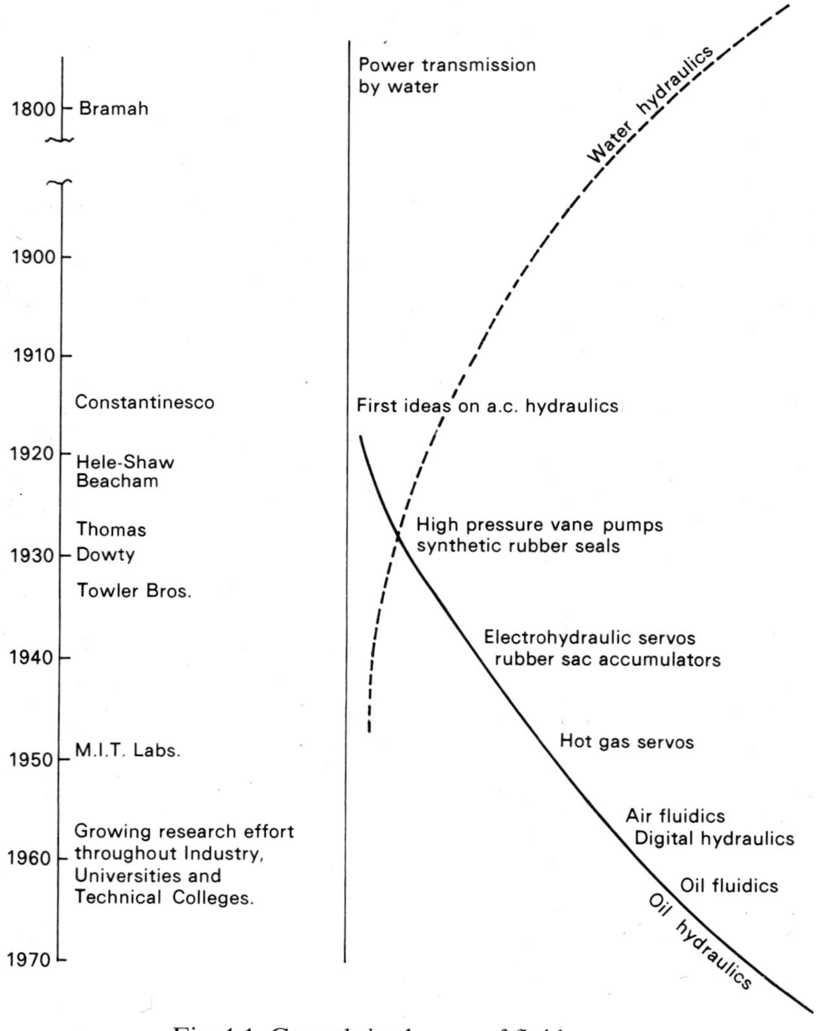

Fig. 1.1. Growth in the use of fluid power

Water pressure was used extensively for the transmission of power until the introduction of electrical power, which was soon shown to be superior to hydraulics for the transmission of power over substantial distances. A time history of the decline of water hydraulics, as electrical power became more popular, and then the gradual increase of interest in oil hydraulics, is shown in Fig. 1.1. The use of air as a power transmission medium has also been used extensively over the years, with particular application in the chemical and process industries.

The manufacture of hydraulic components in quantities was also building up in the 1850s. One firm Tangye Bros., was established in Birmingham to make hydraulic jacks, shearing and riveting machines and other hydraulic equipment. This company supplied the large number of jacks required to launch Brunel's *Great Eastern* steamship into the Thames.

In 1906, an important milestone in fluid power applications was achieved when an oil hydraulic system was developed to replace the electrical system for elevating and controlling warship guns.

Applications of fluid power, both oil and pneumatic, have become virtually limitless. These include systems on machine tools, cars, earth moving and agricultural equipment, aircraft, ships and many other areas too numerous to mention.

1.2. Advantages of fluid power systems

A fluid power device [6] is a component or system which employs a fluid to transmit power from one point to another. The fluid can be, for example, a liquid such as oil or a gas such as air. There are certain distinct differences between a system using a gas and a system using a liquid and, in this section, an attempt is made to draw comparisons between the two basic types of fluid power system and other methods of power transmission.

Fluid power devices lie between electrical and mechanical power transmission devices, in respect to performance. Fluid and electrical power systems transmit more easily over appreciable distances, as compared with their mechanical counterpart, and are often more accurately controllable. Electrical power devices are especially good in this respect at low power levels.

With regard to efficiency over short transmission paths, mechanical power is best, with fluid and electrical power about equal. As far as cost is concerned it is very difficult to generalise, as so much depends on the application and the type of component used. However, very roughly speaking, electrical components are the cheapest at low power levels, while in high power ranges all the components are much the same price.

Table 1.1. is an attempt to summarise the salient features of the various types of systems. The final choice is a difficult one and is often influenced by the rest of the system.

One major advantage of fluid over electrical power is that the former is not so limited by physical limitations of the materials. For example, the saturation limit of steel limits the performance of the electromagnet. This sort of limitation does not apply to hydraulic systems until the material strength limit is reached.

Table 1.1

Property	Best	Good	Fair
Torque/inertia	H	P	M, EM
Power/weight	H, P	—	EM, M
Rigidity in steady state	H	M	F, P, EM
Friction	EM, F	H, M	P
Dirt vulnerability	E, EM, M	—	*H, F, P
Speed of response	E	F, H	P, M, EM
Compactness	E	F, H	P, M, EM
Ability to work in adverse conditions	F	P, M, H	EM, E
Relative cost	M, E, EM	H, P	F

Key: M—mechanical; E—electronic; F—fluidic; EM electromechanical; P—pneumatic; H—hydraulic

In fact, commonly available fluid power devices can supply forces up to twenty times greater than electromagnetic devices.

On the other hand, fluid power devices have the following disadvantages:

Transmission of power by liquids can be messy if proper care is not observed in workmanship and maintenance.

Danger of burst pipes and components containing high pressure fluid, although small, is always present.

With hydraulics there is a danger of fire and explosion in extreme circumstances.

Liquid devices have the following basic advantages over gas systems:

Much stiffer system due to the so called 'incompressible' fluid.

Better speed of response.

Less inherent friction due to the lubrication properties of the liquid.

1.3. Fluids for power transmission

Ideally, fluids used in power transmission [4] and control should be cheap, be non-corrosive, have infinite stiffness, good lubrication properties, store well, be non-toxic, non-inflammable and remain stable with good properties over a wide range of temperatures. This specification appears to be impossible to meet but the use of additives in the case of hydraulic oils has achieved a fair measure of success.

Many different types of fluid have been used for power transmission, ranging from water, mineral oils, vegetable oils, synthetic and organic liquids to even molten metals. Water was the first hydraulic fluid to be used. It is cheap, readily available and can transmit power efficiently. However, it has several disadvantages. For example, it freezes readily, rusts metal parts, boils and is a relatively poor lubricant. Mineral oils are far superior in all these respects. Part of the success of mineral oils is due to the relative ease with which their properties can be changed using additives. Some of the more common type of additives are listed in Table 1.2.

Table 1.2.

Property	Chemical used
Antioxidants	Complex phenols and amines, organic phosphorus and sulphur compounds
Corrosion inhibitors	Organo-metallic compounds, naphthene esters
Rust inhibitors	Amines, amides, soaps, phosphoric esters, organic acids and esters
Anti-foams	Silicones, calcium soaps, sodium alkyl sulphates
Lubrication improvers	Organo-phosphorus, chlorine and sulphur compounds, lead soaps
Pour point depressants	Polyalkylnaphthalenes, polyalkyl phenol esters, polymethacrylates
Viscosity index improvers	Polyisobutylene, polyacrylates

1.3.1 Density

Density is the mass or quantity of matter contained in unit volume of the substance. It is denoted by the symbol ρ (see Appendix I). The mass density of a liquid varies little with temperature and pressure but this is not the case with a gas.

The change of density of a gas with either temperature or pressure can be found from the perfect gas law,

$$P = \rho R T \tag{1.1}$$

where for example in the SI system (Appendix I)

P = pressure Nm^{-2} absolute
ρ = density kg m^{-3}
R = gas constant $\text{J kg}^{-1}\text{ K}^{-1}$
T = temperature °K

For most pneumatic processes the perfect gas law applies up to 3000 lbf in^{-2} ($\simeq$ 20 MNm^{-2}).

1.3.2 Compressibility

Consider the mass flow rates of fluid entering and leaving a system to be denoted by W_1 and W_2. Let M be the mass of fluid contained in the system of volume V and density ρ, then

$$W_1 - W_2 = \frac{\mathrm{d}M}{\mathrm{d}t}$$

$$M = \rho V$$

Hence

$$W_1 - W_2 = \rho\frac{\partial V}{\partial t} + V\frac{\partial \rho}{\partial t}$$

Now

$$W = \rho Q$$

where Q is the volume flow rate. If ρ is assumed to be equal to ρ_2, then

$$Q_1 - Q_2 = \frac{\partial V}{\partial t} + \frac{V\partial \rho}{\rho \partial t} \tag{1.2}$$

Thus the difference in incoming and outgoing volumetric flow rates is the rate of change of the volume V, plus another term due to compressibility. If the compressibility C of the fluid is defined as

$$C = -\frac{1}{V}\frac{\partial V}{\partial P} = -\frac{1}{\rho}\frac{\partial \rho}{\partial P} \tag{1.3}$$

then equation (1.2) becomes

$$Q_1 - Q_2 = \frac{\partial V}{\partial t} + \frac{V}{\rho}\frac{\partial \rho}{\partial P}\frac{\partial P}{\partial t} = \frac{\partial V}{\partial t} - VC\frac{\partial P}{\partial t}$$

indicating how compressibility affects the flow through the system.

The stiffness or compressibility of the fluid is the fractional reduction in volume of fluid per unit increase of applied pressure. It is more usual to express this term as the reciprocal, termed the bulk modulus,

$$N = -V\frac{\partial P}{\partial V} \tag{1.4}$$

The value of bulk modulus for most hydraulic oils is about 2×10^5 lbf in^{-2} above pressures of 1000 lbf in^{-2}, and provided aeration does not exceed 0·2 per cent.

When determining the bulk modulus of a gas it is important to define the conditions of compression. If the temperature is kept constant by allowing heat to flow into or out of the volume of gas being compressed, the resulting quantity is termed isothermal compressibility. This is normally a slow process in which

$$PV = \text{constant}$$

It is easy to show that in this case $N = P$.

If the heat flow is prevented, a different value results, termed the adiabatic compressibility. This is normally a rapid process in which

$$PV^{\gamma} = \text{constant}$$

where $\gamma = C_p/C_v$, the ratio of specific heats. In this case the bulk modulus is $N = \gamma P$, where γ is the ratio of the specific heats (1.4 for air).

It will be seen in later chapters, that the bulk modulus is closely related to the stiffness and speed of response of hydraulic and pneumatic systems. It is desirable to have as high a modulus as possible and this points to the use of high supply pressures.

1.3.3 Viscosity

The dynamic viscosity (μ) of a fluid is a measure of its resistance to shear,

$$\mu = \frac{\text{shear stress}}{\text{shear rate}}$$

This has been expressed in 'poise', having units dyne s cm^{-2}, in 'reyn' having units lbf s in^{-2} and in the SI system in 'poiseuille' with units Nsm^{-2}

$$1 \text{ poise} = 1{\cdot}45 \times 10^{-5} \text{ reyn} = 0{\cdot}1 \text{ poiseuille}$$

In many fluid mechanics equations, the ratio μ/ρ is common. This is termed ν, the kinematic viscosity and may be expressed in 'stoke' with units cm^2 s^{-1}, or in the SI system as m^2 s^{-1}. It is more usual to use the centistoke, cSt, $= 10^{-2}$ stoke

$$1 \text{ cSt} = 1{\cdot}55 \times 10^{-3} \text{ in}^2\,\text{s}^{-1} = 10^{-6} \text{ m}^2\,\text{s}^{-1}$$

Typical values for hydraulic oils are in the range 5–500 cSt, but are of course, very sensitive to temperature.

The rate of change of viscosity with temperature is expressed by a viscosity index. For wide changes in temperature a high viscosity index oil should be used if viscosity variations are to be minimised. It must be pointed out that such an oil will be expensive, because of the chemical additives needed to achieve this state.

When a fluid is flowing it can exhibit two conditions:

Laminar or streamline flow, where the flow pattern is smooth.

Turbulent or irregular flow, where the fluid motion is random.

The flow condition which a fluid experiences is related to the Reynolds number

$$\mathrm{Re} = \frac{vd\rho}{\mu} = \frac{vd}{\nu} \propto \frac{\text{inertia effects}}{\text{viscous effects}}$$

where

v = mean flow velocity
d = a characteristic length (the diameter for pipes)
ρ = mass density
μ = dynamic viscosity
ν = kinematic viscosity

The transition from laminar to turbulent flow in pipes normally takes place when the Reynolds number increases through the range 1200 to 2500, known as the transition region.

1.4. Aeration

How does air get into oil? Air can be dissolved or entrained in hydraulic fluids. Consider first the solubility of air in oil. Henry's law states that air solubility in any given liquid is directly proportional to absolute air pressure of the air above it. In a system reservoir where the pressure is one atmosphere, the air in solution exerts one atmosphere of pressure. The actual amount of air in solution must be determined experimentally and Fig. 1.2 shows the results of some experiments conducted by Leroy *et al* [145]. These results show that mineral oils can dissolve much more air than water glycol or water emulsions. The solubility constant k is defined as the percentage by volume of air dissolved per atmosphere, the air volume being reduced to standard temperature and pressure. Hence, from Fig. 1.2 the solubility constant for a conventional industrial hydraulic oil is approximately 10 and for a water glycol approximately 4. Hayward [146] at the National Engineering Laboratory in Glasgow, Scotland, has produced similar results and has shown that the solubility constant for water is 2, between 6 and 12 for mineral oils, and between 15 and 25 for fuels and silicone fluids. The results of Fig. 1.2 relate to saturated fluids, that is fluids in equilibrium with the excess of undissolved air. Under certain circumstances the fluid may temporarily hold more than this equilibrium amount of air and in this case the fluid is said to be supersaturated. On the other hand, if the liquid holds less than the equilibrium amount of dissolved air the solution is said to be undersaturated and it will

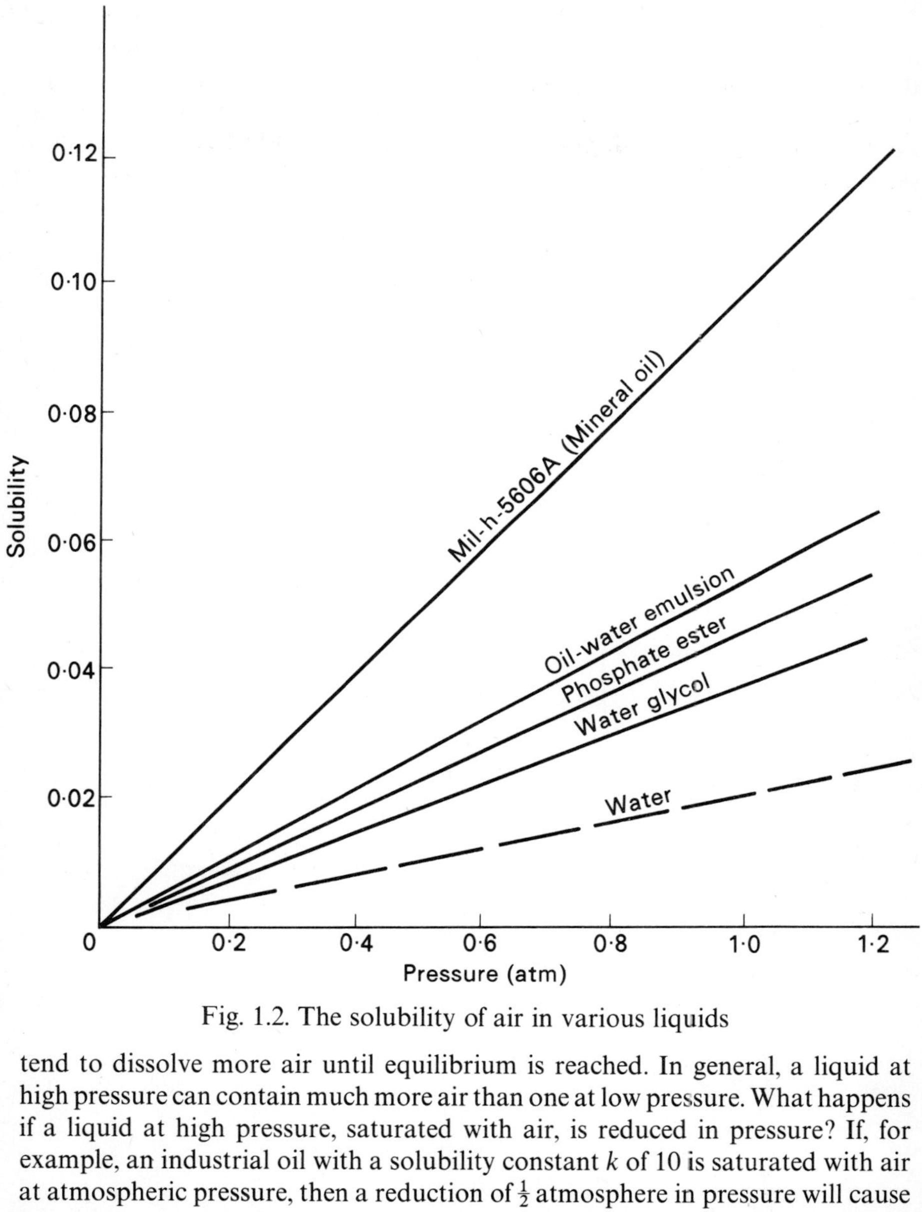

Fig. 1.2. The solubility of air in various liquids

tend to dissolve more air until equilibrium is reached. In general, a liquid at high pressure can contain much more air than one at low pressure. What happens if a liquid at high pressure, saturated with air, is reduced in pressure? If, for example, an industrial oil with a solubility constant k of 10 is saturated with air at atmospheric pressure, then a reduction of $\frac{1}{2}$ atmosphere in pressure will cause supersaturation and 5 per cent of air at standard temperature and pressure will come out of solution. Since the pressure is $\frac{1}{2}$ atmosphere, the actual volume of air will be about 10 per cent of the oil volume. This type of air release is easily seen in a bottle of carbonated beverage. If the top is removed, bubbles of carbon dioxide, which were originally forced into the solution under high pressure, are released and start to stream in the fluid. The gas is released more quickly if the liquid is agitated.

Oil can carry air in solution but in addition it can also carry air by entrainment. If a return line to the reservoir in a hydraulic system is not submerged, the jet of oil entering the liquid surface will carry air with it. This will cause the formation of air bubbles in the oil. Clearly these will cause no harm if they rise to the surface quickly. However if they rise very slowly they may be drawn into the pump intake and this can cause damage, as will be discussed later. Similarly,

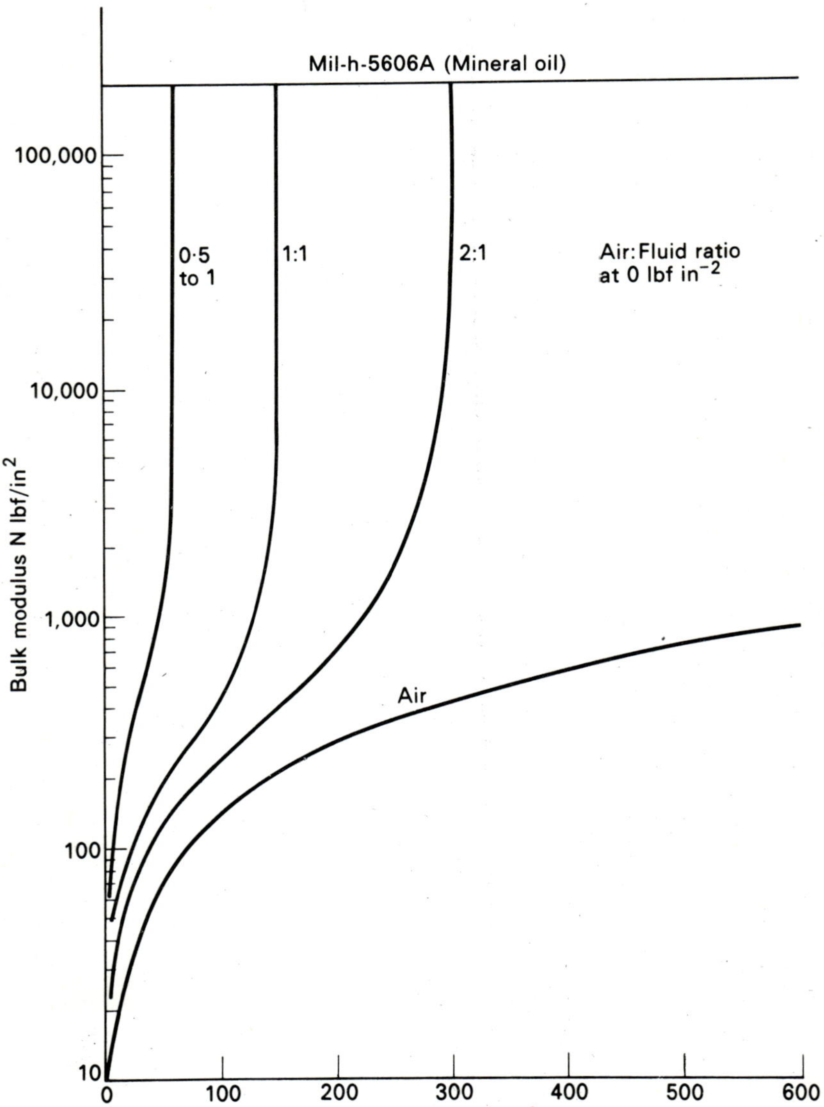

Fig. 1.3. Effect of pressure on the bulk modulus of an air–oil mixture. The liquid is saturated with air at each pressure

a small leak in a suction line can cause the entrainment of large quantities of air. Such leaks are difficult to spot as oil does not leak out, but air leaks in.

How does the presence of entrained and dissolved air affect fluid properties and system performance? Fortunately, air in solution has no measurable effect on fluid properties. Compressibility and viscosity are unaffected. Unfortunately the same cannot be said for entrained air. Hayward has measured the effect of bubble content on the viscosity of bubbly oils. He used a viscometer tube and showed that the effect of air bubbles upon viscosity is relatively small and can be expressed by the empirical equation

$$\mu_\beta/\mu_0 = 1 + 0{\cdot}015\,\beta$$

where β is the percentage bubble content, μ_β is the viscosity of bubbly oil and μ_0 is the viscosity of the oil.

Air bubbles in oil can cause a great reduction in the bulk modulus. Smith *et al* [147] have investigated this phenomenon. Air, of course, is much more compressible than oil and it is to be expected that its presence in free form in the oil will reduce the overall bulk modulus. Figure 1.3 shows the effect of air content on the adiabatic bulk modulus of a common hydraulic oil. Here at each pressure, equilibrium conditions are reached, that is the oil is saturated with air at each pressure. Clearly, as the pressure is increased, the oil dissolves more and more air until. For example, with a 2 to 1 air oil volume ratio, all the air is dissolved at approximately 300 lbf in^{-2}. This agrees with Hayward's results which show that mineral oils have a solubility constant of approximately 10, that is 10 per cent of air dissolved per atmosphere. Hence, to dissolve 200 per cent requires

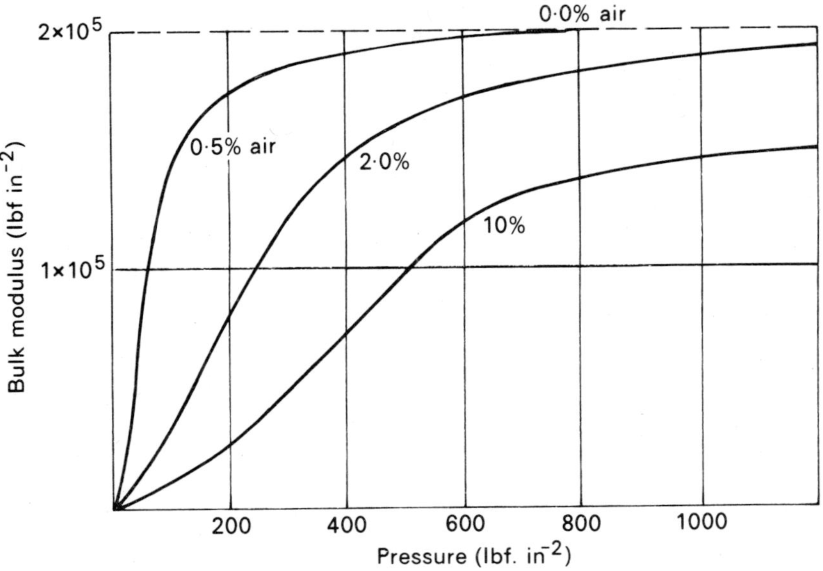

Fig. 1.4. Effect of pressure on the bulk modulus of an air–oil mixture. The air is free

approximately 20 atm, or roughly 300 lbf in^{-2}. If the air does not have time to dissolve, the bulk modulus can be greatly reduced even at very high pressures, Fig. 1.4. Hence, it is clear that it is desirable to keep system pressures as high as possible in order to

Compress the air to give it a higher bulk modulus; and to

Cause the air to dissolve in the oil, in which case it does not affect the bulk modulus.

Air bubbles in the oil can have another indirect effect on its properties. If the bubbles pass through a pump they are compressed and heat is generated. There is still some doubt concerning the actual law of compression. Is it adiabatic or isothermal? Hayward [146] argues that it is nearer isothermal than adiabatic. For an adiabatic compression Schanzlin [148] has shown that the temperature of the air bubbles varies with pressure as shown in Fig. 1.5. For instance,

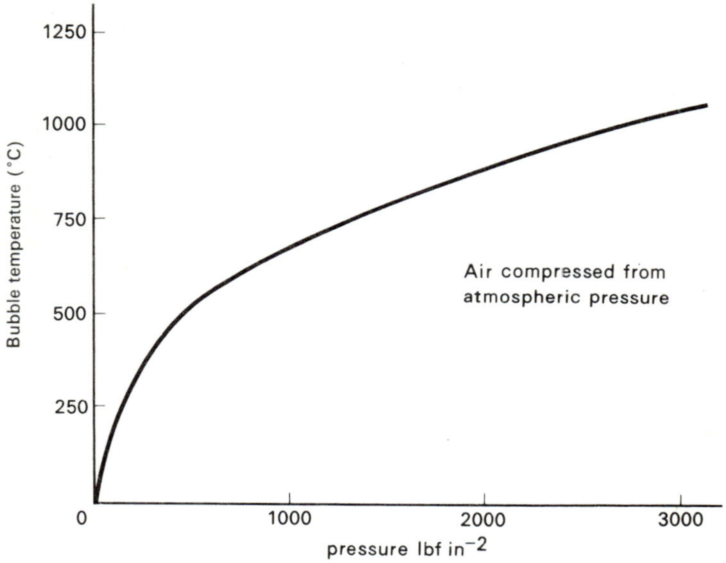

Fig. 1.5. Temperature developed during adiabatic compression

3000 lbf in^{-2} predicts a bubble temperature of 1100°C, although in practice the temperature will be less because of oil going into solution and heat loss to the oil. However, the increase in temperature can cause oil oxidation and can overheat the system and in some cases compressed air has raised the temperature of the reservoir oil by 25 per cent.

How is system performance affected? As far as hydraulic servos are concerned, the reduction in bulk modulus when air is present can have serious results. It is well known that the bulk modulus is a measure of the stiffness of the actuator drive system. If this stiffness is too small, as it may be if air is present, then oscillations caused by the inertia load bouncing on the oil column may be amplified and may reduce stability.

Pump performance can be seriously impaired. Pump delivery is decreased when the amount of dissolved air in the reservoir is increased. The reason for this is clear when it is realised that there is a pressure drop along the pump inlet line. Assume that the oil in the reservoir is saturated with air at atmospheric pressure and bad suction line design results in a pressure drop of 0·3 atm. The volume formed by air coming out of solution will be about 5 per cent of the oil volume, Fig. 1.2. Clearly this amount of air entering the pump will reduce its volumetric efficiency. In addition to this, through bad reservoir design, there may be a substantial quantity of entrained air and this too will contribute to the loss of pump efficiency. Losses in the suction line can be reduced by using a bell-mouthed entrance and by making the diameter of the pipe sufficiently large.

As these bubbles are compressed and passed through the pump they do not go into solution immediately. Hayward [146] has shown that the rate of solution depends on the pressure of the oil and some of his results are shown in Fig. 1.6. These results were taken from a static apparatus. It consisted of a cylinder of bubbly oil on the surface of which was placed a weight. The position of the weight was related to the degree of saturation. Faster rates of solution can be achieved if the oil is agitated, as it is, of course, in a real system and in a static apparatus local saturation of the oil around each bubble occurs fairly quickly

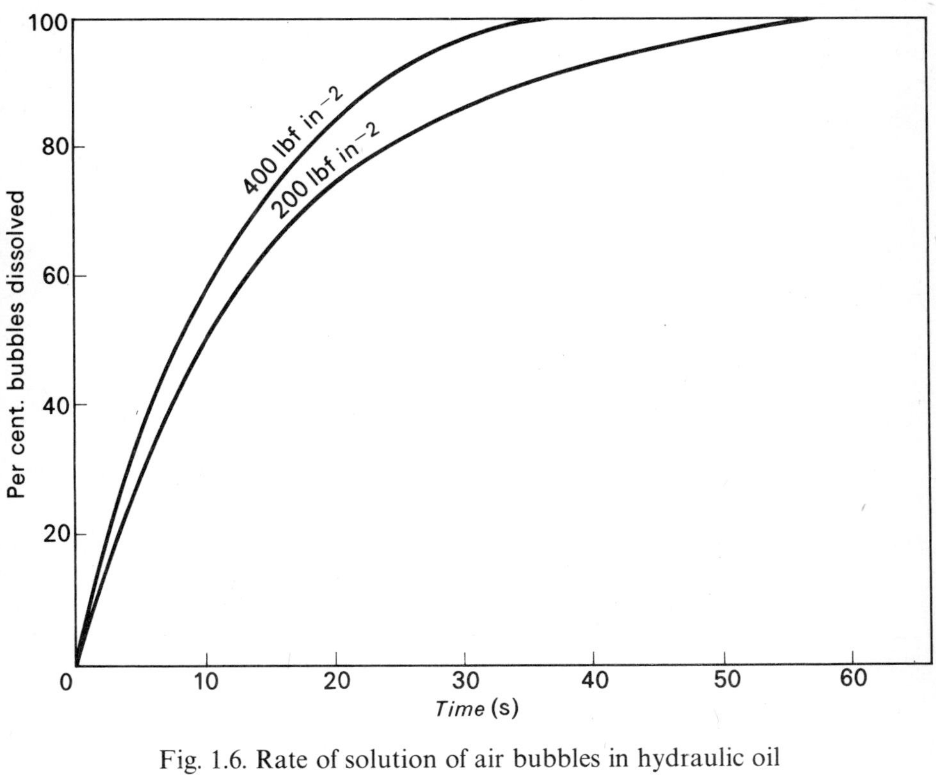

Fig. 1.6. Rate of solution of air bubbles in hydraulic oil

and tends to slow down the rate of solution. From this it is seen that a lower bulk modulus than expected will persist for some distance down the delivery pipe. Both Pigott [149] and Smith *et al* [147] have shown that centrifuge action in pumps can cause the bubbles to coalesce to form larger bubbles. These do not go back into solution readily on the discharge side. In fact under certain conditions, i.e., very large rates of pressure change, these bubbles can be unstable and can collapse implosively on the high pressure side. This is related to the cavitation phenomenon and is sometimes called gaseous cavitation as opposed to vapourous cavitation which will be discussed later. The pressure developed at collapse can be large enough to cause mechanical damage and Schanzlin has shown this to occur on the high pressure outlet of a gear pump.

How can the amount of aeration be reduced? There is not much that can be done about dissolved air. However, some steps can be taken to reduce the amount of entrained air. The majority of the air entrained in a hydraulic circuit is picked up in the sump or reservoir. It is important that the surface of the oil is not agitated by returning oil; therefore all oil return lines must be below the oil level. Return leads from relief valves must also be drowned. In addition to preventing bubble formation in the sump, the design must also prevent air bubbles in the returning fluid from entering the suction line. To this end a separator which encourages bubbles to return to the surface of the oil should be fitted. Tests [146] show that a simple gauze screen of 60 strands to the inch fitted in the reservoir at an angle of 30° stops all bubbles of significant size. The top of such a gauze should be below oil level since if there is any foam on the oil surface it might well penetrate the gauze. Figure 1.7 taken from reference [146] shows a good reservoir design.

Aeration can also arise if the suction line is not immersed deeply enough. In this case a vortex may form and air will be drawn into the system.

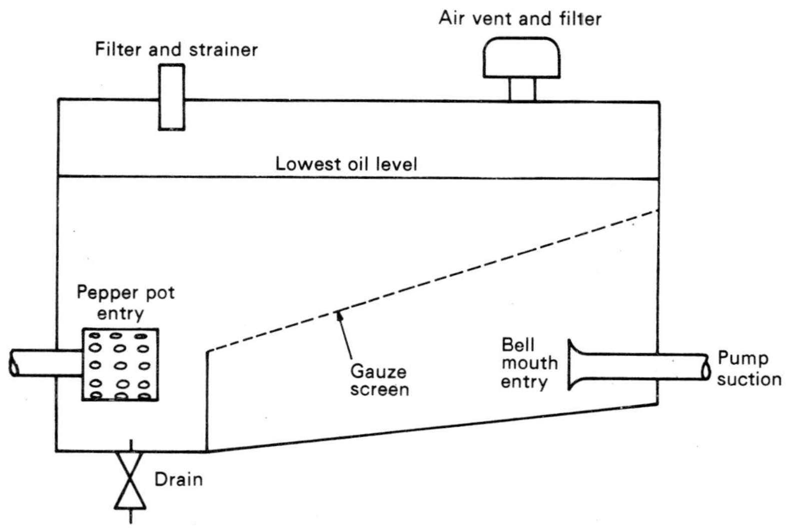

Fig. 1.7. A good reservoir design

1.5. Cavitation

In fluid systems the term cavitation usually refers to the formation and collapse of cavities in the liquid. The cavities may contain air or gas, and if the pressure is reduced far enough hydraulic oil will vapourise and vapour cavities will form. The pressure at which vapourisation commences is called the vapour pressure of the liquid. This vapour pressure is very dependent on the temperature of the oil. The phenomenon of boiling is related to cavitation. At atmospheric pressure water boils at 212°F, i.e., the vapour pressure is 14·696 lbf in^{-2} absolute at 212°F. This vapour pressure is reduced to 0·363 lbf in^{-2} absolute at 70°F. Hence, it is clear that the lower the temperature of a liquid the lower is the vapour pressure and this means that the pressure must be reduced considerably before the liquid boils. The vapour pressures of mineral oils are much lower than that of water and hence cavitation is expected to occur less readily with these fluids. In Fig. 1.8, taken from the paper by Leroy *et al* [145], the vapour pressures of common hydraulic fluids are displayed as a function of temperature. This clearly

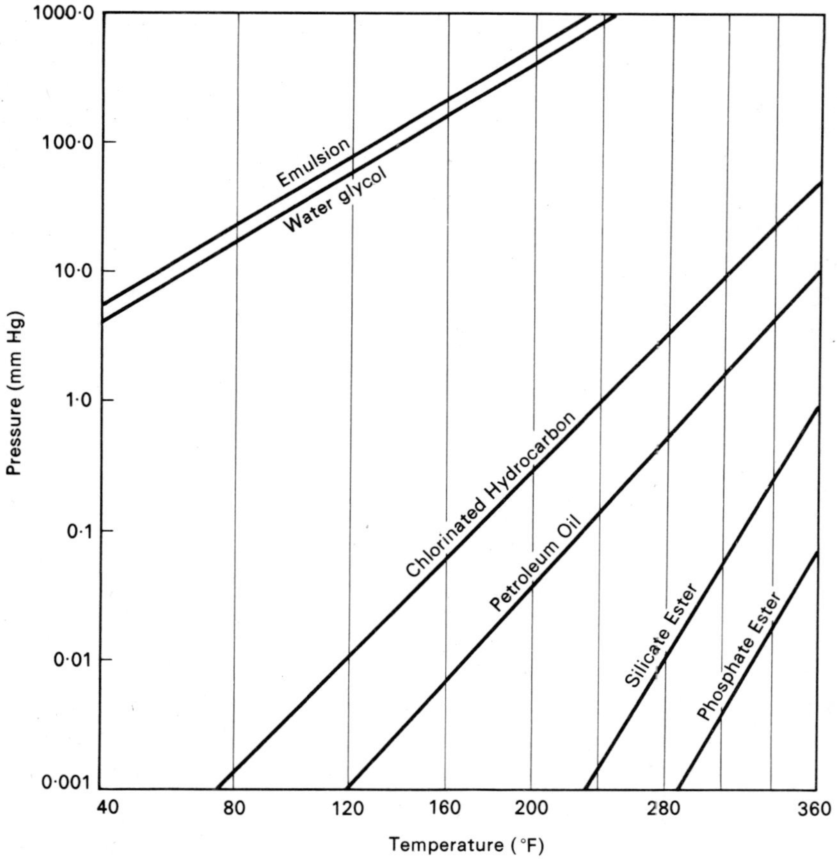

Fig. 1.8. The vapour pressure of hydraulic fluids

shows the increase in vapour pressure as the temperature increases. Those fluids which contain water, such as emulsions and water glycol, have much higher vapour pressures than the other conventional fluids.

It is interesting to note that the formation of cavities in a liquid involves rupturing of the liquid. Since, for instance at 70°F vapour cavities form in water at a pressure of 0·363 lbf in^{-2} absolute, one might argue that the liquid has ruptured at this pressure and hence that its tensile strength is low. This is not so. It has been shown experimentally [150] that the tensible strength of water can be as high as 15 000 lbf in^{-2}. The development of a cavity is now recognised as being associated with a nucleation centre such as a microscopic gas particle. When the liquid is subjected to a tensile stress, cavities do not form as a result of liquid rupture but are caused by the rapid growth of these nuclei [151]. As with aeration it is entrained air and not dissolved air that causes trouble. Experiments have demonstrated that the tensile strength of water saturated with dissolved air can have a very high value. However, conventional hydraulic fluids operating in a conventional system will contain sufficient nuclei to ensure that cavitation will occur when the pressure is reduced to vapour pressure.

The phenomenon of cavitation damage in hydraulic machinery, turbines, pumps and propellers is well known. It has been shown both analytically and experimentally that when cavities collapse as a result of increased hydraulic pressure, very large pressures, sometimes of the order of 100 000 lbf in^{-2}, can be developed. Unfortunately, there is still some controversy about the exact mechanism of the damaging process. The mechanical theory of damage suggests that prolonged exposure to cavity collapse results in brittle fracture and hence wear of the material. Recent work [152] proposes that the strain energy of the material is the controlling property of cavitation erosion. Materials such as aluminium of low strain energy are susceptible to severe cavitation erosion while stainless steel which has a high strain energy withstands cavitation attack very well.

The rate of wear can be greatly accelerated if the cavitation erosion develops under conditions in which corrosion can occur [153, 154]. In such a case the hammering effect of imploding cavities removes the products, such as oxides, of the corrosion attack. At the same time the corrosive effect produces material which is more easily damaged by the hammering than the virgin material.

As with aeration the collapse of the bubbles results in increased temperature of the oil and this causes side effects such as oil oxidation. Resins and slags can form and these will cause deterioration of system performance.

Under what conditions will cavitation arise? It is clear that the fluid must experience a reduction in pressure before cavities will form. This reduction in pressure can come about in several different ways. Since the total energy in a fluid is a combination of dynamic and static pressure heads, a reduction in static pressure will arise if the dynamic pressure is increased. Such a situation arises when a liquid flows through a venturi section. At the point of maximum velocity the static pressure will be minimum, and as the velocity of the fluid is increased a stage will be reached when the minimum pressure drops to the vapour pressure of the liquid. At this point cavitation will occur at the point of maximum velocity. As the cavities move downstream from the point of maximum velocity they will collapse and may well damage the surface of the container.

Thus aeration and cavitation are both undesirable phenomena. They will receive further attention throughout the book [158].

1.6. High temperature systems

Many modern fluid servomechanisms must operate under very high temperature conditions. Examples include nuclear reactors and re-entry vehicles in which controls must be capable of reliable operation at temperatures up to 2000°F. One method of approach is to use hot gases and this is discussed in chapter 9. The problem of finding suitable liquid fluids is a difficult one, since they have not got the high temperature stability of hot gases. The polyphenyls and liquid metals have been used as heat transfer liquids in reactors, and their properties are fairly well known. In the investigation reported in [5] the polyphenols were on the whole found to be superior to the liquid metal alloy of sodium (23%) and potassium (77%). The organic polyphenyl is a fair lubricating fluid and has a bulk modulus of 94 000 lbf in^{-2} at 700°F and 4000 lbf in^{-2} pressure. The liquid metal has a higher modulus, 4×10^5 lbf in^{-2}, but poor lubricity. It is also hazardous, reacting violently with water.

1.7. Sources of Power

The main source of power for hydraulic systems is the hydraulic pump. While the wide range of pumps available may differ in detail design, they are all positive displacement in the sense that for each revolution of the pump shaft, a nearly constant quantity of fluid is delivered, regardless of changes in the pressure which the pump may be developing.

The volume of fluid pumped per revolution can be determined from the geometry of the pump. However, the pump never quite delivers the calculated amount of fluid and the difference is expressed in terms of

$$\text{Volumetric efficiency} = \frac{\text{actual delivery per unit time}}{\text{swept volume per unit time}}$$

A pump's mechanical efficiency mainly depends on the friction forces acting on the moving parts,

$$\text{Mechanical efficiency} = \frac{\text{output work done per rev}}{\text{input work done per rev}}$$

The overall efficiency is the product of the mechanical and volumetric efficiencies. The general characteristics of pumps are summarised in Table 1.3.

Gear pumps are the cheapest and most popular for general medium pressure requirements. They are of simple construction, consisting essentially of two gears in a casing fitted with suitable inlet and outlet ports. The volumetric

Type	Design Arrangement	Max. pressure lbf/in^2	Max. delivery gal/min.	Volumetric efficiency	Overall efficiency	Relative cost
Gear	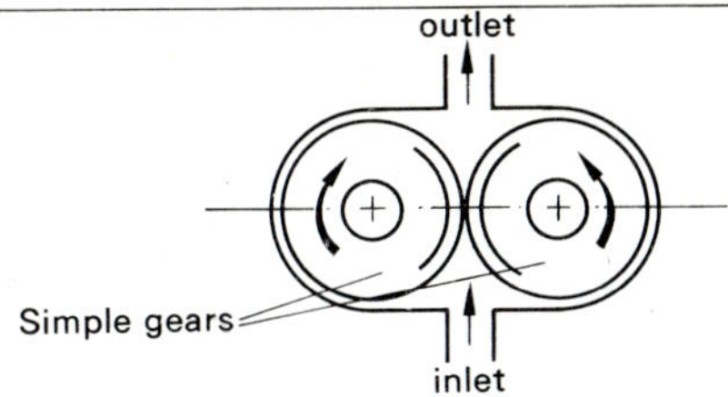	up to 2500	up to 150	93%	83%	cheapest
Vane	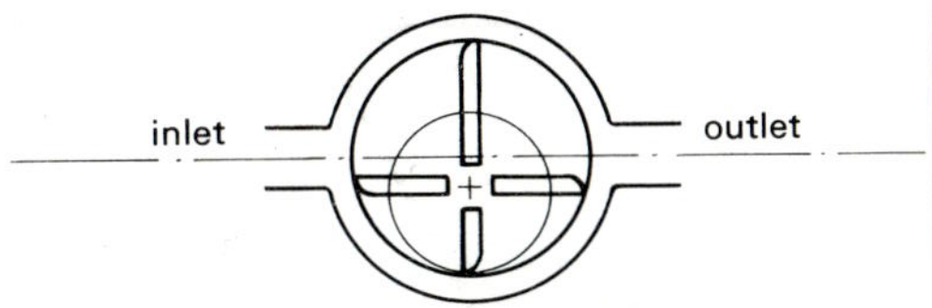	up to 1000	up to 70	85%	82%	cheap
Swash plate (axial piston)	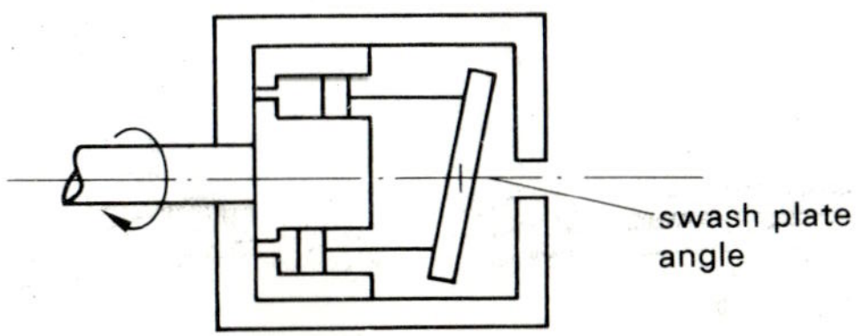	up to 5,000	up to 2,000	97%	94%	more expensive
Reciprocating plunger	Similar to vane pump, but using pistons	up to 10,000	up to 1,000	96%	90%	expensive

efficiency is good under optimum conditions but at low speeds is quite poor. The gear pump is classed as a fixed displacement pump.

The vane pump has a number of spring loaded vanes in a rotor which rotates eccentrically in the pump housing. In some designs centripetal force holds the vanes in contact with the housing. As the space enclosed by the vanes, rotor and housing enlarges, a vacuum is created and liquid is forced into the space by atmospheric pressure. As the space is reduced, the liquid is forced out at the outlet. The volume of oil pumped can be varied by altering the degree of eccentricity between the housing and the rotor.

The axial or swash plate pump lends itself well to closed loop control, since the angle of the swash can be readily varied. This angle controls the stroke of the pistons and hence the delivery. Piston pumps of this type are most suitable for high pressure applications, where sealing becomes a problem but the more complex design and closer tolerances require a much more expensive pump.

Most pneumatic power sources use a compressor. These may vary in size from small industrial bench supplies for air driven tools, up to 10000 lbf in^{-2} units consuming 15,000 hp. Most compressors for pneumatic control systems have deliveries in the region of 200 ft^3 mm^{-1} at pressures up to 100 lbf in^{-2}.

Most units are either reciprocating piston units or vane type. The problem in this case is keeping the units cool, and while air cooling is satisfactory for the small units, water cooling is essential for the big compressor sets.

The standard method of rating compressors or, in fact, any other type of pneumatic equipment, is to convert the flow requirement to volume rate of unpressurised air. Hence, the volume V_r at room temperature and pressure is,

$$V_r = \frac{V_1(P_1 + 14{\cdot}7)}{14{\cdot}7} \text{ ft}^3 \text{ of free air}$$

where V_1 is the volume at pressure P_1 (gauge).

In some cases the compressor rating is given in terms of swept volume or displacement.

$$D = L \times A \times N \text{ ft}^3 \text{ min}^{-1}$$

where
D = displacement
L = stroke (ft)
A = piston area (ft^2)
N = strokes min^{-1}

For example, an air flow of 1 ft^3 min^{-1} at 100 lbf in^{-2} gauge pressure becomes a flow of 7·8 ft^3 min^{-1} at room pressure and temperature and therefore requires a compressor of 7·8 ft^3 min^{-1} free air delivery.

The water content in air supplies is a common cause of trouble in pneumatic systems. It is therefore important to fit the standard filter, pressure regulator and lubricator unit.

1.8. Accumulators and filters

The accumulator is an energy storage unit allowing hydraulic oil systems to use the energy storage capacity of compressed gas. The most common type of accumulator is shown in Fig. 1.9, consisting of a high pressure cylinder containing a rubber bag. The bag is normally charged with an inert gas (normally nitrogen) to about one-third of the final maximum pressure. As oil is pumped into the cylinder around the bag the pressure will rise.

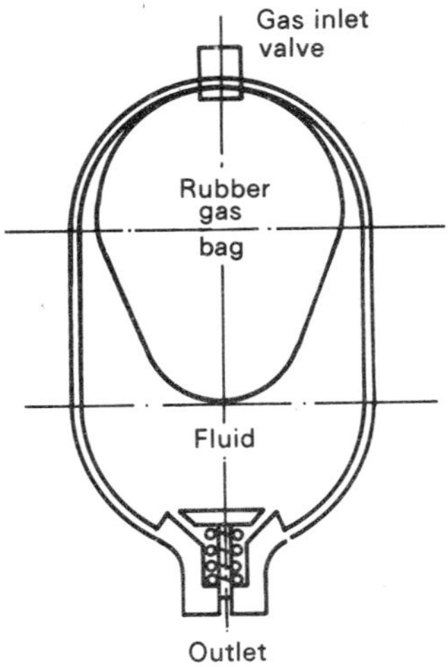

Fig. 1.9. A bag-type accumulator

This type of simple energy storage system can be used as a reserve of power in a pump circuit, in which there are short duration high flow demands. This is much cheaper than a twin pump system. In fact, many missile systems use this approach instead of the conventional pump to supply oil to the control system. The accumulator can be charged from the exhaust gases of the rocket motor.

The accumulator can also be used as a low pass filter, helping to smooth out fluctuations in pressure and flow due to the pump action. The degree of pulsation varies with each type of pump but in general the pulsation amplitude increases with pumping pressure. A more violent pulsation can occur with the rapid shutting of a valve or when components are either stopped or reversed. These effects are often quite audible and the force transmissions are visible in the form of external movement of the piping. Such surges and shocks can create transient pressure forces that far exceed the rated pressures of the components.

Another use of the accumulator in a closed system is for compensation for internal or external leakage or volumetric changes due to thermal expansion and contraction.

The method of determining the accumulator size is based on Boyle's law,

$$P_1V_1 = P_2V_2 = P_3V_3$$

which assumes isothermal operation.

Example

Let $P_1 = 1000 \text{ lbf in}^{-2}$ a minimum oil discharge pressure (gas precharge pressure)

$P_2 = 1500 \text{ lbf in}^{-2}$ a minimum oil working pressure

$P_3 = 3000 \text{ lbf in}^{-2}$ a pump delivery pressure

Let the volume discharged during the working cycle be $61{\cdot}8 \text{ in}^3$ of oil. Then

$$\begin{aligned} V_1 &= \text{? volumetric size of accumulator} \\ V_2 &= V_3 + 61{\cdot}8 \text{ in}^3 \\ V_3 &= V_2 - 61{\cdot}8 \text{ in}^3 \end{aligned}$$

Since,

$$\begin{aligned} P_2V_2 &= P_3V_3 \\ 1500\,(V_3 + 61{\cdot}8) &= 3000\ V_3 \\ \therefore\ V_3 &= 61{\cdot}8 \text{ in}^3 \\ V_2 &= 61{\cdot}8 + 61{\cdot}8 \\ &= 123{\cdot}6 \text{ in}^3 \end{aligned}$$

Now from

$$\begin{aligned} P_1V_1 &= P_2V_2 \\ 1000\ V_1 &= 1500 \times 123{\cdot}6 \\ V_1 &= 185{\cdot}4 \text{ in}^3 \\ &= 0{\cdot}8 \text{ gallons} \end{aligned}$$

A 1 gallon accumulator would therefore be chosen, with a gas precharge pressure of 1000 lbf in^{-2} of nitrogen.

Another important component in hydraulic circuits is the filter. The degree of filtration required will depend on the application; for example, servo valves require a working fluid with contamination levels less than 5–10 micron. Contamination takes the form of many different types of solid material, some introduced during manufacture and assembly, others due to wear of moving parts. These deposits cause friction between moving parts of valves resulting in erosion of the meeting edges and in extreme cases jamming of the valve.

The particle size ranges from 1 micron (10^{-6} m) to greater than 100 micron. Physical meaning of this can be judged from the fact that particles less than 50 micron are not normally visible to the human eye. Although the particle size is quite random it is not difficult to remove particles above 100 micron. Filters are available capable of removing a high percentage of particles down to 5 micron, but these tend to be expensive.

The filter should be installed at the last possible point before the critical component, but when one element is used it is normally placed in the pump outlet.

The filter will cause little pressure drop when new, but as it becomes clogged with contaminants the pressure drop will gradually rise until a rupture could occur. The safest method is to replace the element after the number of hours of service recommended by the manufacturer has been reached.

1.9. Pump sets

Most simple pump set circuits use a fixed capacity type pump running at constant speed and driven by an electric motor. The selection of the pump depends on the quantity of fluid to be circulated and the maximum pressure required.

The speed of the electric motor is fixed at 1470 rev min^{-1}, and can be sized on the horsepower requirements of the pump, where horsepower = 0·00015 × pressure (lbf in^{-2}) × flow (in^3 s^{-1}). The pump pressure and flow rating must take account of losses through the pipework and other components in the circuit. A typical circuit is shown in Fig. 1.10. A three-phase electric motor

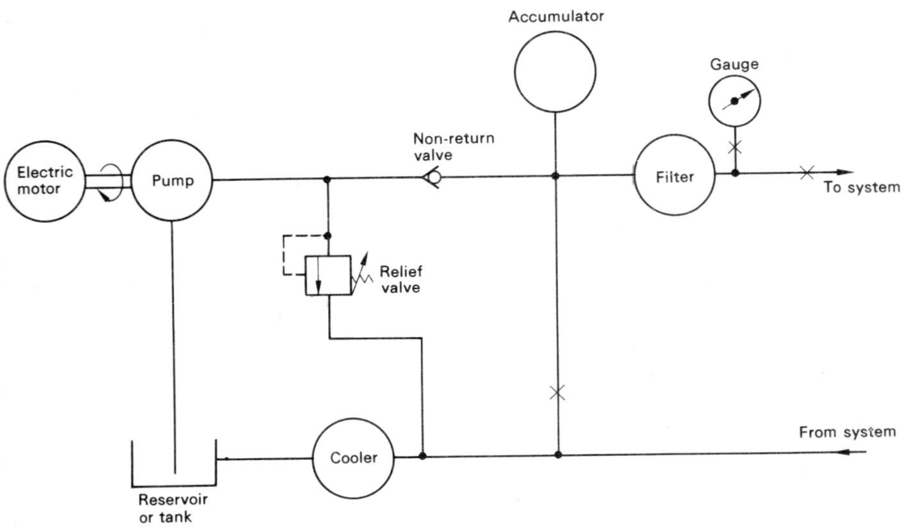

Fig. 1.10. Typical circuit for a pump set

drives a fixed displacement pump, the outlet of which is piped to a relief valve. This allows the working pressure to be pre-set, and any excess pressure is relieved to tank. The non-return valve prevents flow being forced back to the pump and also stiffens the hydraulic circuit. An accumulator is included to smooth the pressure pulses and provides additional flow during short transients. The outflow is passed through a filter to the outlet manifold via a stop cock. It is best to measure the system pressure at this point. A by-pass line with a stop cock to tank is provided for ease of discharge of the accumulator. All the return flow is passed through the cooler and returned to the tank. Where the system arrangement is such that this type of pump set would blow off for long periods, a twin pump off-loading system is more desirable.

The heat generated in a pump set is mainly caused by the flow through the relief valve and this must not be allowed to raise the fluid temperature above 100–115°F. A heat exchanger or cooler will control this condition by force convection cooling with water. It is advisable to let the manufacturer advise on the correct size of unit.

The design of the tank is also important, and should be at least three to six times the rated delivery of the pump. Baffles between the inlet and return lines will direct the oil along a longer path, thus helping to settle out any dirt and aid cooling. As mentioned earlier, a wire gauze screen is useful in preventing air bubbles from recirculating into the system. The inlet line should also include a coarse filter or strainer with a capacity of at least twice the pump capacity. The oil should be returned to tank at the end furthest from the pump inlet pipe, Fig. 1.7.

1.10. Simple concepts of automatic control

A servomechanism is a particular sort of feedback control system, and in fact it is commonly defined as a control system in which the output is mechanical. For example, the hydraulic position control system for positioning a machine tool cutting head is a servomechanism, while a computer is a feedback control system.

Most progressive industrial companies recognise that control system technology is an essential part of efficient operation. Its applications range from simple low cost automation aids, to complex computer controlled process plant. The servomechanism plays an important part in all these applications.

There are many advantages to be gained by using automatic control systems; some areas where their application can be beneficial are listed below.

1. Many processes of modern industry are becoming so fast and complex, or operate in such unsuitable environments, that their control is beyond the capabilities of human operation. For example, the control of nuclear reactor moderator rods.
2. An experienced operator can achieve good, consistent operation for short periods, but his performance will eventually fall off due to monotony and fatigue. A properly designed control system can ensure adequate control over long periods.
3. Some processes are critically dependent on efficiency and accurate control —a result of small profit margins in highly competitive industries or high precision in the specialist industries. The use of automatic control systems will ensure that scrap rate, in either case, is kept to a minimum.

1.11. Feedback control systems [7]

Control systems can be divided into two groups depending on whether or not feedback is used. The concept of feedback is illustrated diagrammatically in Fig. 1.11 where the box represents the component or components in the system, and the lines indicate paths along which information is passed. This method of representation is especially useful in complex systems of components, allowing

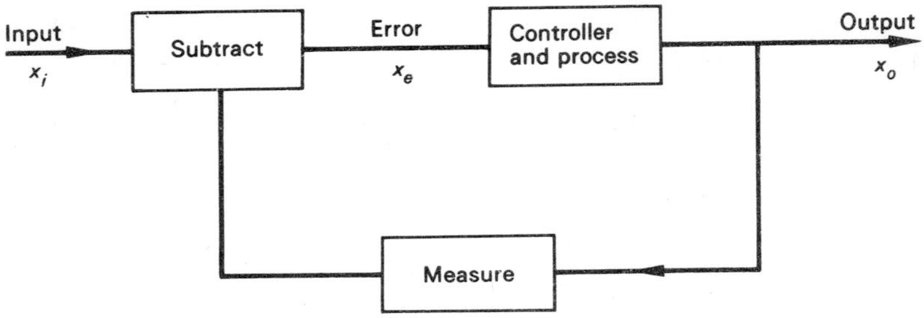

Fig. 1.11. The elements of a feedback loop

signal paths and interconnections to be easily traced. In this representation, the output signal (x_o) is fed back to be compared with the input signal (x_i), the difference between the two being the error signal (x_e). This error is then fed into the system components causing the output to alter. The concept of feedback makes such a system zero error seeking in the sense that the system will continue to operate until the output and input are identical. Only then is the error zero. The basic principle of a feedback system can be expressed by a simple equation,

$$x_e = x_i - x_o$$

Systems in which no feedback is used are called open loop systems. These can also be applied to the control of processes. For example, a spool valve when displaced will allow fluid to enter the hydraulic motor to produce a given motor speed, Fig. 1.12(*a*). The motor speed will be proportional to the valve opening and provided that the load on the motor is not altered it will continue to operate at constant speed. If now the speed of the motor is fed back to the input, any alteration of load at the output would be sensed and a corrective action taken to bring the speed back to the required level. This demonstrates the advantage

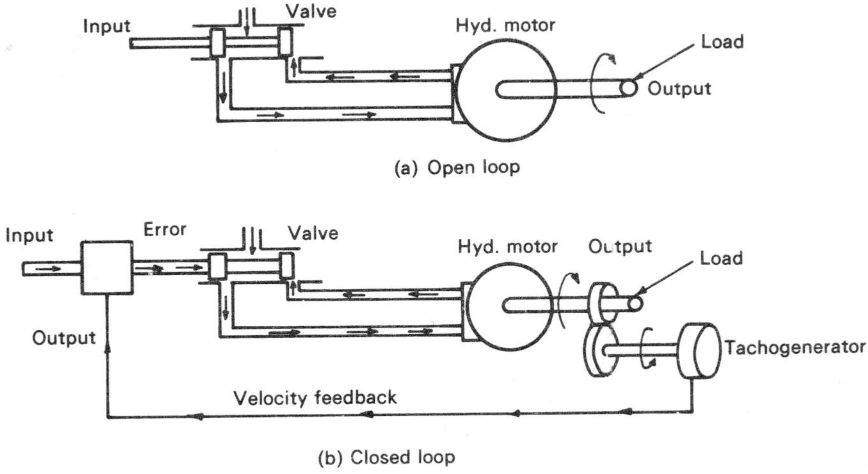

Fig. 1.12. (*a*) An open loop speed controller, (*b*) the closed loop controller

of closed loop feedback control, the input can now see what the output is doing. However, in order to gain this advantage, the system needs to be more complicated. The output must be measured by a tachogenerator and an error measuring device must be included.

1.12. Feedback in practice

At first glance it would appear that a feedback control system operated quite simply, the output behaving exactly as demanded by the input. Unfortunately, this is not the case in practice. There are always lags between input and output since it takes a finite time for information to be transferred through the system. For example in the closed loop speed control in Fig. 1.12(*b*), pressure in the motor lines will not respond instantaneously to change in valve position. Fluid compressibility does not allow this. In addition, the load inertia does not allow the output velocity to instantaneously follow changes in motor torque. These lags result in a degradation of performance as shown in Fig. 1.13. The dashed

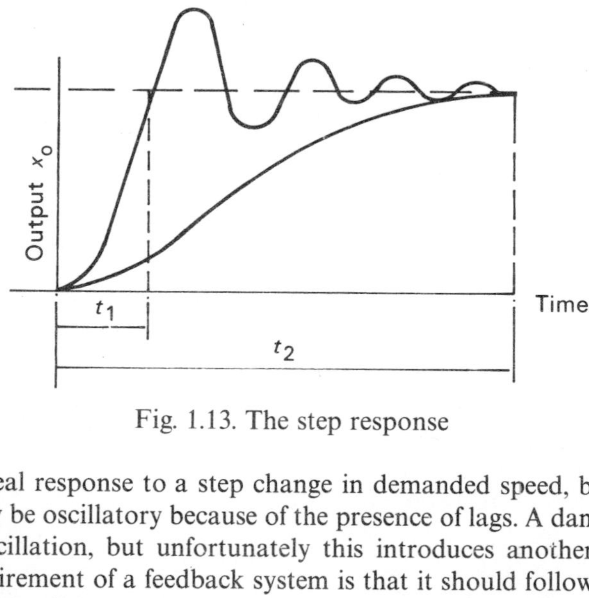

Fig. 1.13. The step response

line is the ideal response to a step change in demanded speed, but the actual response may be oscillatory because of the presence of lags. A damper can help to reduce oscillation, but unfortunately this introduces another problem. A primary requirement of a feedback system is that it should follow the input as accurately as possible—the output should reach the demanded level as quickly as possible and with a minimum of overshoot. In Fig. 1.13 it is seen that these requirements are not compatible in practice. The system with little damping reaches the desired level at time t_1, but shows considerable oscillation; the heavily damped system has a long response time t_2. Hence a compromise must be accepted. In fluid power servos the valve element contributes little damping, and stability often relies upon viscous damping in the actuator and load. If this does not give a satisfactory response, signal shaping techniques are necessary and velocity and acceleration signals can be derived either by means of fluid or electric circuitry.

1.13. Some problems associated with fluid power control systems

Later chapters will discuss the control of pressure, speed, flow rate and position. All of these fluid systems are difficult to analyse because of their inherent non-linear nature. In most hydraulic valves the flow is turbulent and the flow rate is thus proportional to the square root of the pressure drop. The relationship between flow rate and pressure drop is even more complicated for gas flow, and the pneumatic servo thus presents a formidable analytical problem. It will be seen later that the non-linearity results in a variation of system damping and sensitivity with valve displacement, and as a result a servo may be stable for one input and not another.

In addition, most valve ports are circular, resulting in a non-linear variation of valve flow area with valve displacement.

There are several other problems areas. Some are listed below.

There is a lower constraint on actuator pressure. The occurrence of cavitation does not allow the pressure to fall below vapour pressure.

System damping is sensitive to temperature change. As temperature increases, viscosity reduces, and damping caused by the relative motion of actuator parts falls off.

Stiffness of the drive, and hence the stability of the hydraulic servo, are sensitive to changes in bulk modulus. Aeration and cavitation cause problems here.

Valve sensitivity near the null position can be impaired by silting or by cavitation erosion. Filtration can help reduce the former and good design the latter.

Gas powered systems have low stiffness especially at low pressure. As a result it is difficult to achieve a steady low speed feed if friction forces are present.

The characteristics of valve orifices vary with Reynolds number. Hence flow rate through an orifice depends not only on pressure drop, but also on temperature and valve geometry. The next chapter is devoted to a detailed study of this particular area.

Chapter 2

Discharge characteristics of servovalve orifices

2.1. Introduction

An orifice offers resistance to fluid flow and thus can be used to control the flow rate. Control valves employ many different orifice configurations, Fig. 2.1, and their characteristics play an important part in the design of hydraulic control devices.

Figure 2.2 illustrates an ideal sharp edged orifice. At high Reynold's numbers the flow through the orifice forms a jet, the narrowest section being called the vena contracta. Downstream of the vena contracta turbulence causes the jet to spread and diffuse. Consider the energy at section u upstream of the orifice, and at the vena contracta, section vc

$$P_u + \tfrac{1}{2}\rho v_u^2 = P_{vc} + \tfrac{1}{2}\rho v_{vc}^2 + \Delta P_f \tag{2.1}$$

where ΔP_f is a friction loss from section u to vc and ρ is mass density. Continuity gives

$$a_u v_u = a_o v_o = a_{vc} v_{vc} \tag{2.2}$$

The subscript o refers to conditions at the plane of the orifice. The area a_{vc} is related to the orifice area by

$$a_{vc} = C_c a_o \tag{2.3}$$

C_c is a contraction coefficient.

Equations (2.1), (2.2) and (2.3) combine to give

$$v_{vc} = \frac{1}{\sqrt{\{1 - (C_c a_o/a_u)^2\}}} \sqrt{\left[\frac{2}{\rho}\{(P_u - P_{vc}) - \Delta P_f\}\right]} \tag{2.4}$$

This can be further reduced by introducing the velocity coefficient C_v.

$$v_{vc} = \frac{C_v}{\sqrt{\{1 - (C_c a_o/a_u)^2\}}} \sqrt{\left(\frac{2\Delta P}{\rho}\right)} \tag{2.5}$$

where

$$C_v = \sqrt{(1 - \Delta P_f/\Delta P)}$$

and

$$\Delta P = P_u - P_{vc}$$

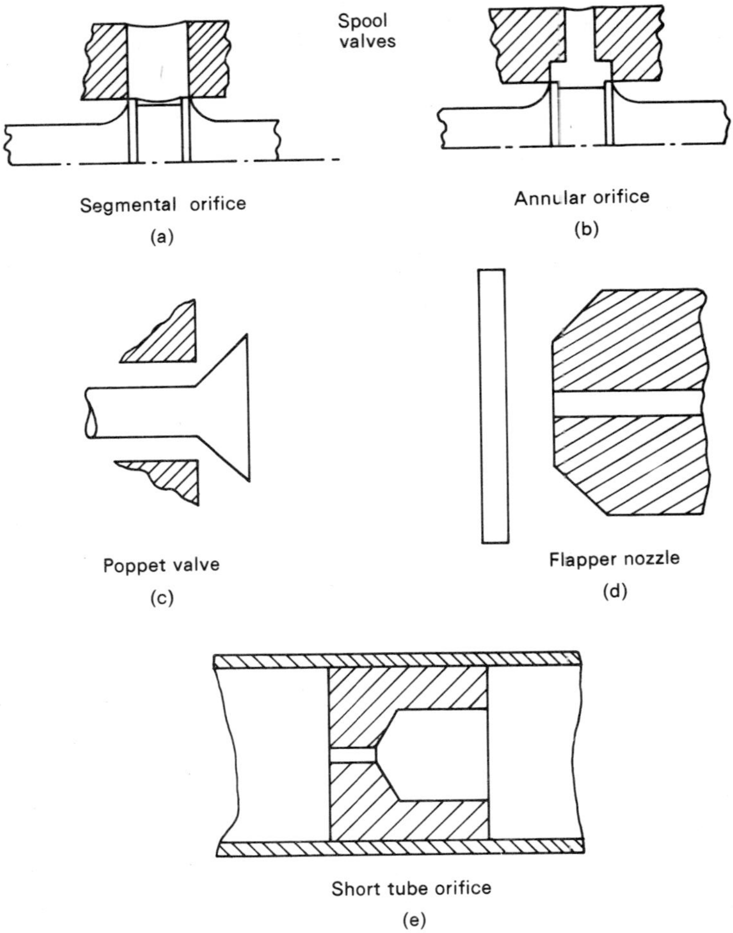

Fig. 2.1. Orifices used in fluid power systems, (*a*) segmental orifice (*b*) annular orifice (*c*) poppet (*d*) flapper nozzle (*e*) short tube

This is the static pressure drop from upstream to vena contracta. The volumetric flow rate $Q = a_{vc}v_{vc} = C_c a_o v_{vc}$
Hence

$$Q = \frac{C_c C_v a_o}{\sqrt{\{1 - (C_c a_o/a_u)^2\}}}\sqrt{\left(\frac{2\Delta P}{\rho}\right)} \tag{2.6}$$

The discharge coefficient is defined as $C_d = C_c C_v$
Hence

$$Q = \frac{C_d a_o}{\sqrt{\{1 - (C_c a_o/a_u)^2\}}}\sqrt{\left(\frac{2\Delta P}{\rho}\right)} \tag{2.7}$$

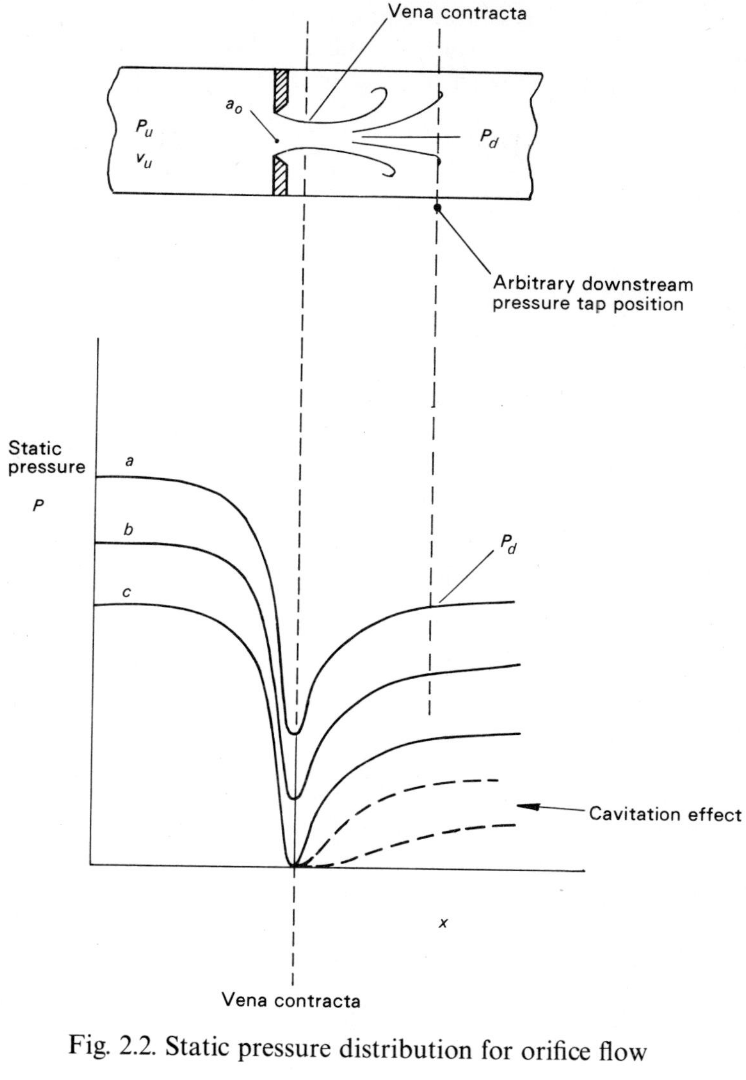

Fig. 2.2. Static pressure distribution for orifice flow

In most orifices employed in fluid power systems, the upstream area is much greater than the orifice area and hence approach velocity effects are negligible. In this case the flow rate is given by

$$Q = C_d a_o \sqrt{\left(\frac{2\Delta P}{\rho}\right)} \tag{2.8}$$

An alternative definition of a discharge coefficient C_d^* is sometimes used.

$$C_d^* = C_c / \sqrt{\{1 - (C_c a_o / a_u)^2\}} \tag{2.9}$$

This includes velocity of approach affects.

The pressure drop across a control orifice can be considered as a system pressure loss; there is less pressure available to drive a load. This is clearer when equation (2.7) is written in the form.

$$\Delta P = \frac{1 - (C_c a_o/a_u)^2}{C^2{}_d} \tfrac{1}{2}\rho v_o^2$$

This formula will sometimes be found in the form

$$\Delta P = K\tfrac{1}{2}\rho v_o{}^2 \qquad (2.10)$$

where K is called a loss coefficient.

2.2. Theoretical determination of the discharge coefficient

In general, the discharge coefficient must be determined experimentally. It will be a function of Reynolds number and orifice geometry. However, in some special cases the coefficient can be predicted analytically and these cases are of particular interest.

von Mises [8] analysed the flow through a general orifice configuration. He made the following assumptions:

1. the flow is steady
2. viscosity effects are negligible
3. the flow is irrotational
4. the flow is two-dimensional
5. gravity effects are negligible.

The use of assumption (2) implies that there is no friction loss from upstream to vena contracta and hence that the velocity coefficient C_v is unity. The Reynolds number is infinite. The flow in a control valve is generally far from being irrotational. However, vortices occurring upstream of the orifice, where the velocity is negligible compared with the velocity at the orifice, have a very small effect. Vortices occurring downstream of the vena contracta need not be considered. Hence assumption (3) is reasonable. All other assumptions can be justified in the case of conventional fluid control valves.

2.3. Spool valve orifices

The contraction coefficients for spool valve orifices, poppet valve orifices and flapper nozzles can be obtained from correct interpretation of von Mises' results. Figure 2.3(*a*) shows a configuration discribed by von Mises as a side opening at the floor of a container. Figure 2.3(*b*) shows how the results also apply to the flow through a spool valve orifice. His results, found by using potential functions and conformal transformation, are shown in Fig. 2.3(*c*), where the contraction coefficient and the jet efflux angle δ (which must be known if momentum forces are to be calculated) are plotted against the non-dimensional valve opening $\bar{x}_1$. C_c is 0·673 at very small openings and decreases to 0·20 at

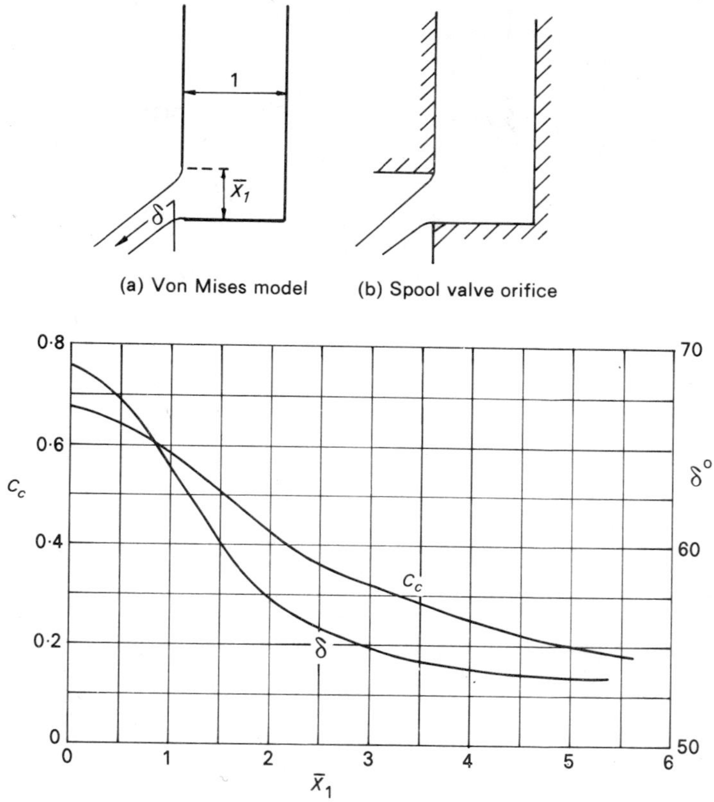

Fig. 2.3. Contraction coefficient and discharge angle for the ideal spool valve orifice (side opening)

$\bar{x}_1 = 5{\cdot}0$. The jet angle δ is 69° at small opening and decreases to 53°20′ at $\bar{x}_1 = 5{\cdot}0$.

Another configuration, the results of which can also be applied to spool valve orifices, is shown in Fig. 2.4. This is described by von Mises as a side lying opening in a horizontal floor. Figure 2.4 shows the effect of non-dimensional orifice opening $\bar{x}_2$ on the contraction coefficient and jet angle. For small openings C_c is 0·673 and at the maximum opening, $\bar{x}_2 = 1{\cdot}0$, C_c is 1·0. The jet angle is 21° at small openings and decreases to zero at maximum opening.

In conventional spool valves, the non-dimensional opening $\bar{x}$ is very small and hence the predicted contraction coeficient is 0·673 and the jet angle is 69°.

The results quoted refer to valves in which there is no clearance between the fixed and moving parts. Of course in practice such clearance must exist and if the orifice opening is very small clearance effects can be expected to be considerable. In this case we refer to von Mises' work on combined floor and side openings. Unfortunately the magnitude of orifice opening and clearance used by

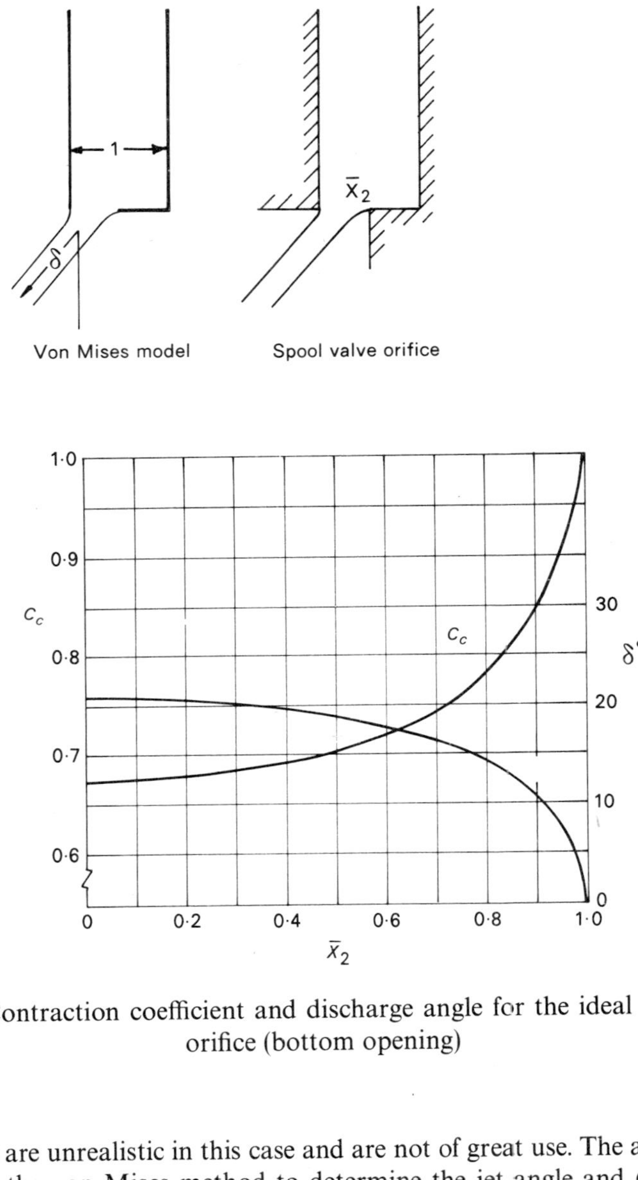

Fig. 2.4. Contraction coefficient and discharge angle for the ideal spool valve orifice (bottom opening)

von Mises are unrealistic in this case and are not of great use. The authors [15] have used the von Mises method to determine the jet angle and C_c for spool valve orifices with realistic openings and clearances and these results are shown in Figs. 2.5 and 2.6. Here $\bar{x}_1$ is the non-dimensional clearance and Fig. 2.6 shows increasing clearance results in increasing C_c. The contraction coefficient is in this case referred to the total orifice area $\surd\bar{x}_1{}^2 + \bar{x}_2{}^2$). Figure 2.5 shows the relationship between jet angle α_j and the orifice angle $\alpha = \tan^{-1}(\bar{x}_2/\bar{x}_1)$. These angles are only equal when the upstream chamber is infinite and $\bar{x}_1 = \bar{x}_2$, i.e. $\alpha = \alpha_j = 45°$.

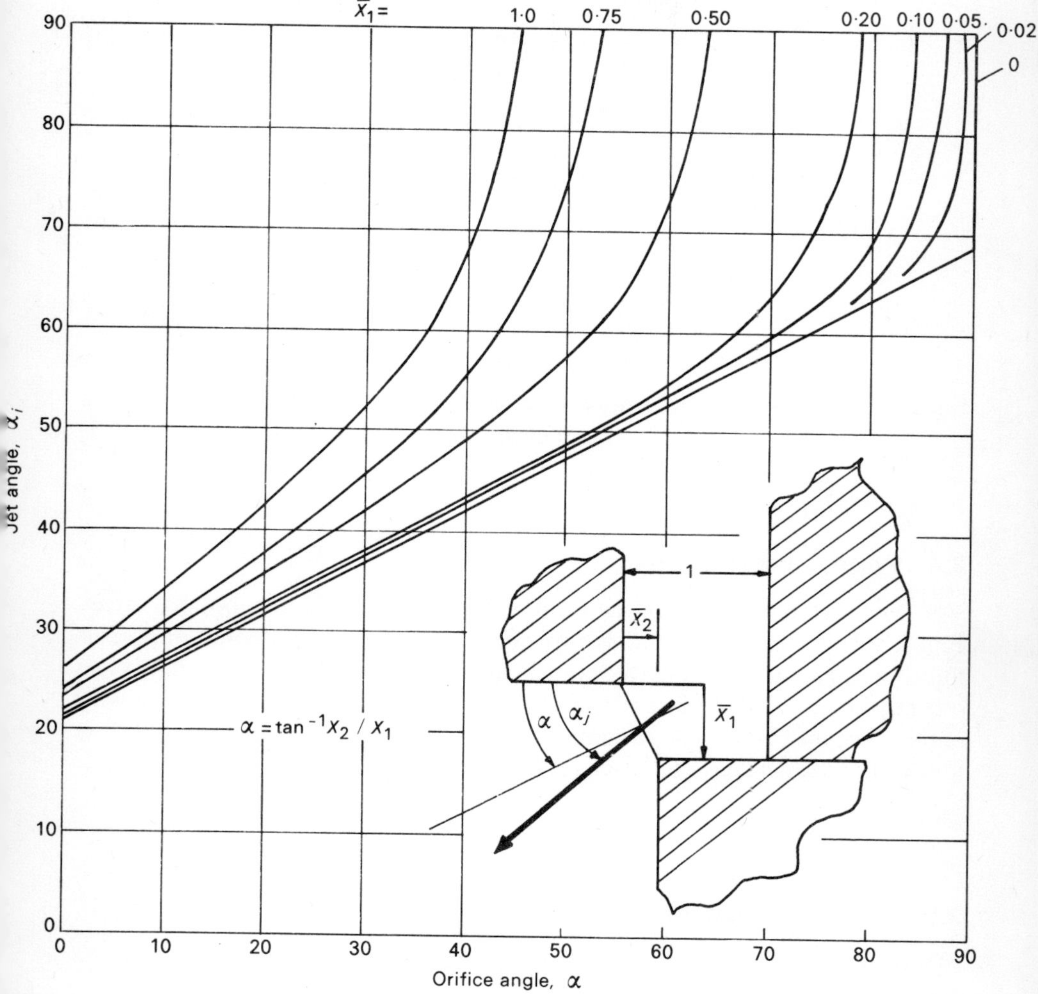

Fig. 2.5. Discharge angle as a function of orifice opening x_2 and clearance x_1

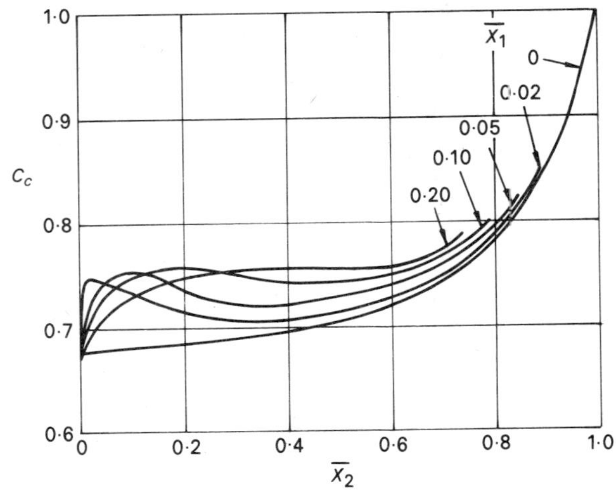

Fig. 2.6. Spool valve orifice contraction coefficient as a function of orifice opening x_2 and clearance x_1

2.4. Poppet valve orifices

Von Mises results for discharge from cones can be applied to poppet valves, Fig. 2.7. This is not immediately obvious when the cone configuration is examined and Fig. 2.7(*b*) has been drawn in an attempt to make this clearer. When one appreciates that the face of the poppet is equivalent to the axis of symmetry of the cone, the rest is easy. It is important to note that this implies that the flow follows the poppet face, i.e. the jet angle and poppet angle are equal.

2.5. Flapper nozzle valves

Flapper nozzle geometry is simulated if the opening $\bar{x}_2$ is negative, Fig. 2.8. The contraction coefficient for various geometries is shown in the figure.

Some points worth noting are:

For a given flapper size (fixed $\bar{x}_2$), C_c reduces as the distance between flapper and nozzles increases.

For a given flapper nozzle separation (fixed $\bar{x}_1$), C_c increases as the flapper gets smaller.

When $\bar{x}_1 > 1$ the orifice area is greater than the supply area and control is lost. In the axisymmetric case the nozzle has unity radius, the supply area is π and the orifice area (curtain area) is $2\pi\bar{x}_1$. Control is lost in this case when $\bar{x}_1 > 0{\cdot}5$.

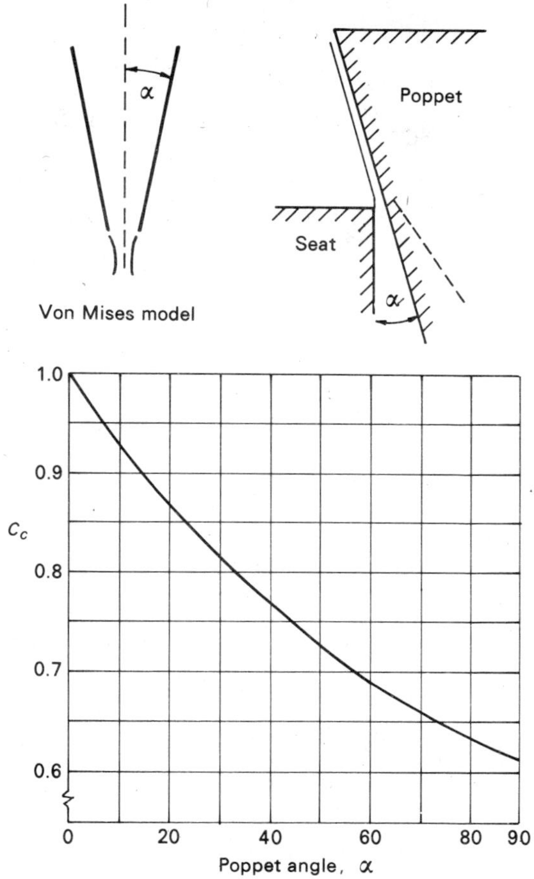

Fig. 2.7. Poppet valve contraction coefficients

2.6. Experimental determination of discharge coefficients

The results of von Mises apply only to an inviscid fluid or in other words to a flow at infinite Reynolds number. The effect of Reynolds number must be determined experimentally. Reynolds number R for an orifice is defined as

$$R = \frac{v_o D_h}{\nu} \tag{2.11}$$

where v_o is the mean velocity through the orifice $= Q/a_o$

D_h is the hydraulic diameter $= \dfrac{4 \times \text{flow section area}}{\text{flow section perimeter}}$

ν is kinematic viscosity $= \mu/\rho = \dfrac{\text{absolute viscosity}}{\text{mass density}}$

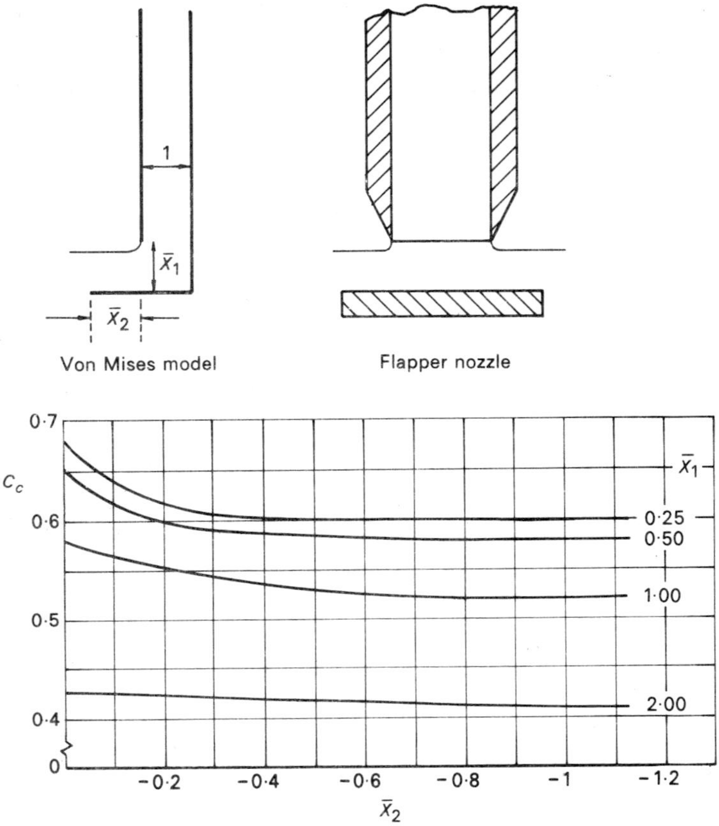

Fig. 2.8. Flapper nozzle contraction coefficients

For a circular orifice, the hydraulic diameter is equal to the actual diameter. For a rectangular slit of width w and length l the hydraulic diameter is

$$D_h = \frac{4wl}{2(w + l)}$$

This is nearly $2w$ for orifices of large aspect ratio l/w.

Reynolds number R is the ratio of inertia to viscous forces acting on the fluid. At high R the inertia forces predominate, the flow is turbulent and the analysis resulting in equation (2.7) is valid. However, when the viscous forces predominate, laminar flow conditions prevail and the analysis becomes invalid. Nevertheless it is general practice to apply this equation in the laminar flow regime also and a typical plot of C_d for a circular orifice in a pipe is shown in Fig. 2.9. Using flow visualisation techniques Johansen [9] showed that at low R (< 10) the outermost streamlines closely followed the orifice profile. As R was increased the flow began to separate from the downstream side until at $R = 150$ a jet formed. This was surrounded by a region of reverse flow. Further increase

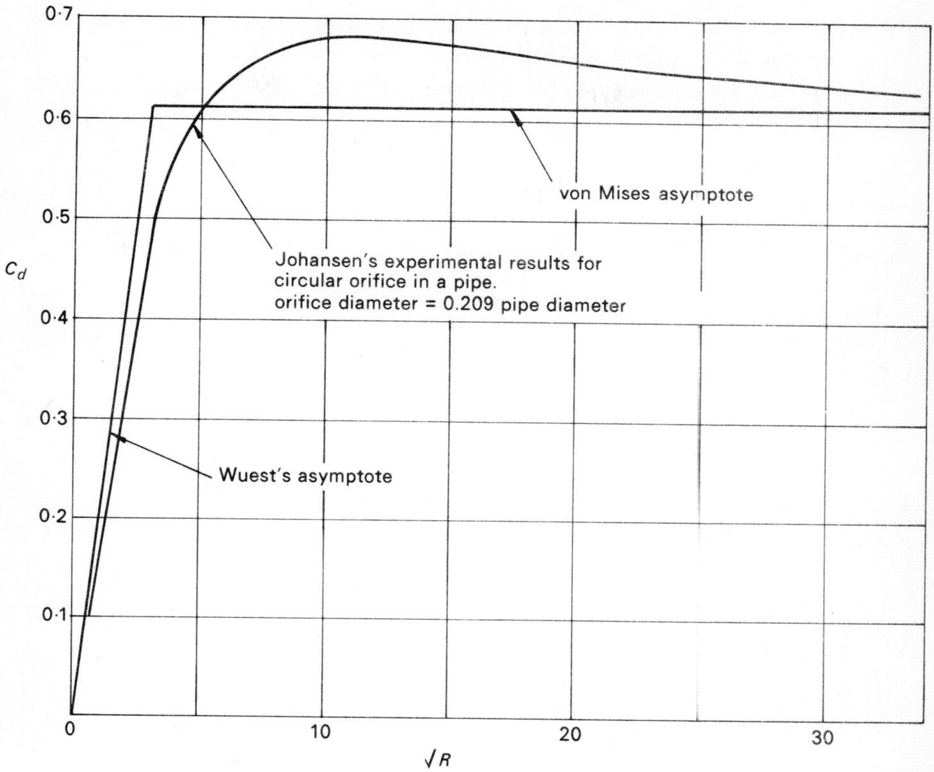

Fig. 2.9. Discharge coefficient of a circular orifice

in R caused the jet contraction to approach a maximum and although the flow outside the jet consisted of a succession of vortex rings, the flow within the jet was not disturbed.

The figure shows that R must exceed approximately 1000 before the discharge coefficient settles to the value predicted by von Mises (0·611). A peak value is reached at $R = 200$ and below this value C_d falls rapidly.

An exact analysis by Wuest [10] of the laminar flow through sharp edged orifices shows that for a circular orifice (diameter d) in an infinite pipe the flow rate is

$$Q = \frac{\pi d^3}{50{\cdot}4\mu} \Delta P \tag{2.12}$$

If this is combined with equations (2.8) and (2.11) it can be seen that for laminar flow

$$C_d = 0{\cdot}2\sqrt{R} = \delta\sqrt{R} \tag{2.13}$$

where δ is the laminar flow constant.

An approximate graph [11] of the variation of C_d with R can now be constructed using the low R asymptote defined by equation (2.13) and the high R value of 0·611 predicted by von Mises. This is shown in Fig. 2.9. The transition Reynolds number R_t is given by

$$R_t = \left(\frac{0{\cdot}611}{\delta}\right)^2 = 9{\cdot}3$$

For a rectangular slit of height b and width w in an infinite plane Wuest showed that $\delta = 0{\cdot}157$.

The majority of variable area orifices of spool servo valves are not circular but take the form of a segment of a circle, Fig. 2.10. The hydraulic diameter in this case is

$$D_h = \frac{4(\theta - \sin\theta)r^2/2}{r(\theta + 2\sin\theta/2)} = \frac{2r(\theta - \sin\theta)}{\theta + 2\sin\theta/2}$$

Results taken from [12] are shown in Fig. 2.10 and curve fitting indicates that δ is about 0·14 in this case.

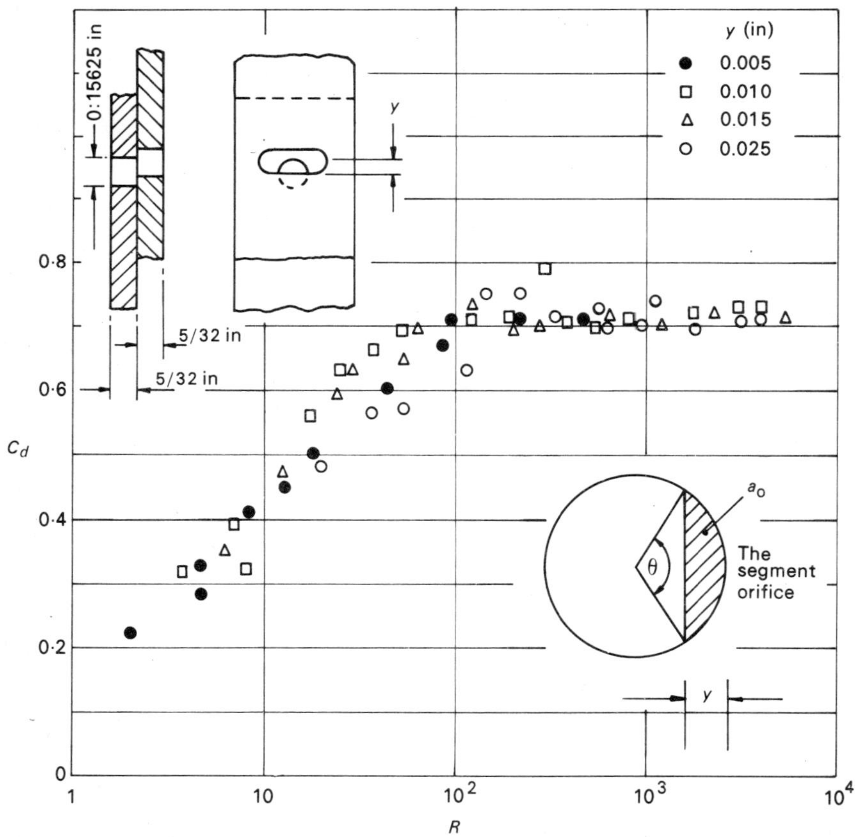

Fig. 2.10. Discharge coefficient of a segmental orifice (after Lewis and Stern [12]).

2.6.2 The flow coefficient

As far as fluid power control valves are concerned it is practically impossible to measure the *discharge coefficient* C_d. Equation (2.8) defines C_d as

$$C_d = Q/\{a_o\sqrt{(2\Delta P/\rho)}\}$$

where ΔP is the pressure drop from upstream to vena contracta. In a real valve it is well nigh impossible to measure the pressure at the vena contracta and it is usual practice to base a coefficient on the pressure drop from upstream to some convenient downstream point. In many cases this coefficient is incorrectly referred to as the discharge coefficient and this can lead to some confusion. Here this will be referred to as a *flow coefficient* C_q, defined as

$$C_q = Q/a_o\sqrt{\{2(P_u - P_d)/\rho\}} \qquad (2.14)$$

where P_d is an arbitrary downstream pressure.
Equations (2.8) and (2.14) combine to give

$$\frac{C_q}{C_d} = \sqrt{1 + \frac{P_d - P_{vc}}{P_u - P_d}} \qquad (2.15)$$

Pressure recovery in the downstream chamber will result in P_d being greater than P_{vc} and hence in C_q being greater than C_d. Clearly the discharge and flow coefficients will be equal only when the downstream static pressure P_d is the same as the static pressure at the vena contracta. This occurs when the downstream chamber is very large compared to the orifice area. The degree of pressure recovery depends on Reynolds number, orifice geometry and downstream chamber size. It can also be affected by cavitation occurrence in the downstream chamber. Figure 2.2 illustrates the differences that can exist between P_{vc} and P_d for an orifice in a pipe. Three static pressure distribution curves are shown, the flow rate being the same in each case. At the vena contracta the pressure falls to a minimum and recovery takes place downstream. The pattern is the same for upstream pressure (*b*), the flow rate being determined by pressure drop from upstream to vena contracta. However, for upstream pressure (*c*) the pressure at the vena contracta has fallen to the fluid vapour pressure, cavities form, pressure recovery is reduced and the flow coefficient falls. If cavitation is very severe so that the vapour cavities extend as far as the downstream tap, then $P_d = P_{vc}$ and $C_q = C_d$. Hence for such an orifice cavitation affects the flow coefficient. Note however that the discharge coefficient is unchanged.

2.6.3 The flow number

Reynolds number has been defined earlier (equation (2.11)). The mean velocity through the orifice must be known before this number can be evaluated and this is usually determined by dividing the measured flow rate by the orifice area. This is all very well if one is conducting a series of experiments on a particular orifice and measured flow rates are readily available. However, if one wishes to predict the flow coefficient from available information concerning the variation of C_q with Reynolds number, the Reynolds number must be evaluated without measurement. How then is the mean velocity estimated? It can clearly be obtained from the relationship

$$v_o = C_c\sqrt{(2\Delta P/\rho)}$$

but unfortunately this cannot be evaluated unless the contraction coefficient is known, which it is not. The problem is not insoluble however, for if one eliminates the flow rate Q from the discharge equation (2.8) and from the Reynolds number equation (2.11), the following relationship evolves.

$$C_d = \left[\frac{\nu}{D_h\sqrt{(2\Delta P/\rho)}}\right]R = \beta R \tag{2.16}$$

Thus lines of constant β can be superimposed on the C_d results. The designer then evaluates β and finds the discharge coefficient from the intersection of this β line and the C_d curve.

In an attempt to avoid this complication some researchers [13, 14] have presented the variation of C_q with a flow number λ where

$$\lambda = (D_h/v)\sqrt{\{2(P_u - P_d)/\rho\}} \tag{2.17}$$

This number can be evaluated without prior knowledge of C_q and the flow coefficient can then easily be found from graphs of C_q and λ.

2.7. Reattached flow in spool valves

One of the authors [14] has measured the flow coefficient of a two-dimensional spool valve orifice, Fig. 2.11. An unusual phenomenon displayed itself in this apparatus. Instead of the liquid passing through the orifice as a free jet at an angle of 69° to the horizontal, it formed a curved path and reattached to the vertical wall of the downstream chamber. A small reattachment bubble, roughly four times as long as the orifice opening, was formed. Because of the curvature of the jet the pressure in the bubble was less than the pressure measured at the downstream tap. The magnitude of the pressure reduction within the bubble depends on the jet momentum and its radius of curvature, and calculation showed that vapour pressure could be reached in this bubble even at large downstream pressure P_d. Indeed, experiment [15, 16] showed that cavitation occurred in the bubble even with back pressures as high as 100 lbf in^{-2}.

It is necessary at this stage to introduce the cavitation number K,

$$K = \frac{P_d - P_v}{P_u - P_d} \tag{2.18}$$

P_v is the vapour pressure of the liquid. K is a measure of the tendency of the system to cavitate or of the degree of cavitation present. Since at any particular region in a flowing liquid, the interplay of static and dynamic pressures may bring about cavitation in the region, then K must be defined in terms of the static

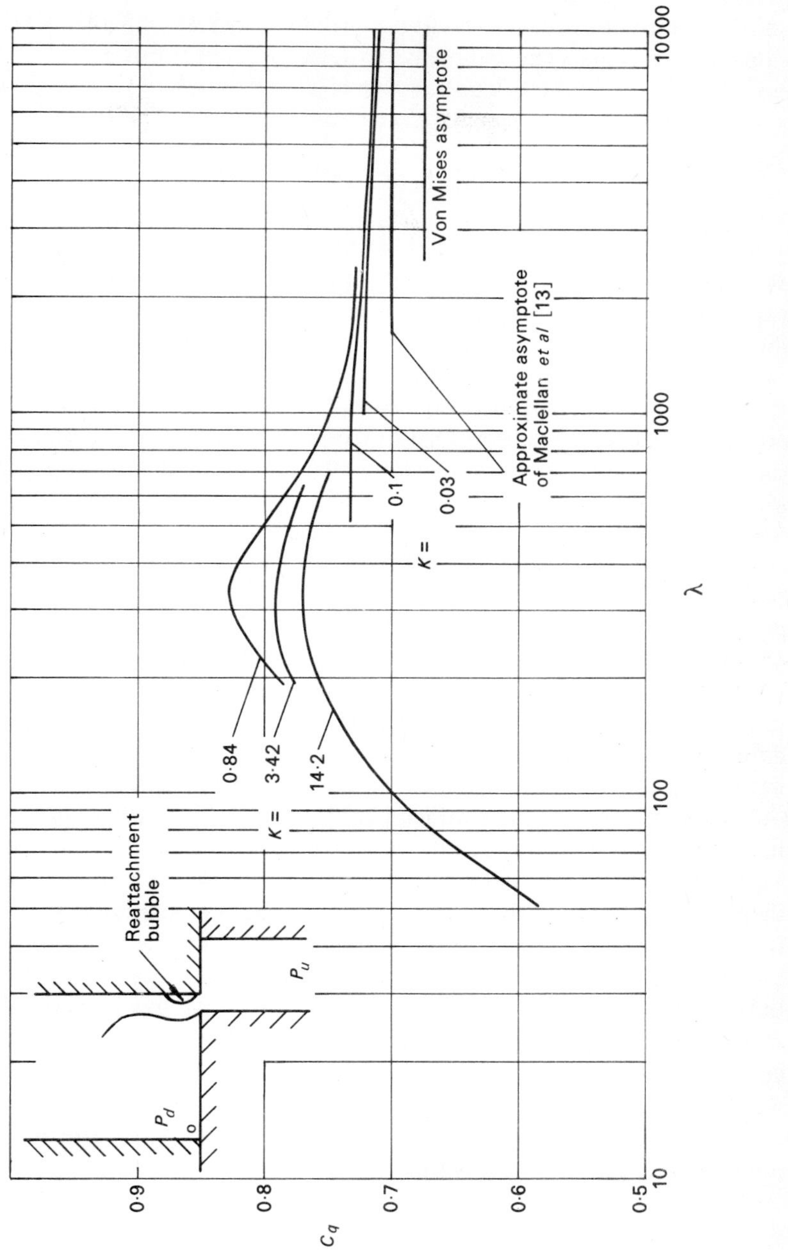

Fig. 2.11. Flow coefficient of two-dimensional spool valve orifice as a function of cavitation number K and flow number λ

pressure, the velocity and the vapour pressure of the fluid. If there is no cavitation and K is gradually reduced, cavitation will first arise when K reaches the critical value K_i, known as the incipient cavitation number. For $K < K_i$ cavitation exists. For $K > K_i$ there is no cavitation. The variation of K_i with flow number λ is shown in Fig. 2.12 where for the two dimensional orifice

$$\text{Flow number } \lambda = \frac{2x\sqrt{\{2(P_u - P_d)/\rho\}}}{\nu}$$

When cavitation occurs in a valve, violent noise is generated as the vapour cavities collapse on entering regions of higher pressure. In one of the authors' experiments a noise pick up was attached to the face of the valve and the resulting oscilloscope traces are shown in Fig. 2.13. At the incipient cavitation condition the noise was emitted in short randomly spaced bursts, while for $K < K_i$ the noise was violent and continuous. The presence of cavitation can lead to erosion of the valve lands or ports and one of the ways to avoid this is to ensure that $K > K_i$. This can be achieved by decreasing the upstream pressure or by increasing the back pressure.

Figure 2.11 shows how the flow coefficient C_q varies with K and λ for reattached flow. It settles out at 0·71 at large λ after reaching a maximum in the range $150 < \lambda < 400$. At large λ the flow coefficient is greater than the value predicted by von Mises for free jet flow (0·673). This is caused by the curved jet and has also been noted by Stenning [17] in his work on pneumatic valves.

When applying these results to a particular valve it is important that the type of flow pattern be known. Is the jet free or does it reattach? Work reported to date on reattached flows has been concerned with two-dimensional devices and models. The reason for this is clear when one appreciates that low pressures can only be maintained in the reattachment bubble by the presence of the front and back plates of the device or model. However, a spool valve is not two-dimensional and although a two-dimensional model may exhibit reattached flow there is no guarantee that a real valve will behave similarly. Fir instance, if there are a number of circular ports in the sleeve, then for the inflow conditions, a number of discrete jets will form and there is no possibility of reattachment bubbles forming under these conditions. However, for flow from spool to sleeve, reattached flow may occur, but this is doubtful since the surface of the port is curved. A reattachment bubble could occur for the inflow condition if the ports were situated in a recess in the sleeve—effectively a rectangular orifice. In this case the fluid forms a continuous sheet after passing into the valve and a pressure differential could exist across this.

Even when the static conditions allow the formation of a reattachment bubble, dynamic conditions may alter this. Foster *et al.* [18] conducted tests on a two-dimensional model spool valve and they found that the flow did not reattach when the spool was oscillated quickly. The reattachment process requires time to establish, this being some 10–30 times the time for a particle to travel the length of the reattachment bubble, depending on the angle between jet and wall. Their results indicate that there is a limiting spool frequency above which re-

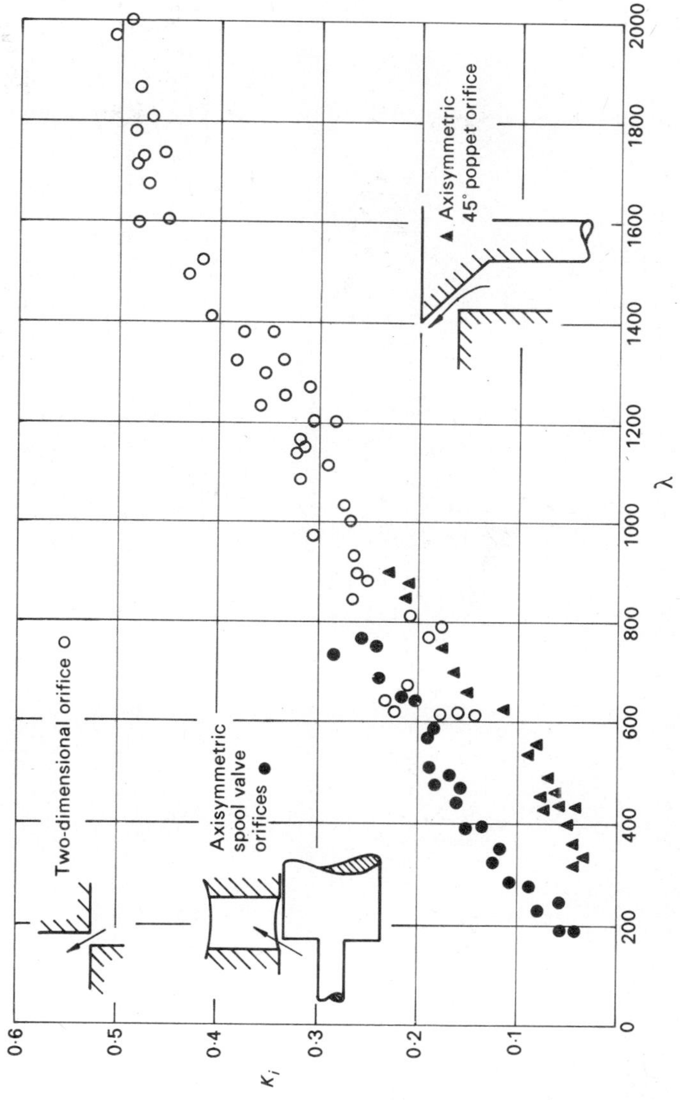

Fig. 2.12. Incipient cavitation numbers

attachment will not occur and that this frequency is a function of orifice opening and pressure drop.

Cavitation in the fluid can also disturb or inhibit reattached flows. There are two types of cavitation to be considered. These are (*a*) cavitation in the mixing region of the turbulent jet and (*b*) cavitation in the reattachment bubble. The existence of the former type was pointed out by Rouse [19]. The latter type has received considerable attention lately by workers in the field of fluidics and results indicate that cavitation in the bubble tends to make it lengthen.

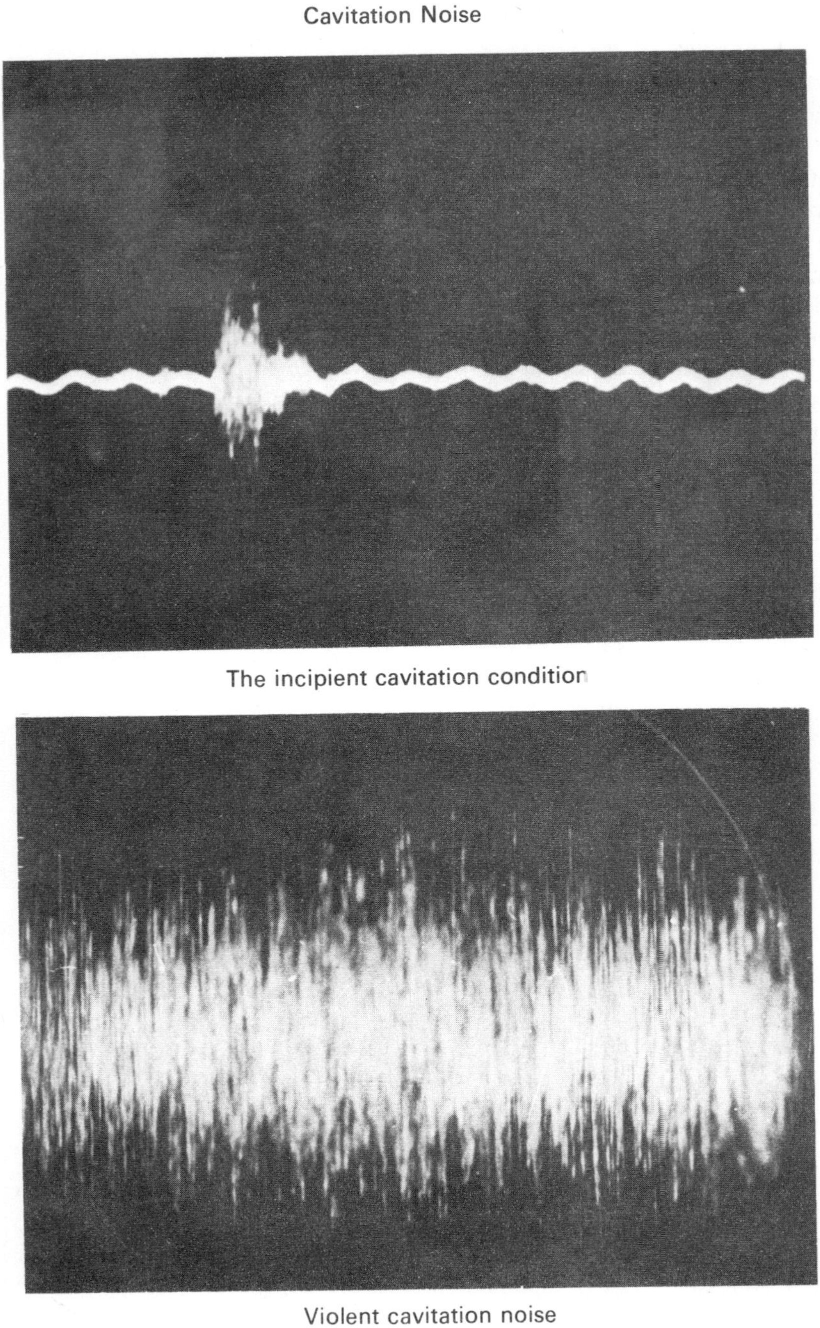

Fig. 2.13. Cavitation noise (*a*) at the incipient cavitation condition (*b*) severe cavitation

2.8. Effects of chamber geometry

In 1956, Turnbull [20] showed photographs of flow patterns in a two-dimensional model valve, Fig. 2.14. His results showed that, for inflow, the flow reattached to the sleeve when the spool displacement x was less than three times the clearance w. When x was greater than $3w$ and also less than $0{\cdot}1a$, the flow reattached to the vertical wall, i.e. the face of the spool piston. Alpay *et al.* [21] have studied clearance effects more closely and showed that as the valve opens the jet breaks away from the sleeve at $x/w = 2{\cdot}32$ and exists as a free jet until $x/w = 3$ when it reattaches to the vertical wall. When the valve is opened even wider a stage is reached when the presence of the horizontal wall, the spool stem, interferes with the establishment of the reattachment bubble. This causes the jet to detach and form a free jet at an angle predicted by von Mises. A hysteresis zone was found to exist. When the valve was opening the reattached jet broke away when $x = 0{\cdot}1a$, but when closing the reverse did not occur until $x = 0{\cdot}075a$.

For the outflow configuration the patterns were:

At small openings and $x < 3w$, the flow reattached to the horizontal land of the spool, Fig. 2.14.

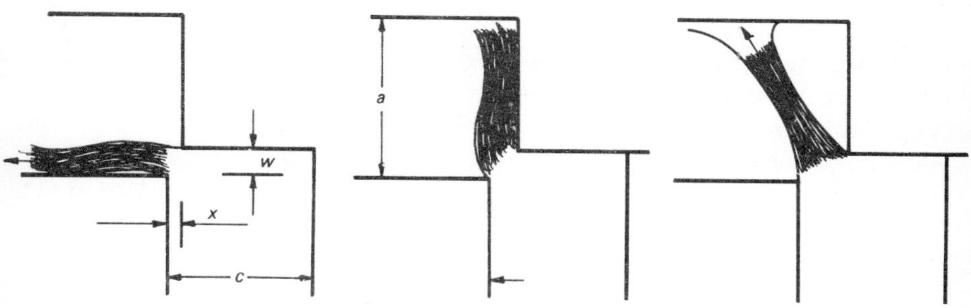

Fig. 2.14 Three possible flow patterns in a spool valve (inflow)

For $x > 3w$ and $x^2 < 0{\cdot}11ac$, the flow reattached to the vertical wall of the cylinder port.

For $x > 3w$ and $x^2 > 0{\cdot}11ac$, the flow existed as a free jet.

In several of these regimes the flow is attached to one wall or the other. This is the Coanda effect which is put to good use nowadays in fluidic elements. This phenomenon relies on the entrainment properties of turbulent jets and it has been pointed out by Glaettli [22] that spontaneous reattachment will not occur if the Reynolds number is less than a critical value. Below this value laminar flow prevails and the speed of reattachment is drastically reduced. It seems reasonable to assume that a similar situation arises in spool valves; there is a critical Reynolds number below which reattachment will not occur spontaneously.

Recent work [23] shows that there is such a critical Reynolds number (≈ 250). It also shows that the flow coefficient for reattached flow is 1·05 times that for jet flow.

2.9. Poppet valves

One of the authors [24] has conducted some tests on a poppet valve. The poppet angle was 45°, the seat was sharp and mineral oil was used at pressures up to 500 lbf in^{-2}. In this case the flow followed the face of the poppet; there was no reattachment bubble and as a result the cavitation number had to be very small before cavitation occurred. This is illustrated in Fig. 2.12. Figure 2.15 is a photograph of cavitation and clearly shows that the flow follows the poppet face. The flow coefficient is defined by equation (2.14):

$$\text{where } a_o = \pi x \sin \alpha \, (D - x \sin \alpha \cos \alpha)$$

D is the poppet seat diameter
x is the poppet lift

$$\lambda = \frac{2x \sin \alpha}{\nu} \sqrt{\frac{2}{\rho}(P_u - P_d)}$$

The variation of flow coefficient with flow number is shown in Fig. 2.16. These results show the same trend as the spool valve results. In this case, because of the large downstream chamber, C_q and C_d are approximately equal. Cavitation effects are slight. The situation is analogous to that described by Fig. 2.2, but in this case pressure recovery is very small since the downstream chamber is large. At the larger flow number $C_q \approx 0{\cdot}77$ which shows reasonable agreement with von Mises value of 0·746. This is not surprising since in the real valve the flow followed the face of the poppet and this was the flow model adopted by von Mises, Fig. 2.7. Work by Stone [25] produced considerable scatter and indicated values of C_q at large λ which were less than the von Mises value. This discrepancy is difficult to explain.

2.10. Flapper nozzle valves

There is little published information on these valves. Attempts to analyse the flow through such orifices indicate that this is a difficult problem, particularly when the edges of the nozzle are not sharp but have a flat. Unfortunately this is usually the case in practice since contact between flapper and nozzle tends to flatten the nozzle. When a flat exists there are two possible flow patterns, Fig. 2.17. In one, the jet flows along the flapper and, in the other, it reattaches to the nozzle tip. Lichtarowicz *et al.* [26] have shown that the flow coefficient can be large when the latter pattern exists, the flapper nozzle gap acting as a diffuser, resulting in good pressure recovery from the vena contracta to the edge of the flapper. Some of their results and those of Feng [27] are shown in Fig. 2.17. The orifice area is taken to be the curtain area πdh and the Reynolds number is $2Q/\pi d\nu$. It might be tentatively deduced from Fig. 2.17 that the flow coefficient is 0·6 if $h/d > 0{\cdot}08$, regardless of the size of the nozzle flat and the flow number. However, at very low flow numbers laminar flow prevails and the coefficient

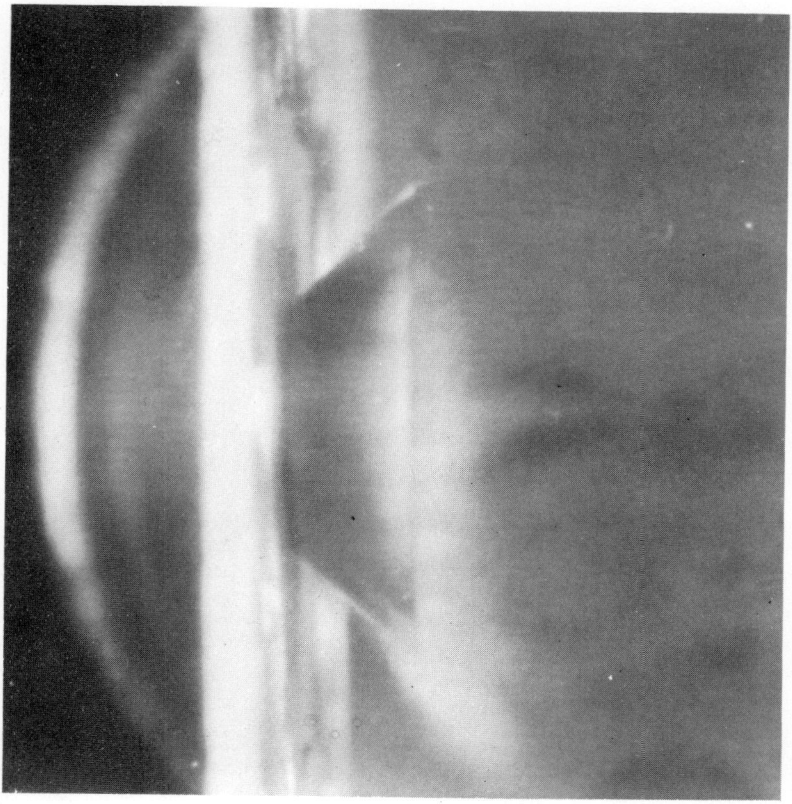

Fig. 2.15. Cavitation in a 45° poppet valve. $K = 0{\cdot}08$, $\lambda = 730$

will be small. Analysis shows that under these conditions the flow rate in the radial gap between flapper and nozzle is

$$Q = \frac{\pi h^3 \Delta P}{6\mu \log (D/d)} \tag{2.19}$$

Equating this to

$$Q = C_q \pi d h \sqrt{\left(\frac{2\Delta P}{\rho}\right)} \tag{2.20}$$

gives

$$C_q = \frac{h^2}{12 d \nu \log (D/d)} \sqrt{\left(\frac{2\Delta P}{\rho}\right)} \tag{2.21}$$

and if Reynolds number is

$$R = \frac{2h\sqrt{(2\Delta P/\rho)}}{\nu}$$

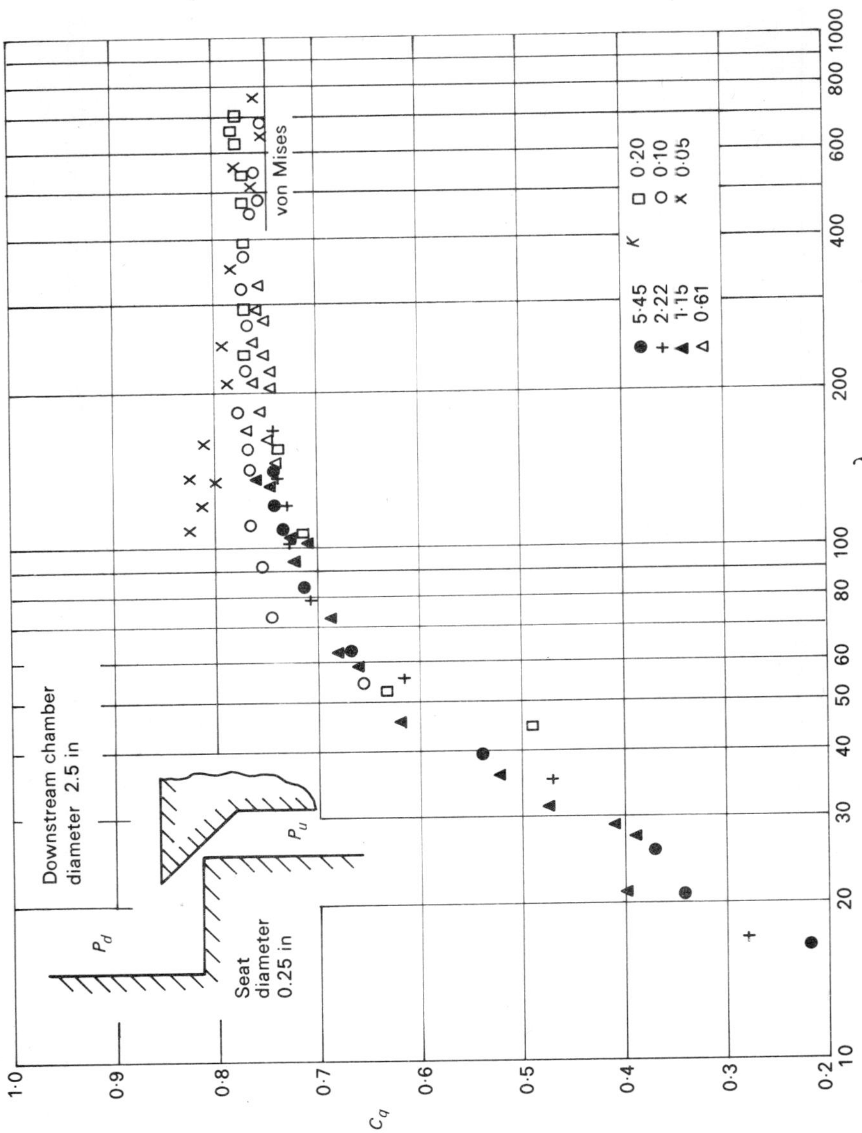

Fig. 2.16. Measured flow coefficient of a 45° poppet valve

then

$$C_q = \frac{(h/d)\,R}{24 \log (D/d)}$$

This formula does not apply if the nozzle is sharp ($D/d = 1$). In this case the von Mises predicted value of 0·65 applies over a wide range of Reynolds numbers and changes in h/d should not result in great changes in C_c since there is now no diffuser effect.

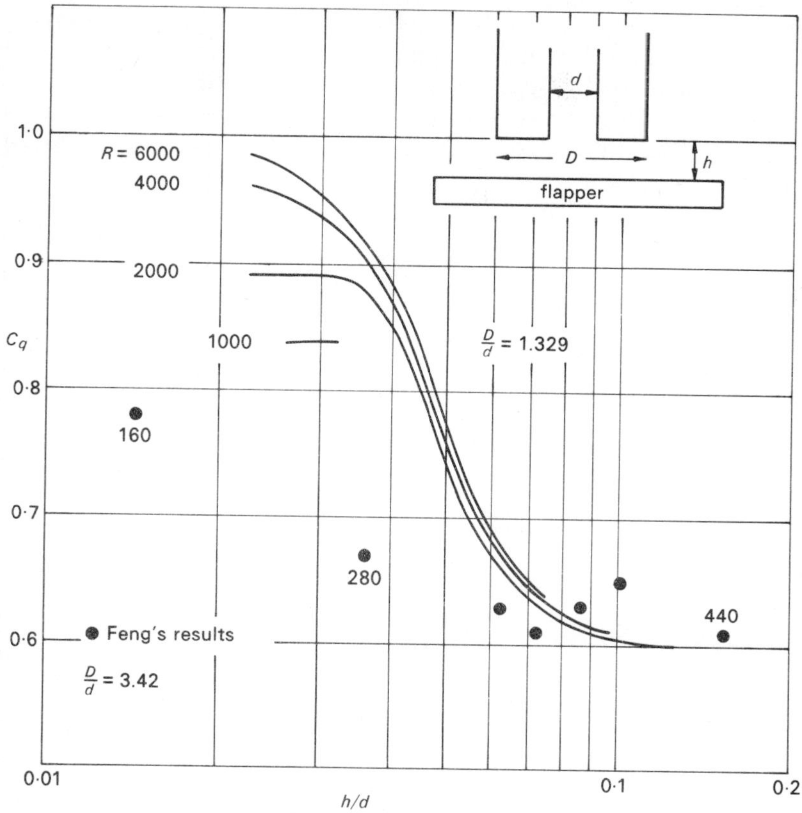

Fig. 2.17. Flow coefficient of a flapper nozzle (after Lichtarowicz and Markland [26] and Feng [27])

2.11. Short tube orifices

An excellent discussion on the discharge characteristics for incompressible non-cavitating flow through this type of orifice can be found in [28]. Shearer [29] points out that this type of orifice, Fig. 2.1(*e*), is usually employed in bridge type hydraulic circuits where the flow rates are very small and where it would be difficult to justify the expensive manufacture of small sharp edged orifices. Again with this type of orifice there is difficulty in deciding whether the measured coefficient is a discharge coefficient or a flow coefficient. In the longer orifices the vena contracta occurs within the tube and use of the pressure drop across the tube gives a flow coefficient. For the short orifices the vena contracta occurs outside the tube and the pressure drop across the tube (with the jet discharging to atmosphere) gives a discharge coefficient. For the sake of generality we will refer to the measured values as flow coefficients. The Reynolds number is defined in this case as

$$R = Qd/a_o v$$

where Q/a_o is the mean velocity through the orifice of diameter d. The flow coefficient is defined, as before, in equation (2.12). Figure 2.18 shows the effects of Reynolds number and of orifice length. The author's conclusions were:

1. Avoid l/d less than 1·5 since C_q varies rapidly with l/d below this value,
2. For $2 < l/d < 10$, C_q is constant for $R_l > 2 \times 10^4$ and is given by $C_{qu} = 0{\cdot}827 - 0{\cdot}0085\ l/d$
3. For $10 < R_l < 2 \times 10^4$, C_q is given by

$$\frac{1}{C_q} = \frac{1}{C_{qu}} + \frac{20(1 + 2{\cdot}25\ l/d)}{R_l} - \frac{0{\cdot}005\ l/d}{1 + 7{\cdot}5(\log 0{\cdot}00015R_l)^2} \tag{2.22.}$$

where

$$R_l = R\frac{\sqrt{(1 - m^2)}}{C_q}$$

and

$$m = a_o/a_1 \quad \text{(See Fig. 2.18)}$$

and a_1 is the upstream pipe area.

2.12. Gas flow through orifices

The orifice equation (2.8) was derived using the assumption that the fluid was incompressible. This is a reasonable assumption for the conventional hydraulic fluids. However, many modern control systems employ gases, particularly air, as the working fluid and in these cases compressibility is an important factor. There are many excellent works dealing with the flow of gases through orifices and so it is unnecessary in the present book to go into compressible fluid dynamics. Suffice it to give the relevant formulae which have been shown to predict mass flow rates fairly accurately. Mass rather than volume flow rate is generally used, since in the case of compressible flow, volume and pressure are interdependent. The mass rate of flow W can be written

$$W = C_d C_m a_o P_u/\sqrt{T_u} \tag{2.23}$$

C_d is the discharge coefficient, a_o the orifice area, P_u the upstream stagnation pressure (absolute), and T_u is the upstream stagnation temperature (absolute). C_m, the mass flow parameter is

$$C_m = \sqrt{\left[\frac{2\gamma}{R(\gamma - 1)}\left\{\left(\frac{P_d}{P_u}\right)^{2/\gamma} - \left(\frac{P_d}{P_u}\right)^{(\gamma+1)/\gamma}\right\}\right]} \tag{2.24}$$

Here P_d is the downstream static pressure (absolute), R is the gas constant (see Appendix I) and γ is the ratio of the specific heats (1·4 for air). Equation (2.24) only applies when

$$\frac{P_d}{P_u} > \left(\frac{2}{\gamma + 1}\right)^{\gamma/(\gamma - 1)} = 0{\cdot}528 \text{ for air}$$

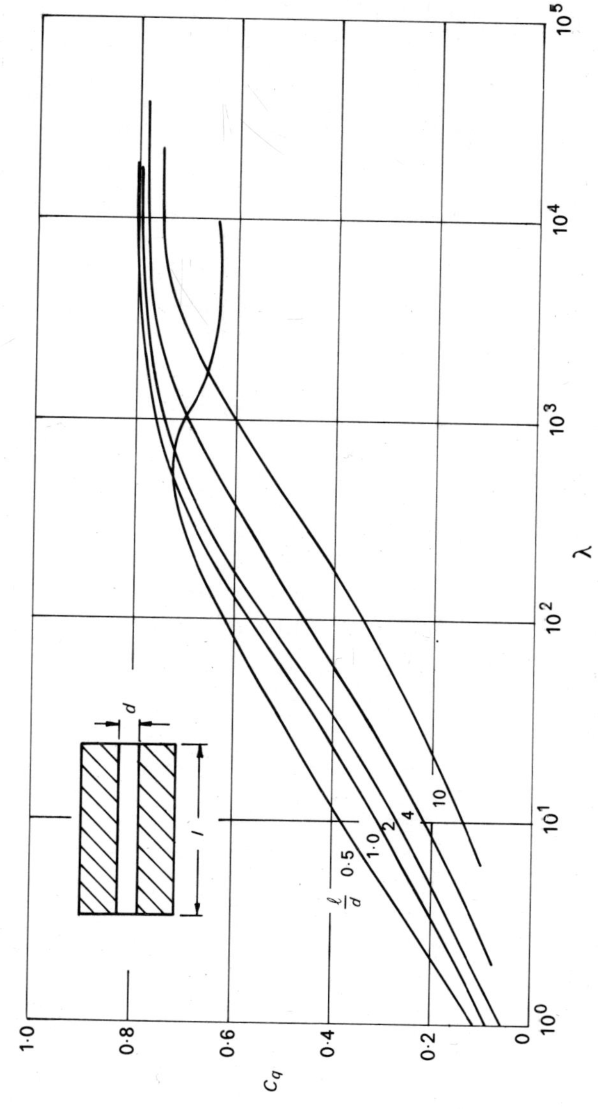

Fig. 2.18. Flow coefficient of tube orifices (after Lichtarowicz *et al.* [28])

When P_d/P_u is less than this value, the flow is choked and C_m becomes

$$C_m = \left(\frac{2}{\gamma + 1}\right)^{1/(\gamma - 1)} \sqrt{\frac{2\gamma}{R(\gamma + 1)}} = 0{\cdot}0166 \text{ for air in the pdl, lb, ft, s, } {}^\circ R \text{ system}$$

$$= 0{\cdot}0405 \text{ for air in the SI system.}$$

The variation of C_m with pressure ratio is shown in Fig. 2.19. An example clarifies the use of equation (2.23). Consider a circular orifice of diameter 0·1 in discharging air from a reservoir charged at 60 lbf in^{-2} gauge to atmosphere. The temperature of air in the reservoir is 70°F. What is the mass flow rate?

The pressure ratio $P_d/P_u = 14{\cdot}7/74{\cdot}7 = 0{\cdot}197$ and Fig. 2.19 shows that the flow is choked and $C_m = 0{\cdot}0166$. If the discharge coefficient is assumed to be 0·8, equation (2.23) gives

$$W = \frac{0{\cdot}8 \times 0{\cdot}0166 \times \pi \times 0{\cdot}1^2 \times 74{\cdot}7 \times 32{\cdot}17}{4\sqrt{530}} = 0{\cdot}0109 \text{ lb s}^{-1}$$

Note than an absolute system of units is adhered to, with mass in pounds and force in poundals.
The volumetric flow rate of free air is

$$Q = W/\rho = \frac{0{\cdot}0109 \text{ lb s}^{-1}}{0{\cdot}075 \text{ lb ft}^{-3}}$$

$$= 0{\cdot}145 \text{ ft}^3 \text{ s}^{-1}$$
$$= 8{\cdot}7 \text{ ft}^3 \text{ min}^{-1}$$

Usually when a gas is flowing through an orifice the Reynolds number is sufficiently large to justify the neglect of Reynolds number effects. This is mainly due to the low kinematic viscosity of air. As a result the main variable of concern

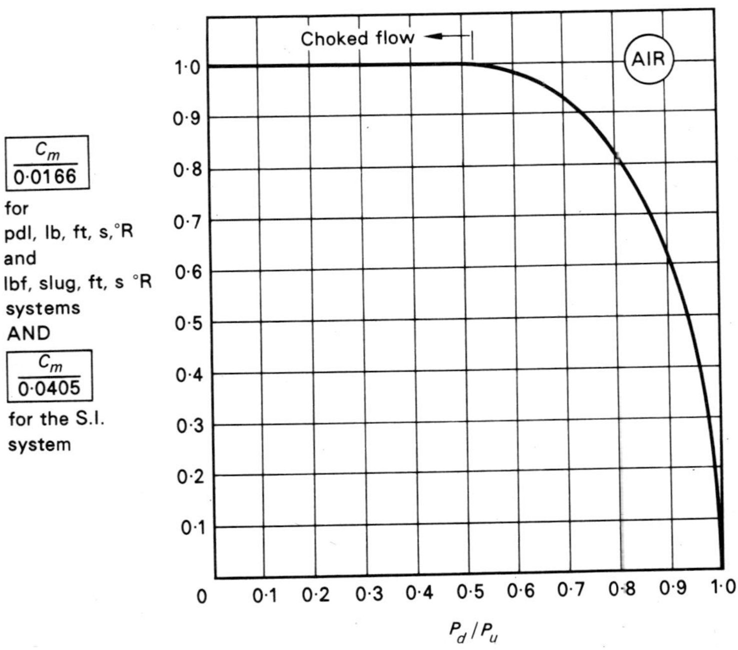

Fig. 2.19. The mass flow parameter for air

is the pressure ratio P_d/P_u and its effects on the discharge coefficient are of interest. Perry [30] has measured C_d for air flow through a sharp edged circular orifice and his results are shown in Fig. 2.20. The coefficient is larger than that for incompressible flow. As the pressure drop reduces ($P_d/P_u \to 1$) compressibility effects become less predominant and C_d tends to the value predicted by von Mises for the incompressible fluid. It should be noted however that at very small pressure drops the flow may be laminar, in which case the C_d will fall off quickly.

Stenning [17] has investigated 'two-dimensional' gas flow through spool valve orifices and he found that the discharge coefficient was greater for reattached flow than for jet flow. Both flow patterns were obtainable when $P_d/P_u > 0{\cdot}3$–$0{\cdot}4$, but the free jet became unstable for pressure ratios less than this value and the jet then reattached.

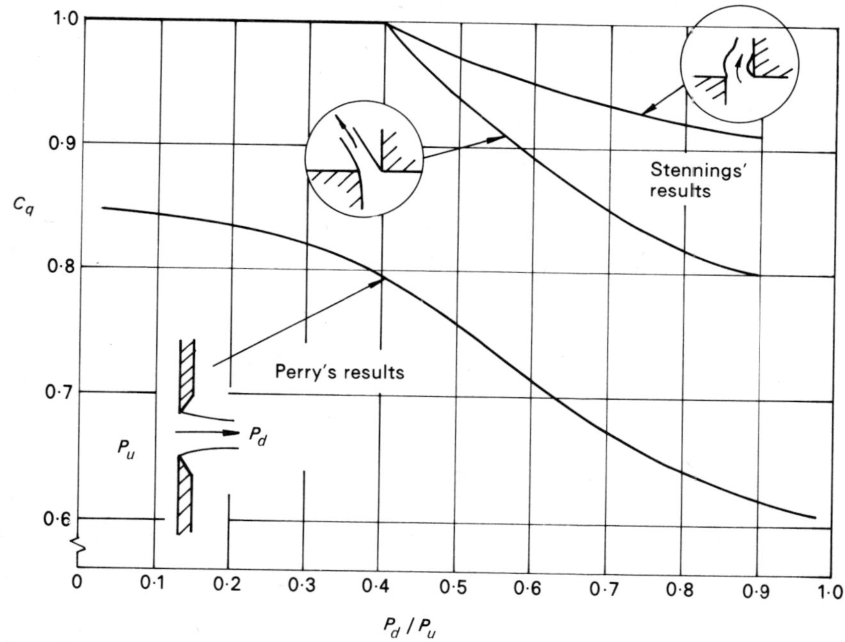

Fig. 2.20. Flow coefficient of pneumatic orifices (after Stenning [17] and Perry [30])

There is very little published data on gas flow through poppet valves. Tsai and Cassidy [31] have measured the characteristics of a 45° poppet. Unfortunately this did not have a sharp seat, so that at small lifts its behaviour is not strictly comparable with the incompressible result of Fig. 2.16. However, the results in Fig. 2.21 do again follow the same trend as those for sharp edged orifices and spools. As the pressure ratio P_d/P_u decreases, the flow coefficient increases. At large lift and pressure ratios near unity (incompressible) the flow coefficient agrees with that discussed earlier for oil flow. At small lifts, the

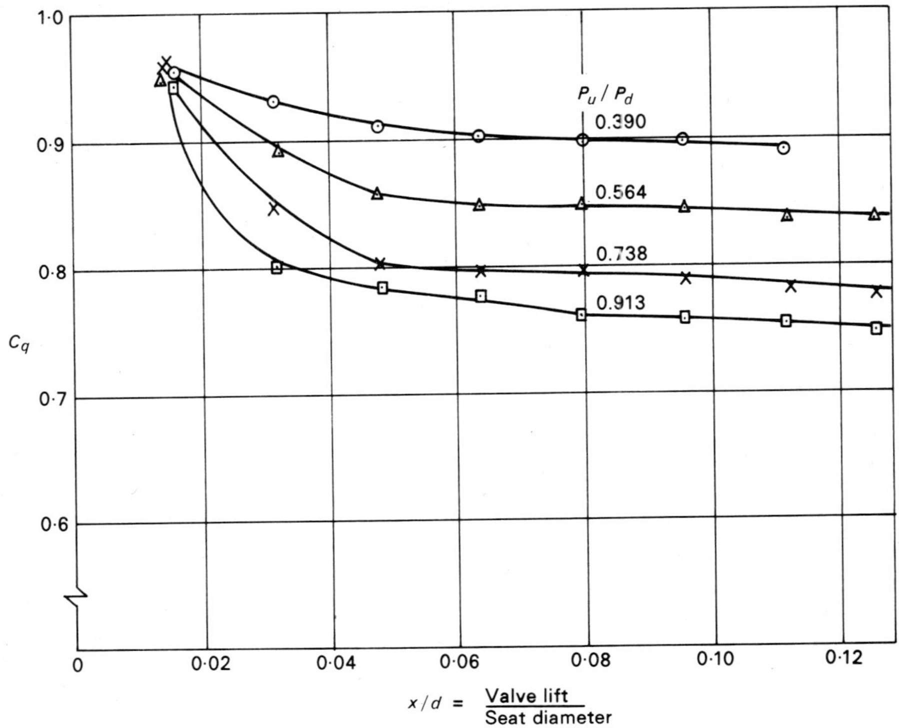

Fig. 2.21. Flow coefficient of pneumatic poppet valve (after Tsai and Cassidy [31])

finite seat area acts as a diffuser resulting in increased flow coefficients. For very small lifts the flow will be laminar and C_q will fall off.

Compressible flow in flapper nozzle valves has not yet been adequately dealt with in the literature. However, the same trend can be expected as for the other orifices discussed above, i.e. C_q should be larger than for oil flow, Fig. 2.8, and the difference should be largest at small pressure ratios. Flow coefficients of unity or more can then be expected at small pressure ratios and at small flapper lifts when the nozzle has a flat [32].

Further information on compressible flow can be found in [33].

Chapter 3

Directional control valves

3.1. Introduction

The direction of a hydraulic motor or cylinder can be changed by reversing the direction of the pump output, but in pneumatic systems directional control is rarely attempted by reversing the direction of the compressor output. The most common means by far for directional control is the use of directional control valves. These consist basically of two or more flow ports so arranged and connected that some external signal can direct the fluid flow to various places in the fluid power system. Such valves have long been classified by the number of flow ports they are intended to control. These are called the 'ways' of the valve; hence the names two-way, three-way and four-way valves.

3.2. Two-way valve

Figure 3.1(*a*) shows a two-way valve with a moving spool as the control element. This is used to vary the flow rate from inlet to outlet, in response to an external signal. A full description of this valve calls it a two-port, two-way spool valve. The number of ports needs to be specified since (see later), in some cases, the number of ports is not equal to the number of ways. In Fig. 3.1(*a*), the spool is mechanically operated and spring returned and a complete description of the valve also needs reference to the means of operation. There are many such means, e.g. roller, solenoid, pilot air, etc.

In addition the particular valve shown is normally closed, i.e. it is closed when there is no applied signal. The number of possible operating positions of the spool must also be specified. Most fluid power valves are used in sequential 'on-off' circuits, in which the flow to or from a cylinder is either on or off. In these cases the valve need only have two (sometimes three) operating positions. A complete description of the valve in Fig. 3.1(*a*) is then—mechanically operated, spring return, normally closed, two-position, two-port, two-way spool valve. Fortunately there is a simple symbol for the valve, Fig. 3.1(*b*). Here the basic envelope is divided into two squares, representing the two positions of the valve. When actuated, the envelope moves to the right relative to the two external connections or ports.

The two-way valves can control a single acting cylinder, Fig. 3.1(*c*). One is used to pressurise and the other to exhaust. In this example it is clear that the valve should offer as little resistance to flow as possible. It is used to change the direction of the flow, not to throttle it. There are, however, cases where on-off or two position control is not good enough, and proportional control must be used. In these cases the valve needs theoretically an infinite number of operating positions, so that the flow rate can assume any value between off and fully on. Throttling by means of variable orifices is needed and the contents of the preceding chapter are relevant. Valves capable of infinite positioning between

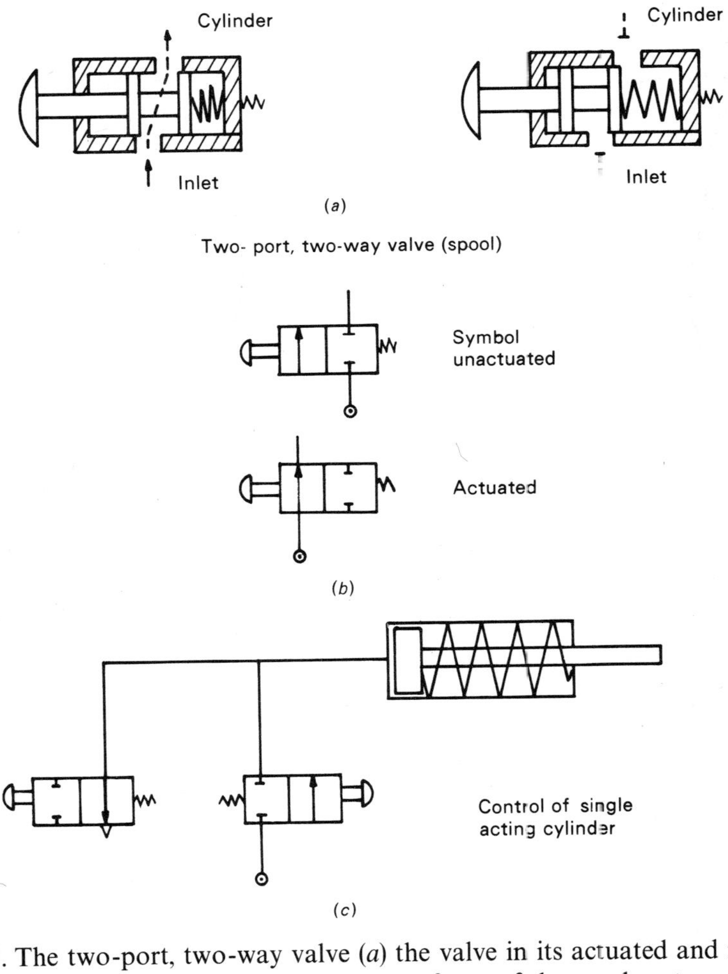

Fig. 3.1. The two-port, two-way valve (*a*) the valve in its actuated and normal positions (*b*) the C.E.T.O.P. symbol (*c*) use of two of these valves to control a single acting cylinder

certain limits are symbolised by the addition of horizontal bars drawn parallel to the envelope.

3.3. Three-way valves

The next on the scale of usefulness is the three-way valve in which there are three functional flow ports. It is used to alternately pressurise and exhaust the working port or cylinder port. Figure 3.2(*a*) shows a mechanically operated,

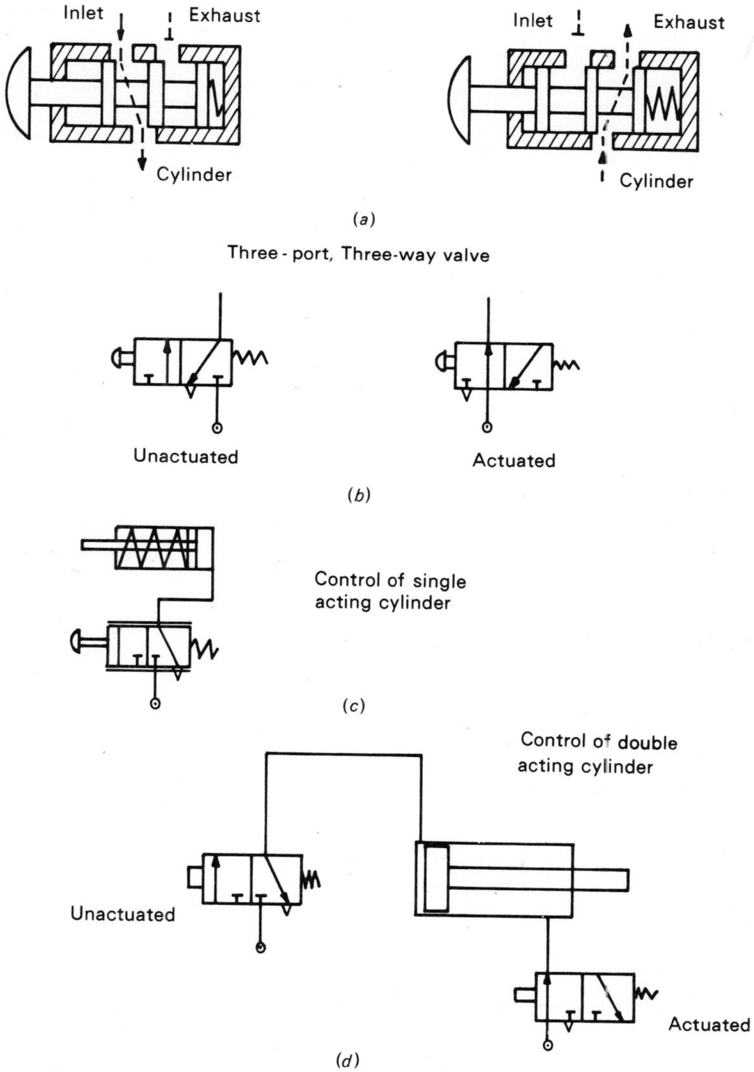

Fig. 3.2. The three-port, three-way valve (*a*) the valve in its actuated and normal positions (*b*) the C.E.T.O.P. symbol (*c*) use of one of these valves to control a single acting cylinder (*d*) use of two valves to control a double acting cylinder

spring return, two-position, three-way, three-port, spool valve. Figure 3.2(*b*) shows the C.E.T.O.P.* symbol. Figure 3.2(*c*) shows how a proportional three-way valve can control the speed of a single acting cylinder. A two-position valve can be used in a similar fashion to drive the cylinder from one extreme position to the other. If the load needs to be stopped at intermediate positions, then a three-position three-way valve is needed. In the third position the fluid is trapped in the cylinder when the load has reached the desired position. A three-way three-position valve and its symbol (air operation) are shown in Fig. 3.3. There

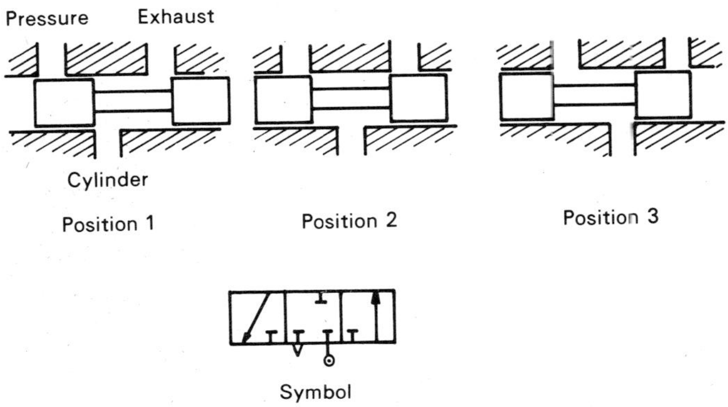

Fig. 3.3. Different mid positions of a three-way valve

are other types of mid-positions but these will be discussed later. Double acting cylinders can be controlled with two three-way valves so arranged that when one valve pressurises one end of the cylinder, the other valve exhausts the other end and vice versa, Fig. 3.2(*d*).

3.4. Four-way valves

The four-way valve combines the functions of two three-way valves. It alternately pressurises or exhausts two working or cylinder ports. The valve shown in Fig. 3.4(*a*) has five ports and is described as a five-port, four-way valve. In some four-way valves, for example, the plate valve and the rotary valve of Fig. 3.5(*a*) and 3.5(*b*), there are only four ports. Only one four-way valve is needed to control a double acting cylinder, Fig. 3.4 (*b*).

As a four-way valve shifts between its extreme positions it passes through the very important neutral position. If the spool can be held in this position for any length of time, the valve is a three-position valve. As with the three-way valve, there are many different possible connections between ports in the mid position. Three such will be discussed here:

1. In the open centre valve, Fig. 3.6(*a*), all lines are interconnected in mid position. Hence in hydraulic systems flow is permitted back to tank at

* Comité European des Transmissions Oleohydrauliques et Pneumatiques

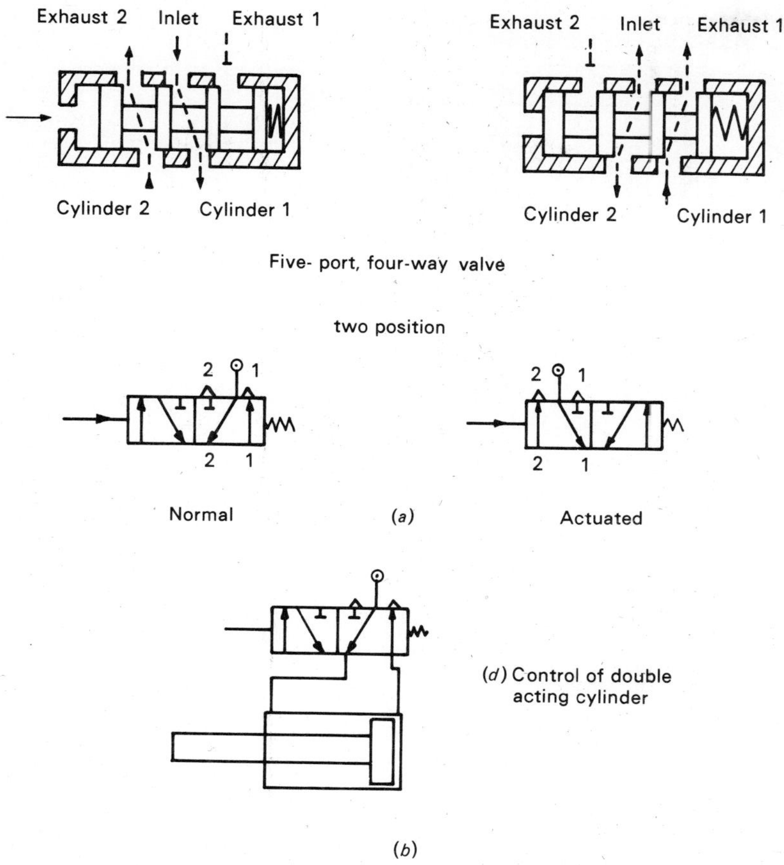

Fig. 3.4. The five-port, four-way valve (*a*) in its actuated and normal positions and the C.E.T.O.P. symbol (*b*) use of the valve to control a double acting cylinder

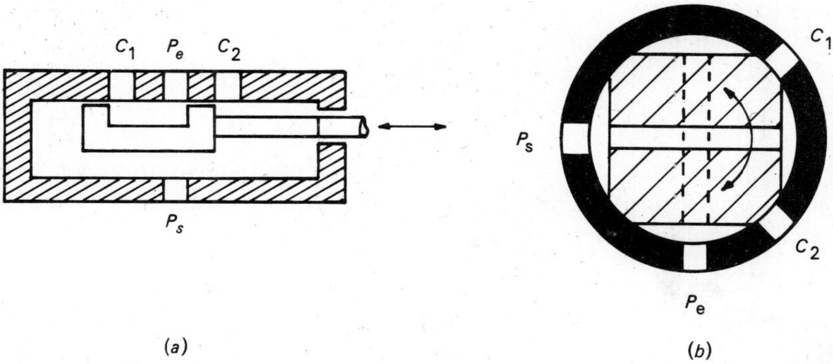

Fig. 3.5. Four-port, four-way valves (*a*) the plate valve (*b*) the rotary valve

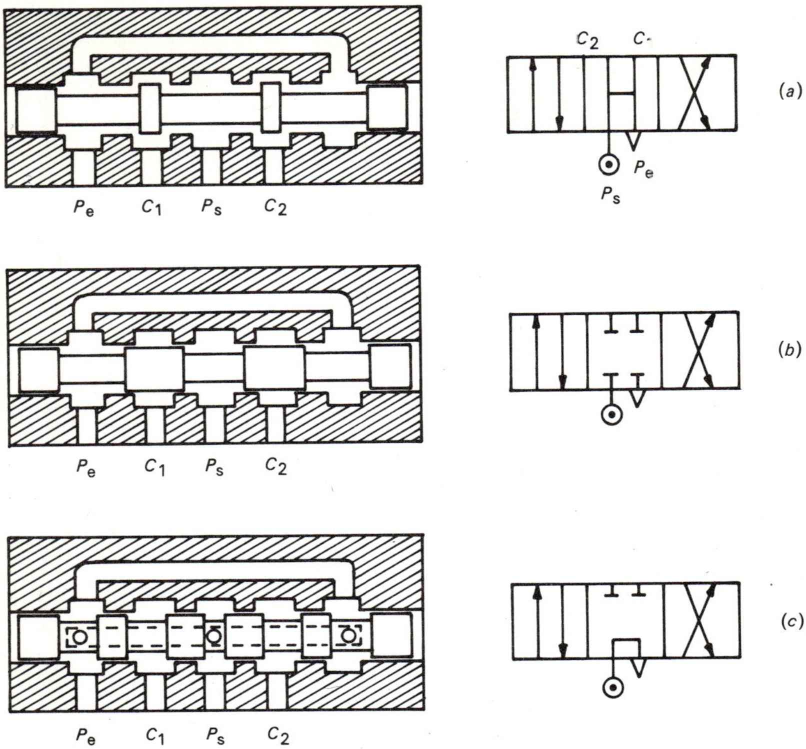

Fig. .3.6. Different mid positions for four-way valves (*a*) open centre (*b*) closed centre (*c*) tandem centre

relatively low pressure, thus eliminating the heat generation associated with closed centre valves and constant delivery pumps.

2. In the closed centre valve, all lines are closed off in the mid position, Fig. 3.6(*b*). As discussed before this enables a cylinder to be held at an intermediate position. There are other reasons for using this type of valve. For instance, if a hydraulic. circuit employs an accumulator, then a closed centre valve will prevent loss of accumulator fluid while the spool moves from one extreme to the other.
3. The tandem centre valve, Fig. 3.6(*c*), combines the advantages of closed and open centre valves. In this case, although the pump is connected to tank in mid position, minimising heat generation, the cylinder ports are blocked holding the load rigidly in position.

3.5. Losses in two- and three-position valves—valve ratings

In chapter 2 data were presented for the calculation of the pressure drop required to provide a specified flow rate through an orifice. This is valuable information for proportional valves in which the main resistance to flow is provided by the orifices, but in two-position valves, which are basically non throttling, orifice resistance does not dominate. In these cases it is a difficult task to determine theoretically the losses in the valve and recourse must be made to experimental determination or to manufacturers' data. Most manufacturers specify the valve flow rating, which is the flow passed by the valve at a given pressure drop. Flow rate at other pressure drops can be calculated using the assumption that flow rate varies with the square root of the pressure drop. For example, a valve with a rating of 10 gal min^{-1} at 60 lbf in^{-2} will give rise to a pressure drop of 86 lbf in^{-2} when passing 12 gal min^{-1}. Valves can also be described in terms of a loss coefficient (chapter 2), giving pressure drop $\Delta P = K\frac{1}{2}\rho v^2$, where v is some defined velocity in the valve.

Most manufacturers of pneumatic valves rate their valves using the K factor or the C_v factor. Weight flow rate can then be calculated for any given pressure drop and pressure ratio. In some cases measured flow capacity charts are published.

3.6. Infinite position valves—Servovalves

When accurate control of position is needed, a servovalve is essential. The two- and three-position valves discussed above find their application in what are essentially open loop systems. In a closed loop system the valve must respond not only to input changes but also to output changes. This is the principle of feedback (as discussed in chapter 1). Most automatic control systems are proportional, in the sense that the controller output is proportional to controller input and experience has shown that such systems can be very accurate. If the controller uses fluid power then the valve must be proportional in its action and infinite position valves are used. Because of their association with servo systems, these valves are called servovalves. Most of these control power by throttling the flow through variable orifices and the data of chapter 2 are relevant when calculating servovalve performance. In what follows the valve supply pressure will be assumed constant, i.e. the constant pressure system. Some systems use constant flow rather than constant pressure, but these will not be considered here.

3.7. Two-way servovalves

A two-way proportional valve is a variable orifice. If P_u is the absolute upstream pressure P_d the absolute downstream pressure, ρ the mass density of the fluid and a the orifice area, then for liquid flow the flow rate is

$$Q = C_q a \sqrt{\{2(P_u - P_d)/\rho\}} \tag{3.1}$$

Here the flow is assumed to be turbulent, although at small pressure drops and areas it may well be laminar.

Defining a reference flow rate Q_r as

$$Q_r = C_{qr} a_r \sqrt{(2P_r/\rho)}$$

leads to

$$\bar{Q} = \bar{a}\sqrt{(p_u - p_d)} = \bar{a}\sqrt{p_u}\sqrt{(1 - p)} \tag{3.2}$$

where

$$p_u = P_u/P_r, \bar{Q} = Q/Q_r, \bar{a} = (C_q a)/(C_{qr} a_r), p = P_d/P_u$$

For example P_r could be the supply pressure.
Since the density is constant the normalised mass flow rate is

$$\overline{W} = \bar{Q}, \text{ with } W_r = \rho Q_r$$

In the case of gas flow [33] the mass flow rate is for $p > p_{crit} = 0{\cdot}528$ for air,

$$W = C_q a \frac{P_u}{\sqrt{T_u}} \left\{ \frac{2\gamma}{R(\gamma - 1)} (p^{2/\gamma} - p^{(\gamma+1)/\gamma}) \right\}^{1/2} \tag{3.3}$$

For $p < p_{crit}$, the flow is choked and

$$W = C_q a \frac{P_u}{\sqrt{T_u}} \left\{ \frac{\gamma}{R} \left(\frac{2}{\gamma + 1} \right)^{(\gamma+1)/(\gamma-1)} \right\}^{1/2} \tag{3.4}$$

Define

$$W_r = C_{qr} a_r \frac{P_r}{\sqrt{T_r}} \left\{ \frac{\gamma}{R} \left(\frac{2}{\gamma + 1} \right)^{(\gamma+1)/(\gamma-1)} \right\}^{1/2} \tag{3.5}$$

$$= 0{\cdot}0166\, C_{qr} a_r P_r/\sqrt{T_r} \text{ for air}$$

using the pdl, lb, ft, s, °R system, or

$$W_r = 0{\cdot}0405\, C_{qr} a_r P_r/\sqrt{T_r} \text{ for air}$$

using the SI system.
Hence the normalised mass flow rate is

$$\overline{W} = \frac{\bar{a} p_u f(p)}{\sqrt{\overline{T}_u}} \tag{3.6}$$

Here $\overline{T}_u = T_u/T_r$ and the function $f(p)$ is unity for $p < p_{crit}$ and

$$f(p) \doteq \left\{ \frac{p^{2/\gamma} - p^{(\gamma+1)/\gamma}}{\left(\frac{\gamma - 1}{2}\right)\left(\frac{2}{\gamma + 1}\right)^{(\gamma+1)/(\gamma-1)}} \right\}^{1/2} \text{ for } p > p_{crit} \tag{3.7}$$

Equations (3.2) and (3.6) and Fig. 3.7 enable the designer to calculate the mass flow rate in hydraulic and pneumatic valves. The pressure flow curve of Fig. 3.7 also enables the designer to decide which size of valve he needs to meet a specific load requirement.

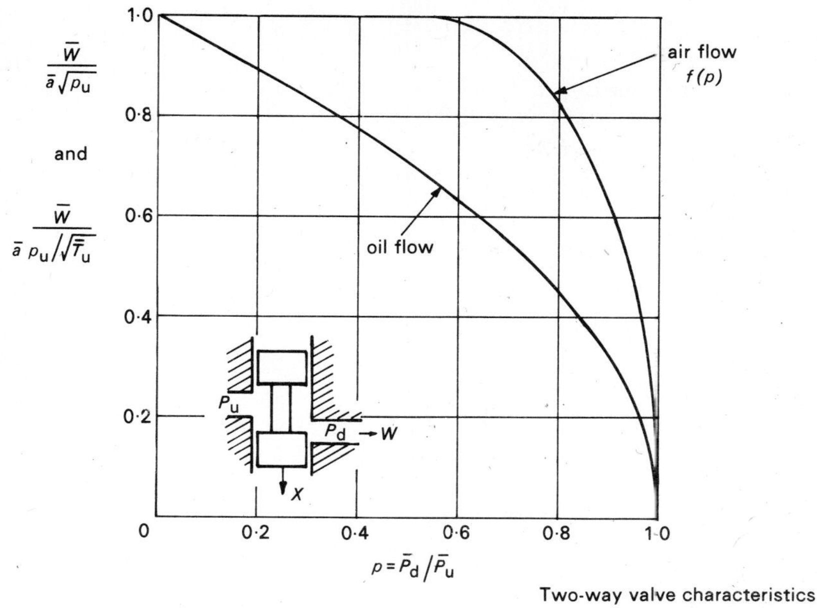

Fig. 3.7. Two-way valve characteristic

Example
Oil flow:

$$\text{If } P_r = P_u = 1000 \text{ lbf in}^{-2} \text{ a}$$
$$a_r = 0{\cdot}01 \text{ in}^2 = \text{a}$$
$$\rho = 62 \text{ lb ft}^{-3}$$
$$C_{qr} = C_q = 0{\cdot}7$$

Then

$$Q_r = \frac{0{\cdot}7 \times 0{\cdot}1}{144}\sqrt{\left(\frac{2 \times 1000 \times 32{\cdot}17 \times 144}{62}\right)} \text{ ft}^3 \text{ s}^{-1}$$
$$= 0{\cdot}188 \text{ ft}^3 \text{ s}^{-1}$$

If the downstream pressure $P_d = 500 \text{ lbf in}^{-2}$ a
Then $p = 500/1000 = 0{\cdot}5$
and since $\bar{a} = 1$ and $p_u = 1$, equation (3.2) or Fig. 3.7 give

$$\bar{Q} = 0{\cdot}707$$

Hence $Q = 0{\cdot}707 \times 0{\cdot}188 = 0{\cdot}133 \text{ ft}^3 \text{ s}^{-1}$
and $W = \rho Q = 62 \times 0{\cdot}133 = 8{\cdot}25 \quad \text{lb s}^{-1}$

Example
Air flow:

$$P_r = P_u = 1 \text{ MNm}^{-2}$$
$$a_r = a = 1 \text{ mm}^2$$
$$C_q = C_{qr} = 0{\cdot}9$$
$$T_r = T = 300°K$$
$$W_r = 0{\cdot}0405 \times 0{\cdot}9 \times 10^{-6} \times 10^6/\sqrt{300}$$
$$= 0{\cdot}002105 \text{ kg s}^{-1}$$

If the downstream pressure is 0·4 MNm^{-2}
Then $p = 0{\cdot}4$
From Fig. 3.7, $f(p) = 1{\cdot}0$, the orifice is choked
Now $\bar{a} = 1$, $p_u = 1$, $\bar{T}_u = 1$
$\bar{W} = f(p) = 1$
and $W = W_r = 0{\cdot}002105$ kg s^{-1}

3.7.1 Maximum hydraulic power transfer

The hydraulic power H delivered by an orifice is the product of the volumetric flow rate and the delivered pressure

$$H = QP_d \tag{3.8}$$

The non-dimensional power is $\bar{H} = \bar{Q}pp_u$ (3.9)

$$\text{where } H_r = Q_r P_r$$

Substituting equation (3.2) in equation (3.9) gives

$$\bar{H} = \bar{a}p_u^{3/2}p\sqrt{(1 - p)} \tag{3.10}$$

Differentiation of equation (3.8) with respect to p shows that maximum power is transferred when $p = 2/3$.

The maximum power is

$$\bar{H}_{max} = 0{\cdot}385\bar{a}_{max}\,(p_u^{3/2})_{max}$$

In the example given above

$$\begin{aligned} H_r = Q_r P_r &= 0{\cdot}188 \times 1000 \times 144 \\ &= 27\,072 \text{ lbf ft s}^{-1} \\ &= 49{\cdot}2 \text{ hp} \\ \bar{H}_{\text{max}} &= 0{\cdot}385 \\ H_{\text{max}} &= 0{\cdot}385 \times 49{\cdot}2 = 19{\cdot}0 \text{ hp} \end{aligned}$$

3.8. Linearisation of valve characteristics

Equations (3.2) and (3.7) are non-linear in p and dynamic analysis of systems employing these valves accordingly involves the solution of non-linear differential equations. This usually requires the use of computers since the equations, except in the simplest of cases, are analytically intractable. Fortunately, experience has shown that in fluid power servos, performance can be fairly accurately predicted using a linearised model, in which changes in the variables are assumed to be small. For example, consider equation (3.2),

$$\bar{Q} = \bar{a}\sqrt{p_u}\sqrt{(1 - p)}$$

This can be expanded as a Taylor's series about some initial point i

$$\bar{Q} = \bar{Q}_i + \left[\frac{\partial \bar{Q}}{\partial \bar{a}}\right]_i \delta\bar{a} + \left[\frac{\partial \bar{Q}}{\partial p_u}\right]_i \delta p_u + \left[\frac{\partial \bar{Q}}{\partial p_d}\right]_i \delta p_d \tag{3.11}$$

+ higher order terms involving $(\delta\bar{a})^2$, $(\delta p_u)^2$, etc.
If small changes are assumed then the higher order terms can be neglected. Hence

$$\delta\bar{Q} = \bar{Q} - \bar{Q}_i = \left[\frac{\partial\bar{Q}}{\partial\bar{a}}\right]_i \delta\bar{a} + \left[\frac{\partial\bar{Q}}{\partial p_u}\right]_i \delta p_u + \left[\frac{\partial\bar{Q}}{\partial p_d}\right]_i \delta p_d \tag{3.12}$$

The partial derivatives are referred to as the valve coefficients. Differentiating equation (3.2) gives

$$[\partial\bar{Q}/\partial\bar{a}]_i = \sqrt{p_{ui}}\sqrt{(1-p_i)} = \bar{Q}_i/\bar{a}_i$$

$$[\partial\bar{Q}/\partial p_u]_i = \frac{\bar{a}_i\sqrt{(1-p_i)}}{2\sqrt{p_{ui}}} + \frac{\bar{a}_i p_i}{2\sqrt{p_{ui}}\sqrt{(1-p_i)}} = \frac{\bar{Q}_i}{2p_{ui}(1-p_i)}$$

$$[\partial\bar{Q}/\delta p_d]_i = \frac{-\bar{a}_i}{2\sqrt{p_{ui}}\sqrt{(1-p_i)}} = \frac{p_i}{2(p_i-1)}\frac{\bar{Q}_i}{p_{di}}$$

For gas flow, equation (3.6) is linearised by putting [33]

$$\overline{\delta W} = \left[\frac{\partial W}{\partial a}\right]_i \delta\bar{a} + \left[\frac{\partial W}{\partial p_u}\right]_i \delta p_u + H^j \left[\frac{\partial W}{\partial p_d}\right]_i \delta p_d + \left[\frac{\partial W}{\partial T_u}\right]_i \delta T_u \tag{3.13}$$

where

$$\left[\frac{\partial\overline{W}}{\partial\bar{a}}\right]_i = \frac{\overline{W}_i}{\bar{a}_i}$$

$$\left[\frac{\partial\overline{W}}{\partial\overline{T}_u}\right]_i = -\frac{1}{2}\frac{\overline{W}_i}{\overline{T}_{ui}}$$

$$\left[\frac{\partial\overline{W}}{\partial p_u}\right]_i = \frac{\overline{W}_i}{p_{ui}}[1+\zeta]$$

with

$$\zeta = \frac{1}{2\gamma}\left\{\frac{(\gamma+1)p_i^{1/\gamma} - 2p_i^{(2-\gamma)/\gamma}}{p_i^{(2-\gamma)/\gamma} - p_i^{1/\gamma}}\right\} = -\frac{p_i}{f(p_i)}\left[\frac{\partial f}{\partial p}\right]_i$$

for $p_i > (p_i)_{\text{crit}}$

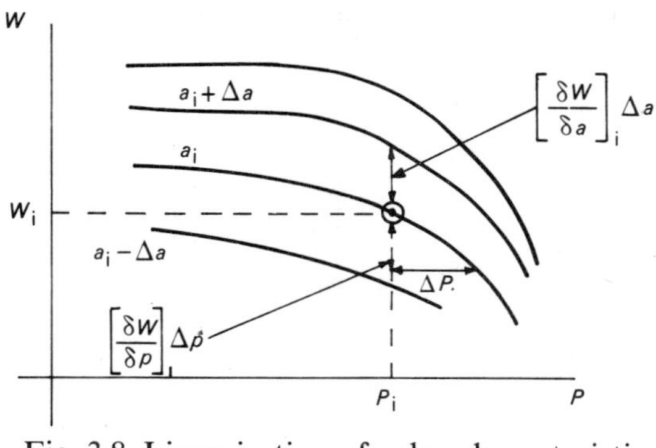

Fig. 3.8. Linearisation of valve characteristic

and $\zeta = 0$ for $p_i < (p_i)_{crit}$

$$\left[\frac{\partial W}{\partial p_d}\right]_i = \frac{\bar{a}_i p_{ui}}{\sqrt{T_{ui}}} \frac{\delta f}{\delta p_d} = -\zeta \frac{W_i}{p_{di}}$$

Valve coefficients can be deduced from experimental results. Figure 3.8 shows a typical variation of measured mass flow rate with back pressure and orifice area. If pressure and area variation are limited to ΔP and Δa respectively then the variation of mass flow about W_i is

$$\Delta W_i = \left[\frac{\partial W}{\partial P}\right]_i \Delta P + \left[\frac{\partial W}{\partial a}\right]_i \Delta a$$

The coefficients are determined as shown.

3.9. The three-way servovalve

Figure 3.9 shows three different three-way valves, driving a single acting cylinder. Each valve has three ports or connections; supply, exhaust and cylinder. Jet

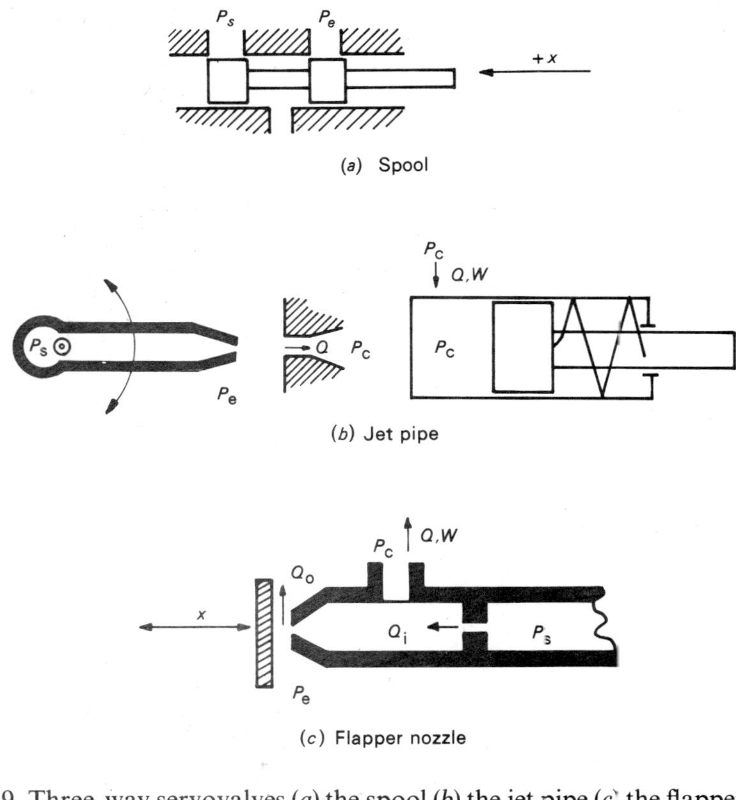

Fig. 3.9. Three-way servovalves (*a*) the spool (*b*) the jet pipe (*c*) the flapper nozzle

pipe and flapper nozzle valves are basically open centre valves. The spool, as shown earlier, can be open or closed centre. Although the spool valve is the most difficult to manufacture on account of the close tolerances required, it is nevertheless the most popular servovalve. It is easy to achieve static pressure balance within this valve and this is an important factor in modern high pressure systems. The valve's behaviour in mid-position can be easily arranged to be open centre, closed centre or critical centre by varying the amount of lap. The spool valve, however, requires a well filtered fluid, otherwise axial friction forces may build up or orifices may become choked. Open centre valves, like the flapper nozzle and the jet pipe valves are relatively insensitive to contaminated fluids and have a self-cleaning tendency. They, however, suffer from high quiescent power loss and are hence usually only found in low power applications such as the first stage of two-stage valves. The relatively large clearances in jet pipe and in flapper nozzle valves make them more easily and cheaply manufactured than spool servovalves. Another important advantage of the jet pipe valve is its low actuating force. These forces can be high in flapper nozzle valves.

3.10. Three-way valve, spool, hydraulic

In most servo systems spool position is affected by the system error and so the variation of orifice area with spool displacement is an important factor. There are three most common porting shapes: rectangular, round and full annulus. If the ports are rectangular, the orifice area increases linearly with spool movement. This is also true of the full annulus port, which is a circumferential slot in the valve sleeve. For circular ports, however, orifice area is not linear with spool movement and the flow equations involve a non-linearity in spool displacement as well as in pressure drop. For the present it is assumed that the valve area displacement characteristic is linear so that, for example, equation (3.1) can be written

$$Q = Kx\sqrt{(P_u - P_d)} \tag{3.14}$$

Referring to the critical centre or line to line valve of Fig. (3.9(*a*). it can be deduced that

For $x \geqslant 0$
$$Q = Kx\varepsilon\sqrt{\{\varepsilon(P_s - P_c)\}} \tag{3.15}$$
$$\varepsilon = \text{sgn}\,(P_s - P_c)$$

For $x \leqslant 0$
$$Q = Kx\phi\sqrt{\{\phi(P_c - P_e)\}} \tag{3.16}$$
$$\phi = \text{sgn}\,(P_c - P_e)$$

The signum terms allow for flow reversal.

Putting $Q_r = Kx_r\sqrt{\{(P_s - P_e)/2\}}$; $P_s^* = P_s - P_e$; $P_c^* = P_c - P_e$; $p_c^* = P_c^*/P_s^*$.

$\therefore$ For $\bar{x} \geqslant 0$
$$\bar{Q} = \varepsilon\bar{x}\sqrt{\{2\varepsilon(1 - p_c^*)\}} \tag{3.17}$$

For $\bar{x} \leqslant 0$
$$\bar{Q} = \phi\bar{x}\sqrt{(2\phi p_c^*)} \tag{3.18}$$

The cylinder pressure P_c will not in practice fall below the oil vapour pressure P_v and the lower limit on p_c^* is then p_v^*. The valve characteristics are plotted in Fig. 3.10, where p_v^* has been taken as $-0{\cdot}111$ (as would be the case for $P_s = 1000$ lbf in^{-2} a, $P_e = 100$ lbf in^{-2} a, $P_v = 0$ lbf in^{-2} a.

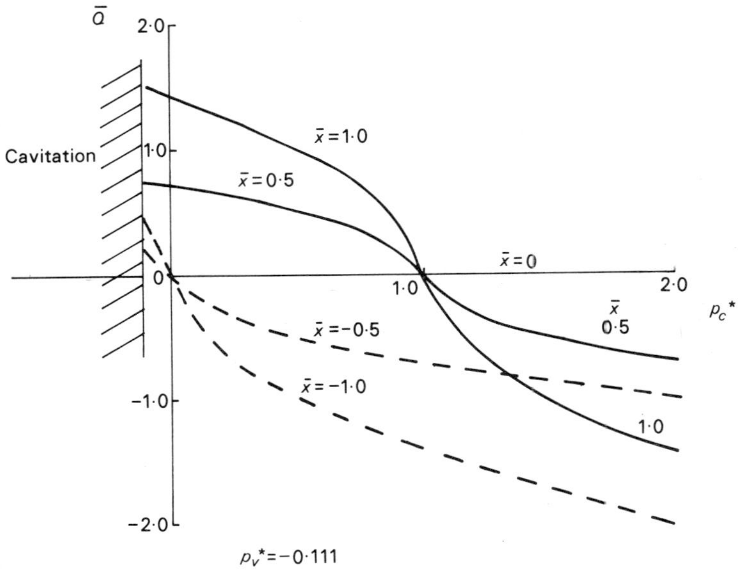

Fig. 3.10. Three-way hydraulic spool valve characteristics

The most important operating point for a servovalve is the null point. Bearing in mind that spool displacement is usually a function of system error in a hydraulic servo, then it is clear that if the servo is a good one, errors will be small and the spool position should depart very little from the null position for most of its working life. Hence it is usual to linearise the valve flow about the null point. In the case discussed here, the null position is $x = 0$. For a three-way valve, the null cylinder pressure is usually chosen as $P_s^*/2$, so that there are equal potential load accelerations and decelerations. Equations (3.17) and (3.18) give, for $\bar{x} = 0$, $p_c^* = 0{\cdot}5$, the normalised flow gain $\partial\bar{Q}/\partial\bar{x} = 1$ or flow gain $\partial Q/\partial x = Q_r/x_r = K\sqrt{(P_s^*/2)}$ and $\partial\bar{Q}/\partial p_c^* = 0$ or the flow pressure coefficient $\partial Q/\partial P_c^* = 0$.

Another useful coefficient is the pressure sensitivity $\partial P_c^*/\partial x$, which relates to the stiffness of the valve load system. This can be determined easily if the other two coefficients are known, since the theory of partial differentials gives

$$\frac{\partial P_c^*}{\partial x} = -\frac{\partial Q}{\partial x}\bigg/\frac{\partial Q}{\partial P_c^*}$$

In this case $\partial P_c^*/\partial x = \infty$. In practice, large values of the pressure sensitivity can be achieved (10^6 lbf in^{-1}).

3.10.1 Underlapped three-way valve, hydraulic

If the underlap U, see Fig. 3.11, is the same on supply and exhaust ports, the flow rate Q is

$$Q = K(x + V)\varepsilon\sqrt{\{\varepsilon(P_s - P_c)\}} + K(x - V)\theta\sqrt{\{\theta(P_c - P_e)\}} \tag{3.19}$$

or

$$\bar{Q} = (\bar{x} + 1)\varepsilon\sqrt{\{2(1 - p_c^*)} + (\bar{x} - 1)\phi\sqrt{(2\phi p_c^*)} \tag{3.20}$$

with

$$Q_r = KV\sqrt{(P_s^*/2)}$$

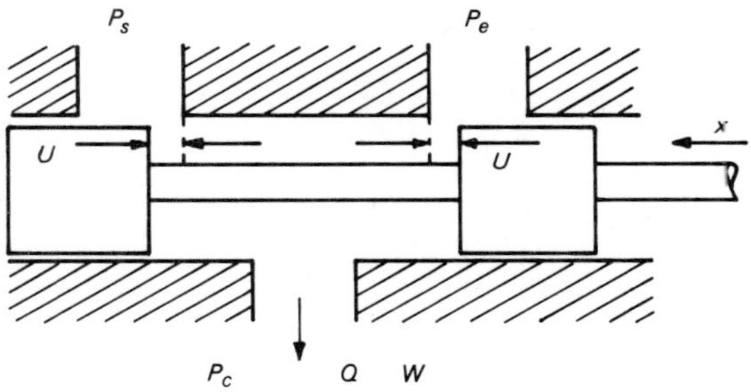

Fig. 3.11. An underlapped three-way valve

The null coefficients are

$$\frac{\partial Q}{\partial x} = 2 \quad \text{or} \quad \frac{\partial Q}{\partial x} = 2\frac{Q_r}{V} = 2K\sqrt{(P_s^*/2)}$$

$$\frac{\partial Q}{\partial p_c^*} = -2 \quad \text{or} \quad \frac{\partial Q}{\partial P_c^*} = -2\frac{Q_r}{P_s^*} = -KV\sqrt{(2/P_s^*)}$$

$$\frac{\partial P_c^*}{\partial x} = P_s^*/V$$

Note that the flow gain is twice that for the critical centre valve. The flow pressure coefficient, which is important as far as servo stability is concerned, is no longer zero. On the other hand the stiffness has fallen. This is an open centre valve and suffers the disadvantage of a null leakage flow rate $\bar{Q}_l = 1{\cdot}0$ or $Q_l = Q_r$.

3.11. Three-way valve, pneumatic

Referring again to Fig. 3.9(a), it is seen that for positive x, the upstream pressure P_u is supply pressure P_s and the downstream pressure P_d is cylinder pressure P_c. In developing a reference mass flow rate W_r (equation (3.5)) it is convenient to take P_s and T_s as reference pressure and temperature respectively. Hence for positive x, $P_u = P_s = P_r$, $P_d = P_c$, $p = P_c/P_s = p_c$.

Hence from equation (3.6)

$$\bar{W} = \bar{x} f(p_c) \tag{3.22}$$

For negative x, $P_u = P_c$, $P_d = P_e$ and $p = P_e/P_c$. It can also be assumed with reasonable accuracy that $T_u = T_s = T_d$.
Hence from equation (3.6)

$$\bar{W} = \bar{x}(P_c/P_s) f(P_e/P_c) \tag{3.22}$$

or

$$\bar{W} = \bar{x} p_c f(p_e/p_c) \tag{3.23}$$

Equations (3.21) and (3.23) are plotted in Fig. 3.12, for the case $p_e = 0{\cdot}1$. Here for negative x,

$$\bar{W} = \bar{x} p_c f\left[\frac{0{\cdot}1}{p_c}\right] \tag{3.24}$$

The graphs are easily plotted by taking values of p_c, referring to Fig. 3.7 for the appropriate values of the function f and calculating W.

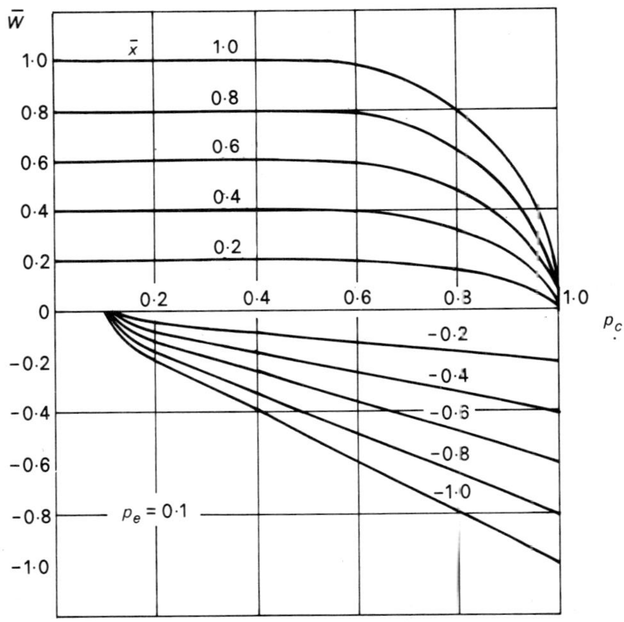

Fig. 3.12. Three-way pneumatic spool valve characteristics

For positive $\bar{x}$ the valve coefficients are evaluated as described earlier, equation (3.13). For negative x, Fig. 3.12 shows that the flow is linear with p_c over most of the range ($f(p_e/p_c) = 1$). Indeed it is reasonable for small p_e to assume linearity over the full range in which case $\bar{W} = \bar{x} p_c$.

3.12. The flapper nozzle valve, hydraulic

In this case, Fig. 3.9(*c*)

$$Q = Q_i - Q_o$$
$$Q = C_{qi}a_i\sqrt{\{2(P_s^* - P_c^*)/\rho\}} - C_q a\sqrt{(2P_c^*/\rho)} \quad (3.25)$$

where the outlet orifice area $a = \pi dx$, d being the diameter of the nozzle.

Define
$$Q_r = C_{qi}a_i\sqrt{(P_s^*/\rho)}$$
Hence,
$$\bar{Q} = \sqrt{\{2(1 - p_c^*)\}} - \bar{a}\sqrt{(2p_c^*)} \quad (3.26)$$

with $\bar{a} = (C_q a)/(C_{qi}a_i) = (C_q\pi dx)/(C_{qi}a_i)$

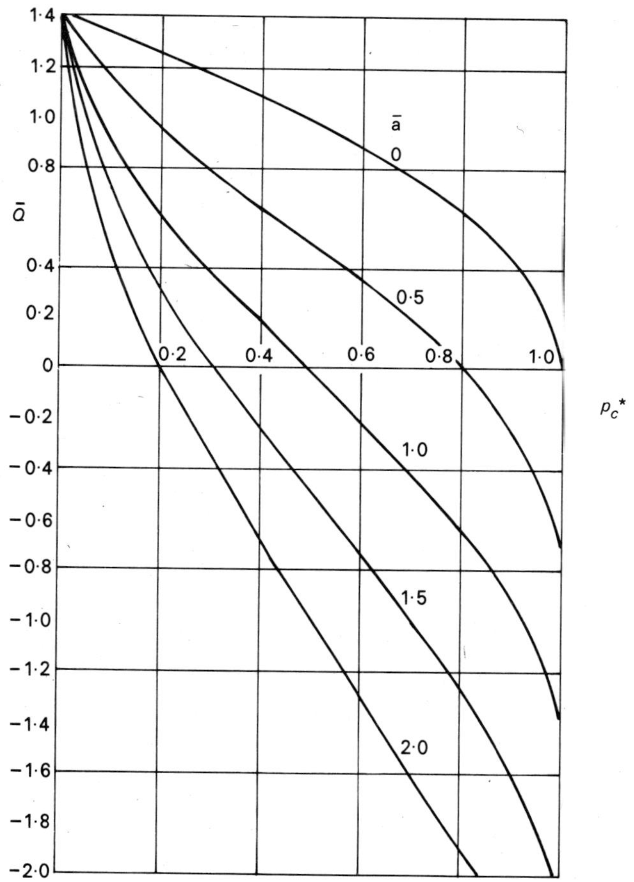

Fig. 3.13. Three-way hydraulic flapper nozzle characteristics

Equation (3.26) is plotted in Fig. 3.13. Flapper nozzle valves are usually designed with null point $\bar{a} = 1$ and null point $p_c^* = 0{\cdot}5$. This gives maximum null pressure sensitivity. The null coefficients are

$$\frac{\partial \bar{Q}}{\partial \bar{a}} = -1 \quad \text{or flow gain } \frac{\partial Q}{\partial x} = -C_q \pi d \sqrt{(P_s^*/\rho)}$$

$$\frac{\partial \bar{Q}}{\partial p_c^*} = -2 \quad \text{or } \frac{\partial Q}{\partial P_c^*} = -2\frac{Q_r}{P_s^*} = -\frac{2C_{qi}a_i}{\sqrt{(\rho P_s^*)}}$$

Pressure sensitivity $\partial P_c^*/\partial x = -P_s^*/x_i$ where x_i is null x. The leakage flow at null is $\bar{Q}_l = 1$ or $Q_l = Q_r = C_{qi}a_i\sqrt{(P_s^*/\rho)}$.

3.13. Flapper nozzle valve, pneumatic

The mass flow rate to the load is

$$W = W_i - W_o \tag{3.27}$$

Defining, $W_r = 0{\cdot}0405\, C_{qi}a_iP_s\sqrt{T_s}$ (S.I. system eg., 3.5)

$$\bar{W} = \bar{W}_i - \bar{W}_o \tag{3.28}$$

Then

$$\bar{W}_i = f(P_c/P_s) = f(p_c)$$
$$\bar{W}_o = \bar{a}p_c f(p_e/p_c)$$

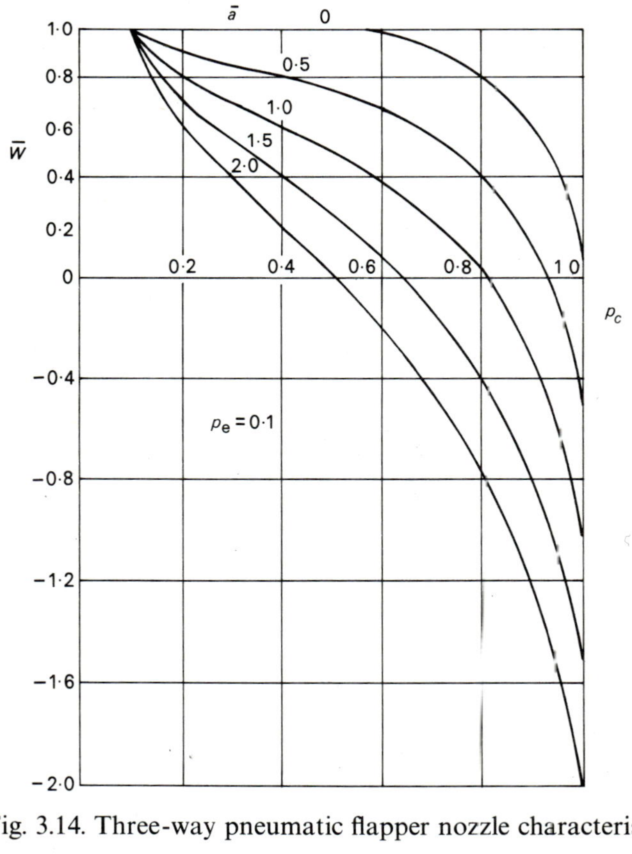

Fig. 3.14. Three-way pneumatic flapper nozzle characteristics

Equation (3.28) is plotted in Fig. 3.14 for the case $p_e = 0{\cdot}1$. Note that if $p_c = 0{\cdot}5$ is chosen as the null pressure, then the null orifice area ratio is 2·0 as compared to unity for the hydraulic case.

3.14. The three-way jet pipe valve

The jet pipe presents a much more difficult analytical problem than the other valves and hence experimental data is invaluable. Several researchers have shown that there are some similarities between the characteristics of hydraulic and pneumatic jet pipe valves. For example experiment has shown that the optimum receiver diameter to nozzle diameter ratio d_r/d_n, Fig. 3.15, for both

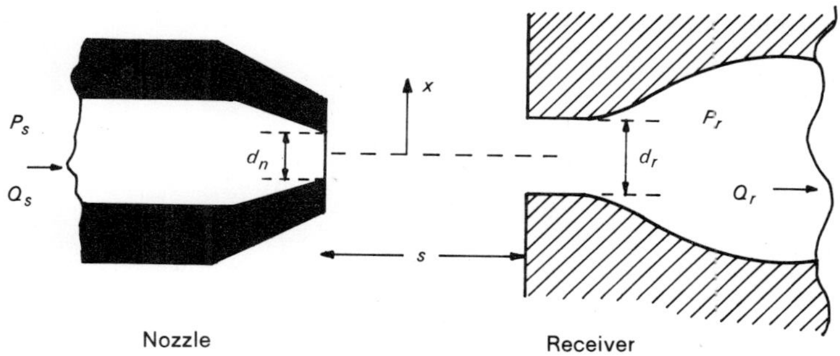

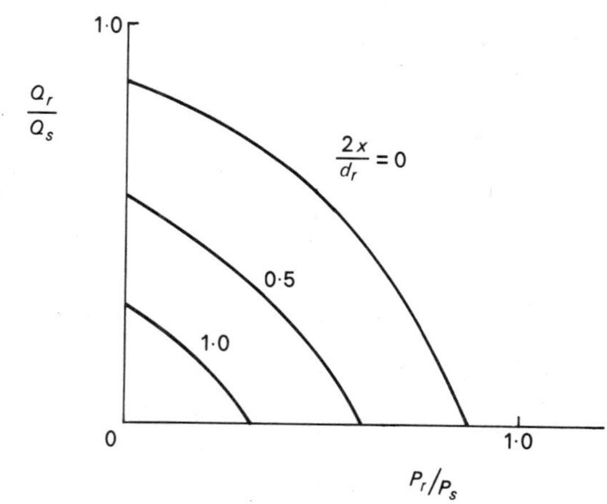

Fig. 3.15. The three-way jet pipe valve

oil and air systems is 1·4, maximum power transfer being the criterion for optimum performance. Nozzle to receiver spacing is more critical in pneumatic valves than in hydraulic valves, due largely to the rapid expansion of high pressure air jets. Alpay [34], working at pressure of 100 and 200 lbf in^{-2} g, showed that the ratio s/d_n could be varied between 0·5 and 5·6 without causing any noticeable change in the performance of a hydraulic jet pipe valve. On the other hand, Reid [35], working at pressures roughly 50–500 lbf in^{-2} g, concluded that nozzle to receiver spacing in pneumatic jet pipe valves should be as small as possible. Figure 3.15 shows qualitatively the flow pressure characteristics of a three-way jet pipe valve. When the load is blocked so that the load flow rate Q_r is zero, it is possible to achieve 90 per cent recovery of the supply pressure. At zero load pressure (very rarely achieved in practice) recovery of 30 per cent of supply flow is possible.

3.15. Four-way servovalves

A four-way servovalve can control the position of a double acting cylinder. Some examples are shown in Fig. 3.16. A four-way valve is equivalent to two

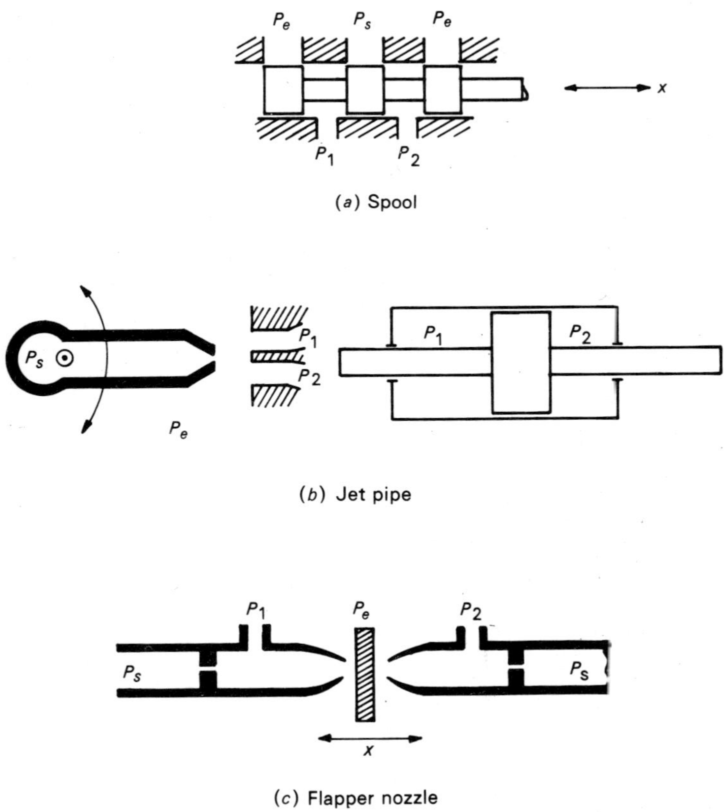

Fig. 3.16. Four-way servovalves (*a*) spool (*b*) jet pipe (*c*) flapper nozzle

three-way valves and hence its characteristics can be deduced from those of a three-way valve. The following example concerning a spool valve illustrates this point.

3.15.1 Four way spool valve, hydraulic

A critical centre valve is shown in Fig. 3.17. It is equivalent to two three-way valves. For positive x,

$$\begin{aligned} \bar{Q}_1 &= \varepsilon_1 \bar{x} \sqrt{\{2\varepsilon_1(1 - p_1^*)\}}, \quad & \varepsilon_1 &= \operatorname{sgn}(1 - p_1^*) \\ \bar{Q}_2 &= -\phi_2 \bar{x} \sqrt{(2\phi_2 p_2^*)}, \quad & \phi_2 &= \operatorname{sgn} p_2^* \end{aligned} \tag{3.29}$$

For negative x

$$\begin{aligned} \bar{Q}_1 &= \phi_1 \bar{x} \sqrt{(2\phi_1 p_1^*)}, \quad & \phi_1 &= \operatorname{sgn} p_1^* \\ \bar{Q}_2 &= -\varepsilon_2 \bar{x} \sqrt{\{2\varepsilon_2(1 - p_2^*)\}}, \quad & \varepsilon_2 &= \operatorname{sgn}(1 - p_2^*) \end{aligned}$$

where $\qquad Q_r = K x_r C_{qr} \sqrt{(P_s^*/2)}$

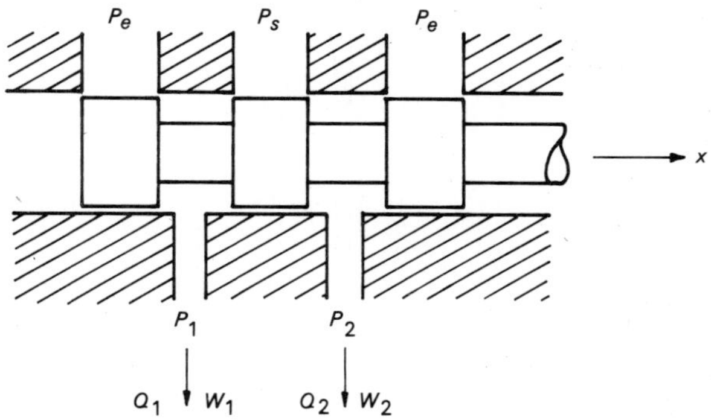

Fig. 3.17. A critical centre four-way servovalve

The load determines the relationship between flow rates $\bar{Q}_1$ and $\bar{Q}_2$. Here steady state conditions are assumed so that for orifice loads and symmetrical double acting cylinders, $\bar{Q}_1 = -\bar{Q}_2$. Hence for positive x

$$\varepsilon_1(1 - p_1^*) = \phi_2 p_2^*$$

or

$$p_1^* + p_2^* = 1 \tag{3.30}$$

Define a load pressure $p_l = p_1^* - p_2^*$, and then for positive x the flow rate is

$$\bar{Q} = \varepsilon_1 \bar{x} \sqrt{\{\varepsilon_1(1 - p_l)\}} \tag{3.31}$$

with $\varepsilon_1 = \operatorname{sgn}(1 - p_1^*) = \operatorname{sgn}(1 - p_l)$.
The lower limit on p_1^* is p_v^* and hence since $p_l = 2p_1^* - 1$, the lower limit on p_l is $(2p_v^* - 1)$.

For negative x it is easy to show that

$$\bar{Q} = \phi_1 \bar{x} \sqrt{\{\phi_1(1 + p_l)\}} \tag{3.32}$$

with $\phi_1 = \operatorname{sgn}(1 + p_l)$.

In this case the upper limit on p_l is $(1 - 2p_v^*)$, at which point pressure P_2 falls to vapour pressure P_v.

Equations (3.31) and (3.32) are plotted in Fig. 3.18 for $x = \pm 1$, $x = 0$ and

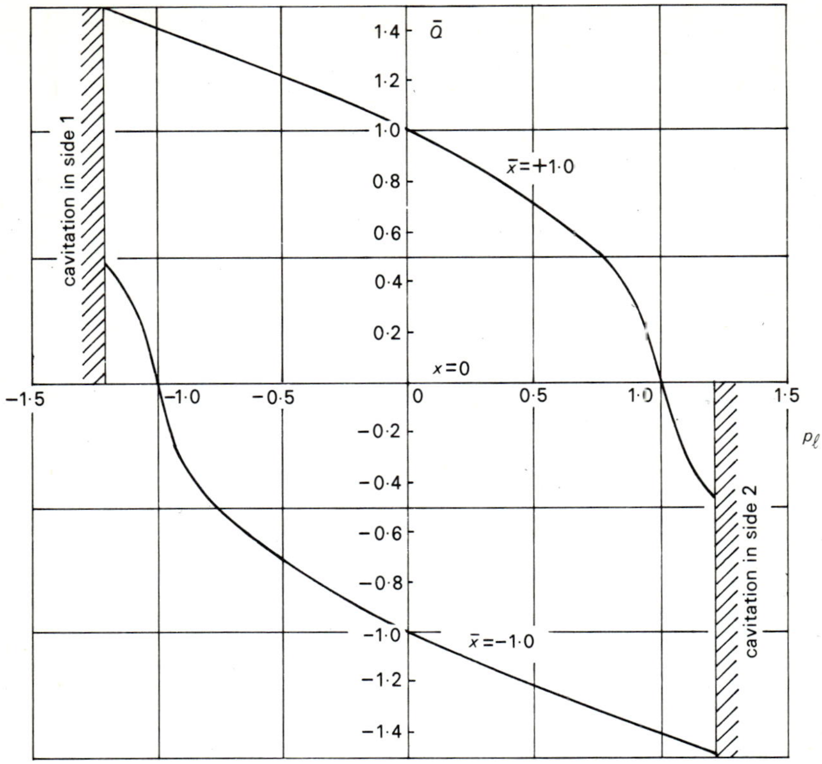

Fig. 3.18. Characteristics of a hydraulic critical centre four-way spool valve

$p_v{}^* = -0{\cdot}111$ (e.g. $P_s = 1000$ lbf in^{-2} a, $P_e = 100$ lbf in^{-2} a, $P_v = 0$ lbf in^{-2} a). When $|p_l| > 1{\cdot}222$, cavitation occurs in the cylinder. The null coefficients, evaluated at $p_l = 0, \bar{x} = 0$ are $\partial Q/\partial x = K\sqrt{(P_s{}^*/2)}$, $\partial Q/\partial P = 0$ and $\partial P_l/\partial x = \infty$. Note that the piston rod velocity $u = Q/A$, where A is the effective piston area.

3.15.2 Four-way spool valve, pneumatic

Put $P_r = P_s$ and $T_r = T_s$. Figure 3.17 shows that the mass flow rate W_1 passes from an upstream pressure $P_u = P_s$ to a downstream pressure $P_d = P_1$. Equation (3.6) then gives

$$\overline{W}_1 = \bar{a}f(P_1/P_s) = \bar{a}f(p_1) \tag{3.33}$$

W_2 can be calculated, knowing that the upstream pressure is P_2 and the downstream pressure is P_e. When normalised this becomes

$$\overline{W}_2 = -\bar{a}p_2 f(p_e/p_2) \tag{3.34}$$

Again the load determines the relationship between W_1 and W_2. An orifice load requires in the steady state that $W_1 = W_2$ or that $f(p_1) = p_2 f(p_e/p_2)$. This can be solved by trial and error using Fig. 3.7 and will give pressures P_1 and P_2 for various flow rates.

For a double acting cylinder moving at steady velocity the volumetric flow rate into one side of the cylinder equals the volumetric flow rate displaced out of the other side. Hence

$$\frac{W}{\rho_1} = \frac{-W_2}{\rho_2} \text{ or } \frac{W_1}{W_2} = \frac{-\rho_1}{\rho_2} = \frac{-P_1}{P_2} \tag{3.35}$$

Equations (3.33)–(3.35) combine to give

$$\frac{1}{p_1} f(p_1) = f(p_e/p_2) \tag{3.36}$$

This too must be solved by trial and error. (A neat graphical solution has been given by Blackburn *et al.* [4].) Having solved equation (3.36) for p_1 and p_2, the mass flow rates can be calculated. It is also useful to know the piston rod

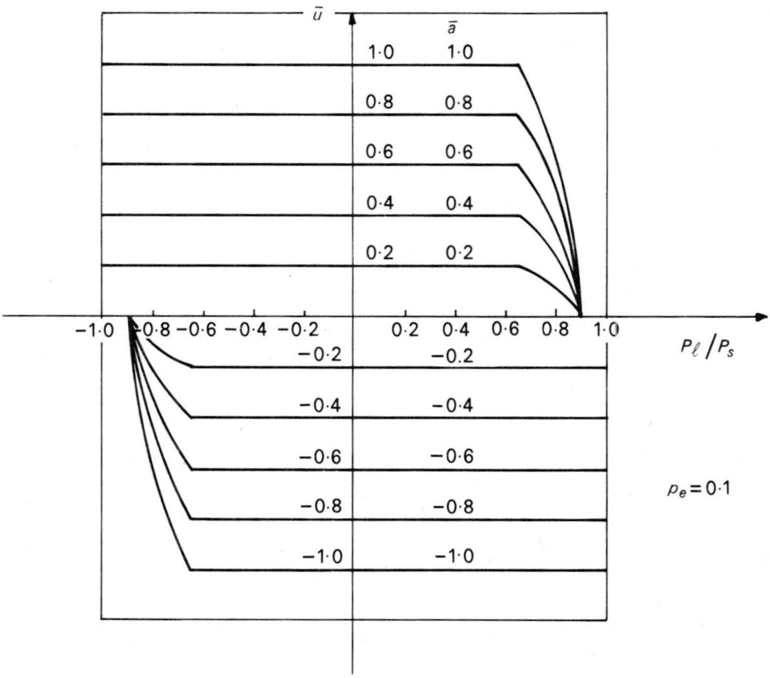

Fig. 3.19. Characteristics of a pneumatic critical centre four-way spool valve

velocity u at any load pressure p_l. This is found as follows:

Velocity $$u = W_1/(\rho_1 A)$$

Define $$u_r = W_r/(\rho_r A)$$

Hence $$\bar{u} = u/u_r = W_1(\rho_r/\rho_1) = \overline{W}_1(P_s/P_1) = W_1/p_1 \qquad (3.37)$$

or $$\bar{u} = \bar{a}f(p_1)/p_1$$

The flow characteristics are plotted in Fig. 3.19.

3.16. Some comments on the use of valve characteristics

The above discussion of valve characteristics may seem at first reading to be somewhat abstruse and of little practical value. This is not so. The characteristics enable a designer to:

Determine the maximum flow rate that the valve can pass. This fixes the maximum velocity of the load.

Determine the valve coefficients. These must be known if a linearised analysis is to be carried out (chapter 6).

Determine the maximum power that the valve can deliver.

More details of their use will be given in chapter 10.

3.17. The electric field valve

In 1939, Willis M. Winslow [36] invented a novel form of hydraulic clutch which depended upon the electrostatic fluid principle. He found that when two metal plates, separated by a film of dirty oil, are charged, the viscosity of the oil increases so much that the plates can lock together. If the circuit is broken the viscosity again reduces and the plates move freely. The phenomenon will not show itself if the oil is pure and Winslow found that a mixture of cornstarch and transformer oil gave best results. It seems that the starch particles link together in chains when they are electrically charged. The speed of response is surprising. Changes in viscosity take place so quickly that the effect has been used to drive a loudspeaker. However, the most promising application in the early days was in automotive clutches. By varying the current to the clutch, it can be locked fast or allowed to slip to any desired degree. Very little electric power is needed to drive such a system. High voltages may be needed but the amount of current flow is small and depends on the insulating properties of the oil–starch mixture.

Recently [37], Boeing have developed a hydraulic vibrator which uses the Winslow effect. Their electric valve is shown in Fig. 3.20(*a*). It consists of five cylindrical shells forming a group of six parallel 0·030 in gaps. A control voltage is applied to alternate shells, the other shells and the inner and outer boundary walls being grounded. This voltage produces an electrostatic field in the gaps between the shells. The flow characteristics for the device, Fig. 3.20(*b*), show that at zero flow the fluid can sustain a considerable pressure drop; the fluid can sustain a considerable shear stress before slippage occurs.

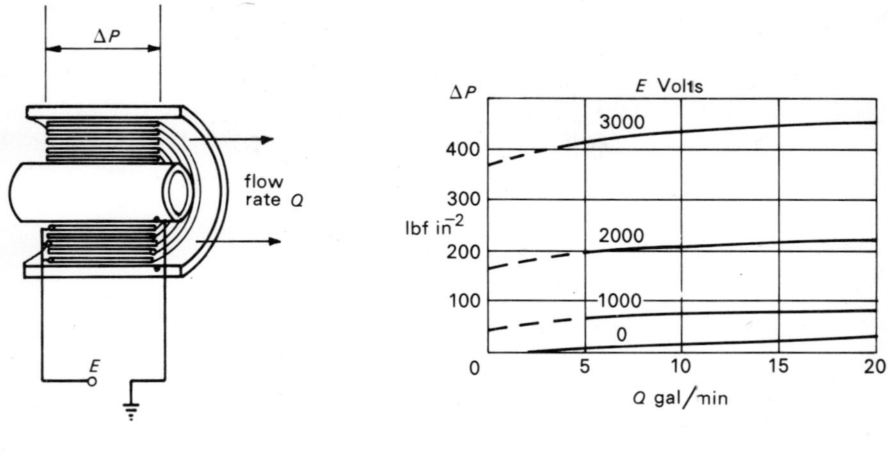

(a) Electric field valve (b) Flow characteristics

Fig. 3.20. The electric field valve (*a*) the valve and (*b*) its characteristics (after [37])

Four valves of the type shown in Fig. 3.20(*a*) are combined to drive the vibrator, Fig. 3.21 (four two-way valves are equivalent to one four-way valve). Valves *A* and *D* are attached to the cylinder and valves *B* and *C* are mounted on the output shaft where they also act as pistons. The cylinder is supplied from a fixed displacement pump which provides a constant flow source. The control voltages E_1 and E_2 are sine waves of opposite phase. Reference to Fig. 3.21 shows that when E_1 is a maximum, valves *A* and *C* try to stop the flow, but since the supply is constant flow, pressure builds up in the chamber between *A*

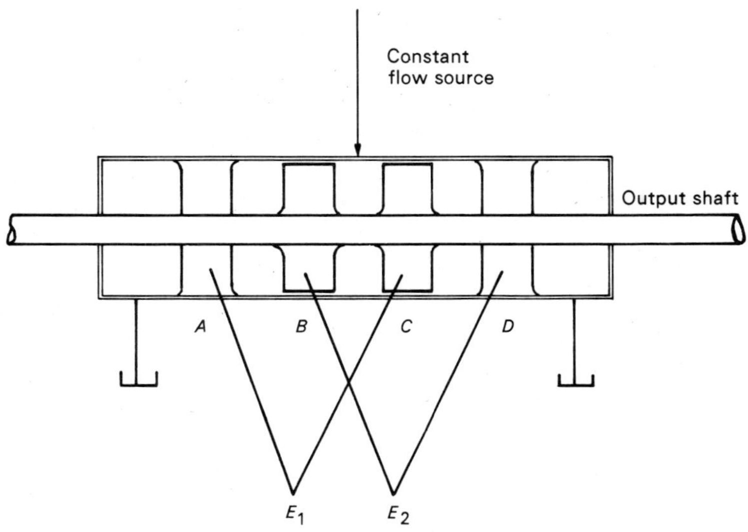

Fig. 3.21. A four-way electric field valve (after [37])

and *C*. The pressure acting on the left face of *C* moves the output shaft to the right.

The vibrator has produced 3500 pounds static force and 2200 pounds peak oscillating force at 1000 Hz, which is much better performance than conventional hydraulic oscillators. For step inputs the response time is fast, peak load acceleration being reached in less than a millisecond.

Chapter 4

Flow forces in valves

4.1. Introduction

It is important that the force required to open or close a valve be kept to a minimum. This is particularly true for single stage valves where large operating forces can create a need for a large and expensive torque motor. Newton's second law states that force is proportional to rate of change of momentum. Clearly there must be changes of momentum when a fluid is throttled and diverted by a control valve and these must be accompanied by momentum forces. Newton's law as applied to fluid flow can be written as follows. For steady state conditions, force equals the net rate of outflow of momentum from a control volume. An example may make this clearer. Consider the curved pipe of varying cross-section shown in Fig. 4.1. The rate of outflow of momentum in the X direction is $\rho Q v_2 \cos\alpha$ where ρ is the mass density, Q the flow rate, and v_2 is the mean velocity at station 2. The rate of inflow of momentum at station 1 is $\rho Q v_1$. Hence the force on the fluid in the X direction is

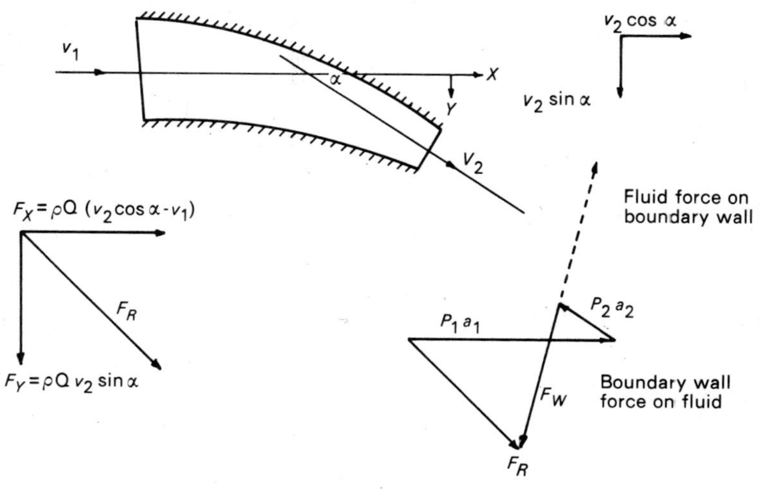

Fig. 4.1. Force acting on a pipe as a result of momentum changes

$$F_X = \rho Q[v_2 \cos \alpha - v_1] \tag{4.1}$$

Similarly in the Y direction the force on the fluid is

$$F_Y = \rho Q v_2 \sin \alpha \tag{4.2}$$

In Fig. 4.1 these two forces are combined to give F_R. This is the resultant force that must be applied to this fluid volume in order to produce the given changes in momentum. Three forces combine to produce the required force F_R. These are (1) the pressure force $P_1 a_1$ (2) the pressure force $P_2 a_2$ and (3) the force F_W exerted by the pipe walls on the fluid. Wall friction is neglected. Vectorial addition of these forces is illustrated in the figure and this determines the unknown force F_W. The reaction of the fluid on the boundary walls is equal and opposite to F_W. In servovalves this reaction force is sometimes known as the Bernoulli force but it is more commonly referred to as the steady state flow force.

4.2. Steady state flow forces in spool valves

Figure 4.2 shows a flow chamber in a spool valve [38]. Chamber A is pressurised and the jet of fluid enters chamber B at an angle θ which has been predicted theoretically by von Mises [8]: for a sharp edged orifice with no clearance

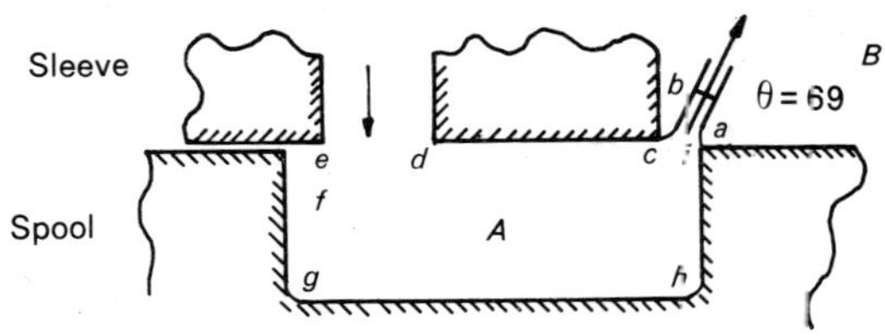

Fig. 4.2. A spool valve chamber

between spool and sleeve, $\theta = 69°$ (see chapter 2). The force on the valve is found by evaluating the momentum changes in the control volume *a b c d e f g h i*. Section *ab* is at the vena contracta of the jet. The following assumptions are made

The fluid is inviscid.
The flow is two-dimensional.
The flow is irrotational.

In a conventional valve the rate of inflow of momentum through the section *de* is much less than the rate of outflow through *ab*. The net rate of outflow of momentum in the X direction is $pQv \cos \theta$ where the fluid velocity at *ab* is

$$v = C_v \sqrt{(2\Delta P/\rho)} \tag{4.3}$$

C_v is a velocity coefficient and ΔP is the pressure drop across the orifice.

The flow rate Q is

$$Q = aC_d \sqrt{(2\Delta P/\rho)} \tag{4.4}$$

where a is the orifice area and C_d is the discharge coefficient. Hence the force applied to the fluid in the X direction is

$$F_X = \rho Q v \cos\theta = 2C_d C_v a \Delta P \cos 69^\circ$$

If $C_v = 1$ this reduces to

$$F_X = 0{\cdot}72 C_d a \Delta P \tag{4.5}$$

Of course in an actual four-way valve there are two orifices in series and the total force is twice that given by equation (4.5). Note that this is the force that must be applied to the fluid in the X direction. The fluid applies an equal and opposite force to the boundary walls. This tends to close the valve. The valve shown in Fig. 4.2 is pressure balanced (i.e. when the orifice is closed there is no force acting on the valve), and in this case the only force acting on the spool is the flow force. If the upstream land fg were removed then the resultant opening force caused by the pressure on land hi could easily swamp the closing flow force.

The efflux angle θ is affected by clearance and by roundness of the orifice edges. Figure 2.5 shows how chamber size [15], valve opening and clearance, affect the jet efflux angle.

It is worthwhile to include an example at this stage in order to give an indication of the magnitude of the forces involved. If the spool displacement is 0·010 in, and the pressure drop 1000 lbf in^{-2}, we get, assuming $C_d = 0{\cdot}7$ and that the ports are circumferential in a sleeve diameter $\frac{1}{4}$ in:

$$F_X = (4 \times 0{\cdot}7 \times \pi/4 \times 0{\cdot}01 \times 1000 \times 0{\cdot}36) = 7{\cdot}9 \text{ lbf}$$

4.3. Compensation of flow forces in spool valves

These forces can be sufficiently large to limit the employment of single stage valves to low power applications. Recent papers on servo dynamics show that flow forces reduce system gain and stability. Thus, methods of eliminating flow forces are of interest. All of these methods aim at changing the net rate of efflux of momentum from the control volume shown in Fig. 4.2.

4.3.1 Method 1. Radial hole orifices

This is illustrated in Fig. 4.3. The valve ports are in this case two small holes drilled in the sleeve in diametrically opposite positions. This ensures that there is no lateral force on the spool, i.e. the momentum components $\rho QU \sin\theta$ cancel out. Initially as the holes are being uncovered the jet angle is 69°, but as the opening increases the angle increases until at full opening it becomes 90°. At this point the net axial rate of efflux of momentum is zero and there is no flow force, Fig. 4.3(*b*). However, Clark [39] points out that the abrupt change in the force gradient can lead to stability problems.

This can be improved if each of the radial holes is replaced by a series of smaller holes. These are arranged in a spiral so that one begins to open before its neighbour is completely closed, Fig. 4.4(*a*). The desired valve area displacement characteristic determines the size of these holes. Clark has found that an

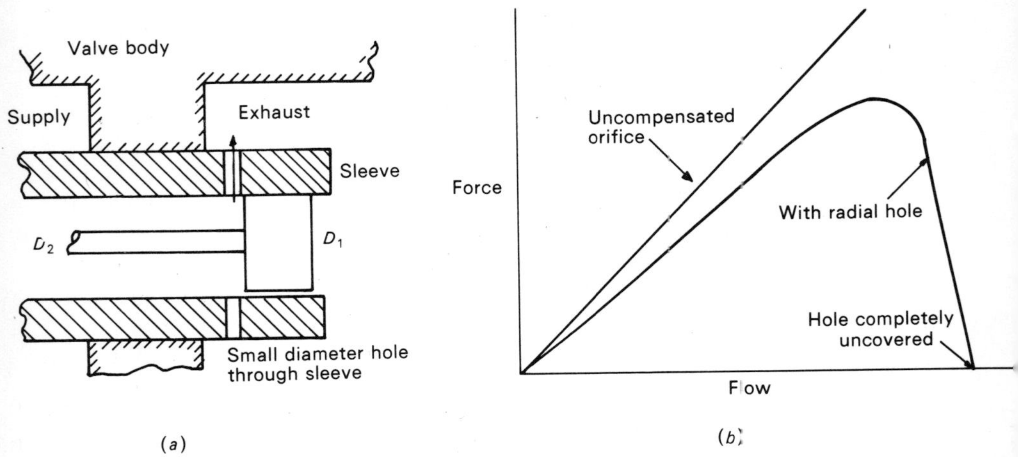

Fig. 4.3. Flow force compensation by means of radial hole orifices (*a*) sleeve with small radial holes (*b*) effect on flow force

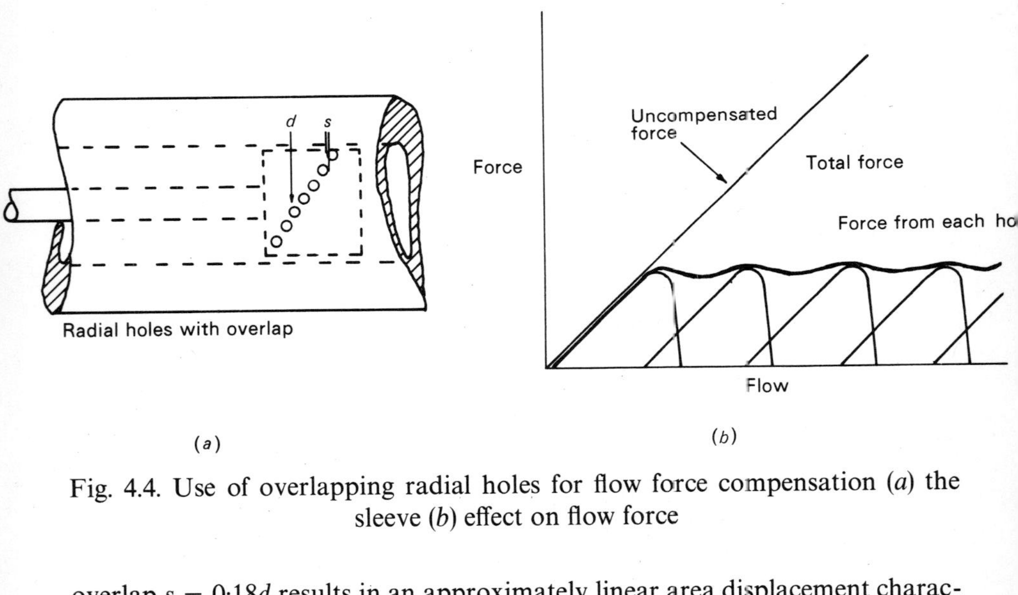

Fig. 4.4. Use of overlapping radial holes for flow force compensation (*a*) the sleeve (*b*) effect on flow force

overlap $s = 0{\cdot}18d$ results in an approximately linear area displacement characteristic. Figure 4.4(*b*) shows how the flow force contributions from each hole can be made to add up to give a constant flow force, independent of flow rate.

4.3.2 Method 2. Angled port compensation

The flow force exists because there is a net rate of efflux of momentum from the control volume. In Fig. 4.2 the momentum at the outlet or orifice was the only contributor since there was no axial component of the inlet momentum. However if the inlet port is angled as in Fig. 4.5(*a*) an axial component of inlet

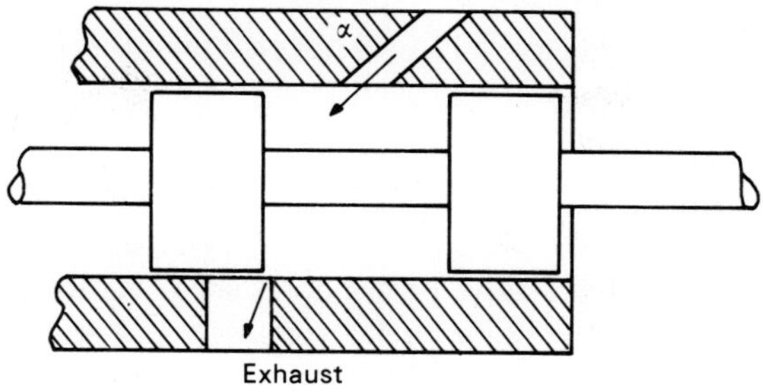

Angled port compensation

(a)

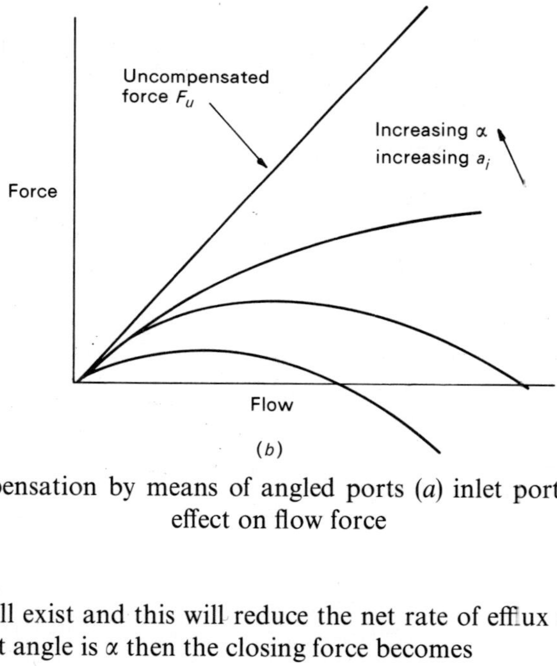

(b)

Fig. 4.5. Compensation by means of angled ports (a) inlet port at angle α (b) effect on flow force

momentum will exist and this will reduce the net rate of efflux of momentum. If the inlet port angle is α then the closing force becomes

$$F = \rho Q v \cos\theta - \rho Q v_i \cos\alpha \tag{4.6}$$

where v_i is the mean velocity at the inlet port.

Hence $$F = \rho Q v[\cos\theta - (v_i/v)\cos\alpha] \tag{4.7}$$

Now $$C_c a v = C_{ci} a_i v_i \tag{4.8}$$

where C_c is a contraction coefficient and the 'i' refers to the inlet port.

Take $$C_c = 0{\cdot}65 \text{ and } C_{ci} = 0{\cdot}90 \tag{4.9}$$

Hence $$v_i/v = 0{\cdot}72(a/a_i) \tag{4.9}$$

Substitution in equation (4.7) gives

$$F = \rho Q v \cos\theta\,[1 - 0{\cdot}72a\cos\alpha/(a_i\cos\theta)] \tag{4.10}$$

If $\theta = 69°$ this gives

$$(F/F_u) = 1 - 2(a/a_i)\cos\alpha \tag{4.11}$$

where F_u is the uncompensated force. Clearly the flow force is zero when $\cos\alpha = 0{\cdot}5(a_i/a)$ or when $a = 0{\cdot}5a_i/\cos\alpha$. For instance if $\alpha = 45°$ zero force exists when $a = 0{\cdot}71a_i$. Since valve orifice areas are generally small this implies that the inlet area should be made pretty small too and so this method causes an additional restriction in the valve. This can be reduced by making the inclination of the inlet port to the longitudinal axis as small as possible, since this makes for larger ports [40].

4.3.3 Method 3. Pressure drop compensation [13]

In the analysis of momentum effects frictional forces in the valve chamber were neglected. However, if the annular space between spool and sleeve is small, velocity in the valve chamber will be high and a friction loss, proportional to v^2, will

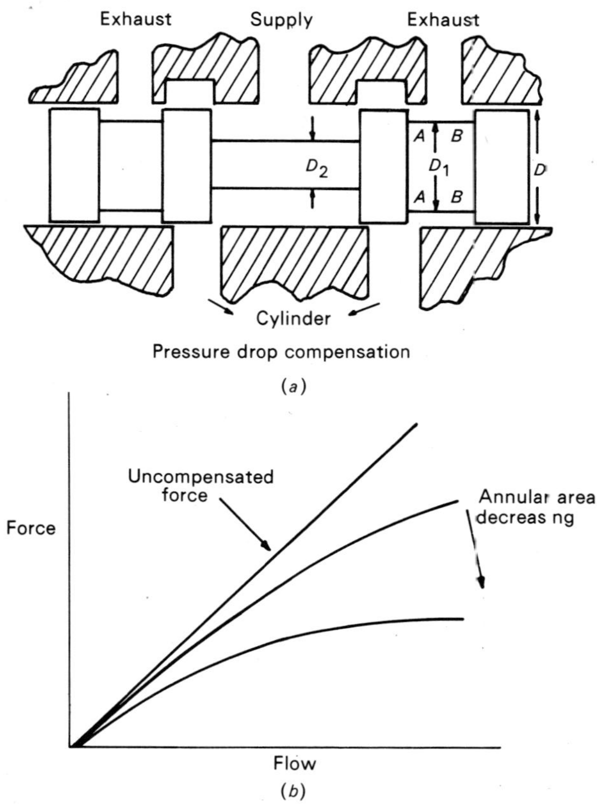

Fig. 4.6. Use of valve chamber pressure drop for flow force compensation (*a*) the valve with a thick stem on the exhaust side (*b*) the effect on the flow force

occur. Referring to Fig. 4.6(*a*) this will result in the pressure in the exhaust side of the valve at A being greater than that at B. This gives an opening force which opposes the closing force discussed above. The smaller the annular area the greater the opening force. However, this cannot be carried too far since it causes choking of the flow. Lee [38] points out that this is serious if the ratio of the wetted areas is less than 1 :4. That is, if

$$(D^2 - D_1^{\,2})/(D^2 - D_2^{\,2}) < 1/4$$

This method of compensation is sometimes referred to as the 'thick stem' method.

4.3.4 Method 4. Spool profiling or spool shaping

Earlier in the discussion we noted that there are two orifices in series in a four-way valve and each contributes to the opening force. Lee and Blackburn [38] have devised a method for changing the sign of the force caused by flow from control port to exhaust port, Fig. 4.7(*a*). This is accomplished by machining tapers on parts of the valve stem and recesses in the sleeve. This is costly but it is effective and in addition it usually results in a slight lowering of the hydraulic restriction of the valve [62].

Figure 4.7(*a*) shows a negative force port. The stem is shaped to look like a turbine bucket and the recess in the sleeve makes the downstream chamber larger. The jet enters at angle θ_1 and leaves at angle θ_2 and hence if the cross-section of the flow remains constant, the closing force would be

$$F = \rho Q v\,(\cos\theta_1 - \cos\theta_2) \tag{4.12}$$

If $\cos\theta_2 > \cos\theta_1$, i.e. $\theta_2 < \theta_1$, the force will tend to open the valve. In addition to this an eddy is formed in the chamber causing an additional flow to enter the chamber at angle θ_3. This also tends to open the valve. The total force is almost impossible to predict analytically. However, the method is very effective as can be seen from Lee's results in Fig. 4.7(*b*).

4.4. Transient flow forces in spool valves

The forces so far discussed are static and their value can be calculated for a given static valve opening and flow rate. However during unsteady or transient conditions the flow rate is not steady and the fluid accelerations involved give rise to additional flow forces. Consider the fluid slug, of length L_1 and cross-sectional area A, in the valve chamber shown in Fig. 4.8. Newton's law gives

$$F = \text{mass} \times \text{acceleration}$$

or

$$F = \rho L_1 A \frac{\mathrm{d}}{\mathrm{d}t}\left[\frac{Q_1}{A}\right] = \rho L_1 \frac{\mathrm{d}Q_1}{\mathrm{d}t} \tag{4.13}$$

Here the fluid is assumed to be incompressible so that the flow rate through the chamber equals the flow rate through the orifice. In addition the slug is

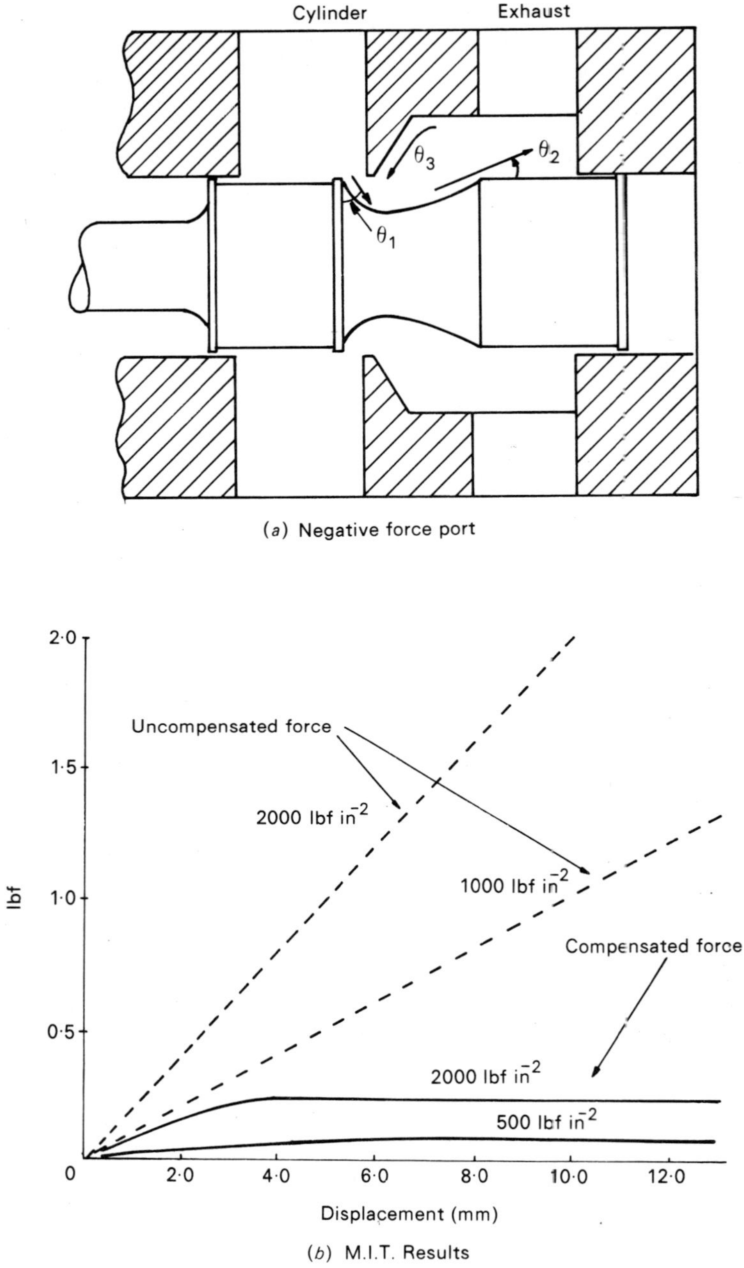

Fig. 4.7. Use of valve shaping to reduce flow force (*a*) a negative force port (*b*) effect on flow force (after Lee and Blackburn [38])

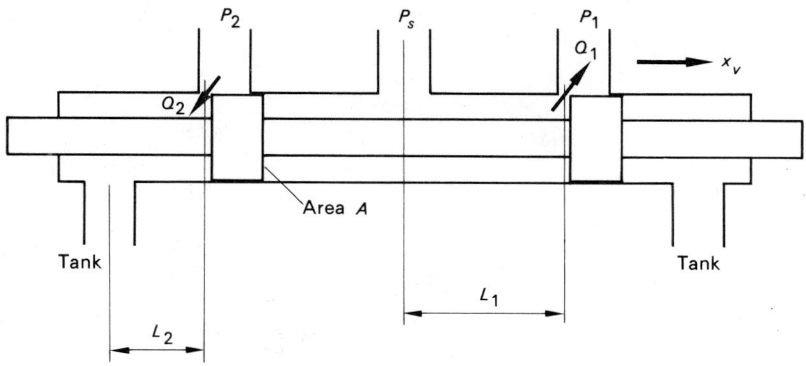

Fig. 4.8. Supply and exhaust damping lengths L_1 and L_2

assumed to move at velocity Q_1/A. Note that F is the force *required* to produce the rate of change of momentum. The reaction F_r on the spool is equal and opposite to F. For example if the flow Q_1 is increasing, the accelerating force F must be in the positive x_v direction, but the reaction F_r on the spool is opposite and tends to close the valve. F_r is thus stabilising. If, however, the same analysis is applied to the outflow Q_2, it is seen that as Q_2 increases, the F_r is destabilising and tends to open the valve further. Thus, in general, flow out of the valve chamber through an orifice is stabilising and flow into the valve chamber is destabilising. The total closing force on the spool in Fig. 4.8 is

$$F_{rt} = \rho L_1 \frac{\mathrm{d}Q_1}{\mathrm{d}t} - \rho L_2 \frac{\mathrm{d}Q_2}{\mathrm{d}t} \tag{4.14}$$

But $Q_1 = Q_2$

$$F_{rt} = \rho(L_1 - L_2)\frac{\mathrm{d}Q_1}{\mathrm{d}t} \tag{4.15}$$

Hence for stability $(L_1 - L_2)$, called the damping length L_d, must be positive [41]. If the damping length is negative, instability can arise and the valve may 'sing'.

Since

$$Q_1 = C_q a \sqrt{\left\{\frac{2}{\rho}(P_s - P_1)\right\}}$$

where $a = kx_v$

Then

$$F_{rt} = \rho L_d k C_q \sqrt{\frac{2}{\rho}}\left\{\frac{\mathrm{d}x_v}{\mathrm{d}t}\sqrt{(P_s - P_1)} + \frac{x_v}{2\sqrt{(P_s - P_1)}}\frac{\mathrm{d}(P_s - P_1)}{\mathrm{d}t}\right\} \tag{4.16}$$

Normally the pressure rate term is small in comparison to the other term so that the transient flow force can be written

$$F_{rt} = \beta\sqrt{\Delta P}\,\frac{dx_v}{dt} \tag{4.17}$$

where $\beta = \rho L_d k C_q \sqrt{(2/\rho)}$
and ΔP = valve orifice pressure drop

4.5. Flow forces in poppet valves

A pressure balanced poppet valve is shown in Fig. 4.9(*a*). It has been shown by Schrenk [42], Fig. 4.9(*b*), that different flow patterns can exist in the downstream chamber and work carried out by one of the present authors [43] showed that three of the patterns could be established in a large two-dimensional poppet valve, using water as the fluid. Flow forces on this model were determined by measuring the pressure distribution around the poppet profile. The non-dimensionalised results are shown in Fig. 4.10. The theoretical result was based on the simple momentum result

$$F = \rho Q v \cos\theta$$

where the jet angle was taken to be the same as the poppet angle, 45° in this case. Forces are higher than the theoretical when Type I flow exists since in this case, as the flow bends after it leaves the orifice, it induces high velocities and hence low pressures along the poppet face. The slight fall off of the measured forces at large flow rates may be due to the neglected inlet momentum effects. Stone's [25] results for a high pressure poppet also showed quite good agreement with theory.

In poppet valves both the size and shape of the downstream chamber and the shape of that part of the poppet in the downstream chamber have a strong influence on the flow force. For instance, if the chamber is very small a tight vortex may form above the poppet, reducing static pressures there and hence reducing the closing force, Fig. 4.10(*b*).

In many ways the poppet valve is easier to compensate than the spool valve. The poppet angle dictates the magnitude of the flow force. If it can be made 90°, the flow force is zero. If a 90° angle is impractical then a vane or appendage can be attached to the poppet in the downstream chamber in such a way that it collects the jet momentum and cancels the closing force. Figures 4.11(*a*) and 4.11(*b*) show how this can be done in practice [44]. In Fig. 4.11(*c*) pressure in the downstream chamber causes an opening force which reduces the closing force. This opening force increases with the square of the flow rate, whilst the closing force is linearly proportional to flow rate and hence it is not possible to achieve perfect compensation for all flow rates. None of the valves in Fig. 4.11 is pressure balanced, but this can easily be achieved by adding an upstream land as in Fig. 4.9(*a*).

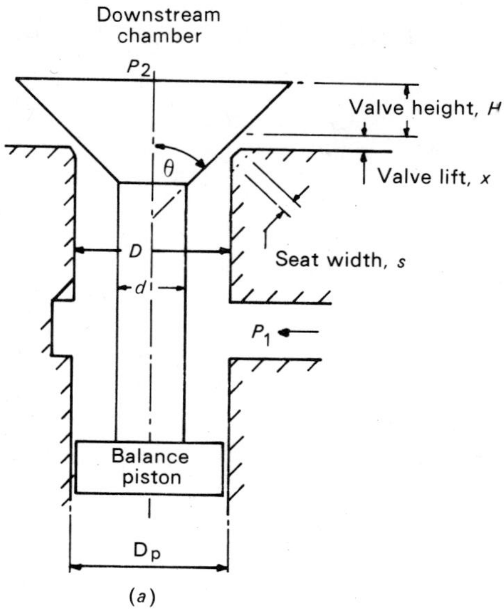

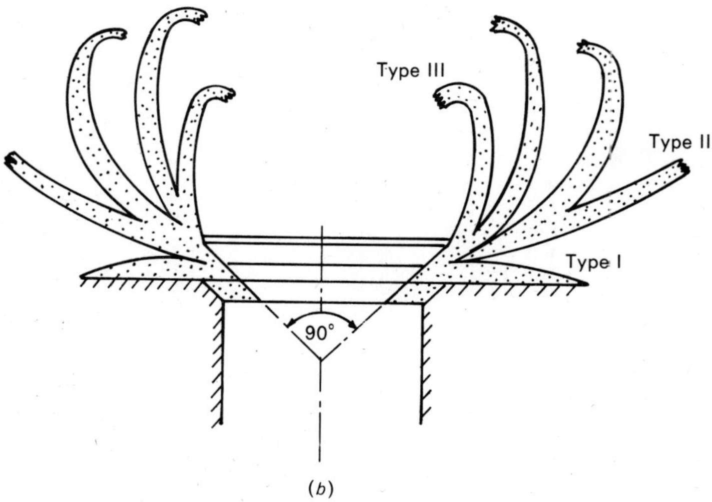

Fig. 4.9. The poppet valve (*a*) a balanced poppet valve (*b*) possible flow patterns

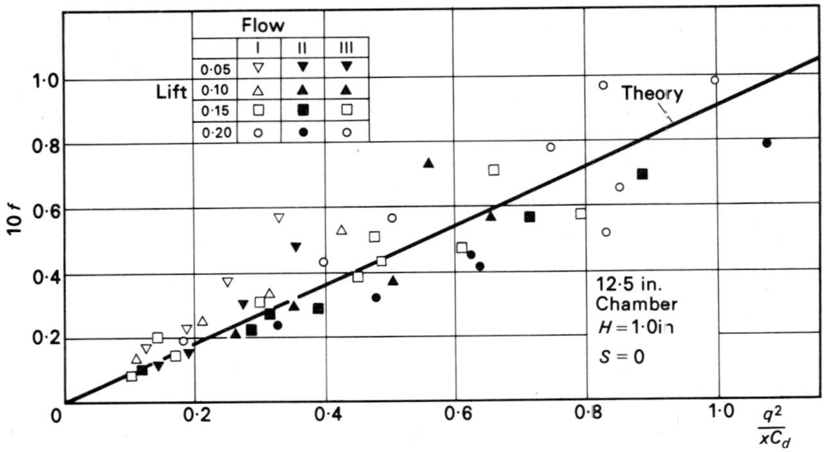

(a) Flow forces on poppet valves

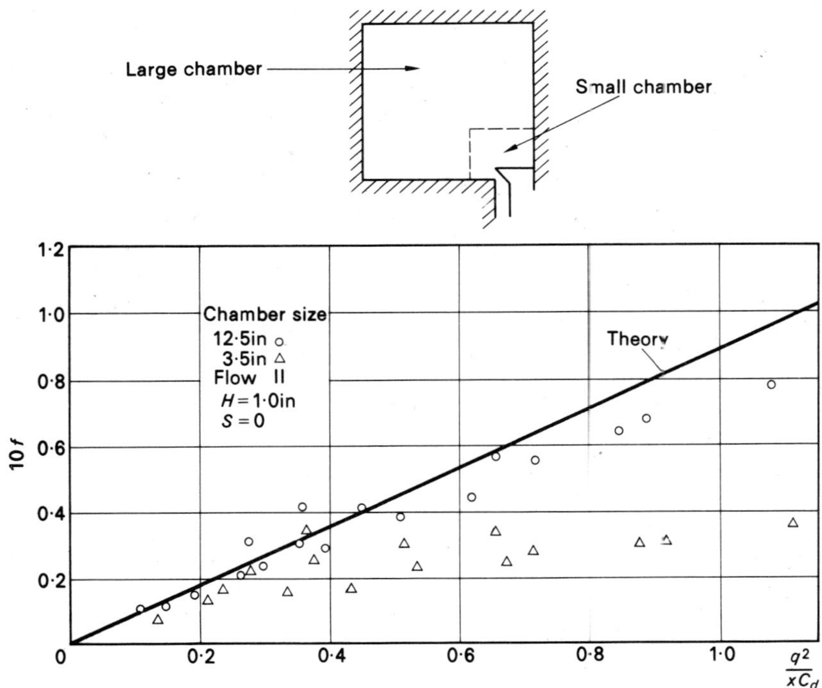

(b) Effect of chamber size on flow force

Fig. 4.10. Measured poppet valve flow forces (*a*) effect of flow pattern (*b*) effect of downstream chamber size

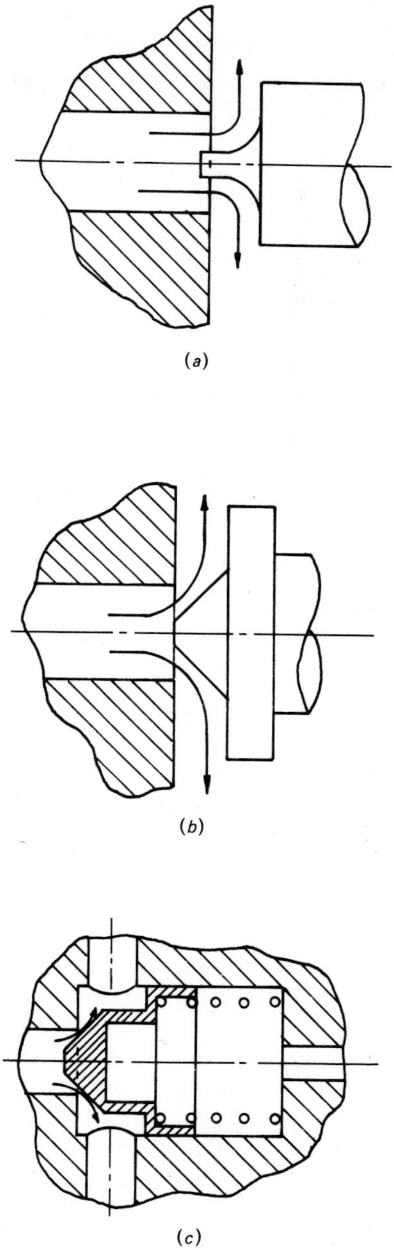

Fig. 4.11. Flow force compensation in the poppet valve (*a*) the 90° poppet (*b*) another means of increasing the poppet angle (*c*) the downstream chamber provides an opening force

4.6. Forces on flapper nozzle valves

A flapper nozzle valve is shown in Fig. 4.12(*a*). The orifice is circular of radius r and has a sharp edge. If the flapper displacement x is small compared to r (less than $r/2$), the curtain area $2\pi rx$ is less than the orifice area and determines the flow rate. The flow rate is given by the orifice equation

$$Q = 2\pi r x C_d \sqrt{(2\Delta P/\rho)} \tag{4.18}$$

where $C_d \approx 0{\cdot}6$.

The total force on the flapper can be found using the analysis of J. F. Blackburn [4]. With the flapper close to the nozzle, Fig. 4.12(*b*), consider three sections of the flow, one far upstream of the nozzle, one at the nozzle and one at the vena contracta. Bernoulli's equation applied at sections 1 and 2 gives

$$P_1 + \tfrac{1}{2}\rho v_1{}^2 = P_s = P_2 + \tfrac{1}{2}\rho v_2{}^2 \tag{4.19}$$

Continuity gives

$$v_1 \pi r^2 = C_d v_2 2\pi r x \tag{4.20}$$

Momentum consideration give

$$F = P_1 \pi r^2 + \rho Q v_1 \tag{4.21}$$

Hence we get

$$F = \pi r^2 P_s[1 + (2C_d x/r)^2] \tag{4.22}$$

This equation holds where $x < r/2$ and the curtain area is the controlling area. In this region the force on the flapper is not far removed from the simple static pressure force $\pi r^2 P_s$. Feng's measured forces [27] show excellent agreement with theoretical predictions, Fig. 4.12(*c*) Equation (4.22) indicates that the flapper force increases with displacement and this effect is not immediately apparent from Fig. 4.12(*c*). The reason for this is clear when it is appreciated that in an actual valve, such as that shown in the figure, the chamber pressure is a function of the flapper displacement and P_s must be replaced by P_c in equation (4.22). Hence when the valve opening is large and the above mentioned increase in force is expected, the chamber pressure is low, the force is small and any force increase above $P_c \pi r^2$ is difficult to see from Feng's results.

Pawlak [45] has examined the forces in a pneumatic flapper nozzle valve. Assuming isentropic flow he calculated results for a sharp edged nozzle, Fig. 4.13(*a*), and these show the same trend as those for an incompressible fluid.

When the nozzle has a flat edge, Fig. 4.13(*b*), analysis becomes much more complicated. If the fluid is incompressible and the flapper displacement is very small, the well developed theory of hydrostatic bearings can be used to determine the flow force. If the flow is laminar an exact solution of the Navier–Stoke's equation gives

$$F = \frac{\pi[(r + w)^2 - r^2]\Delta P}{2 \log_e[1 + (w/r)]} \tag{4.23}$$

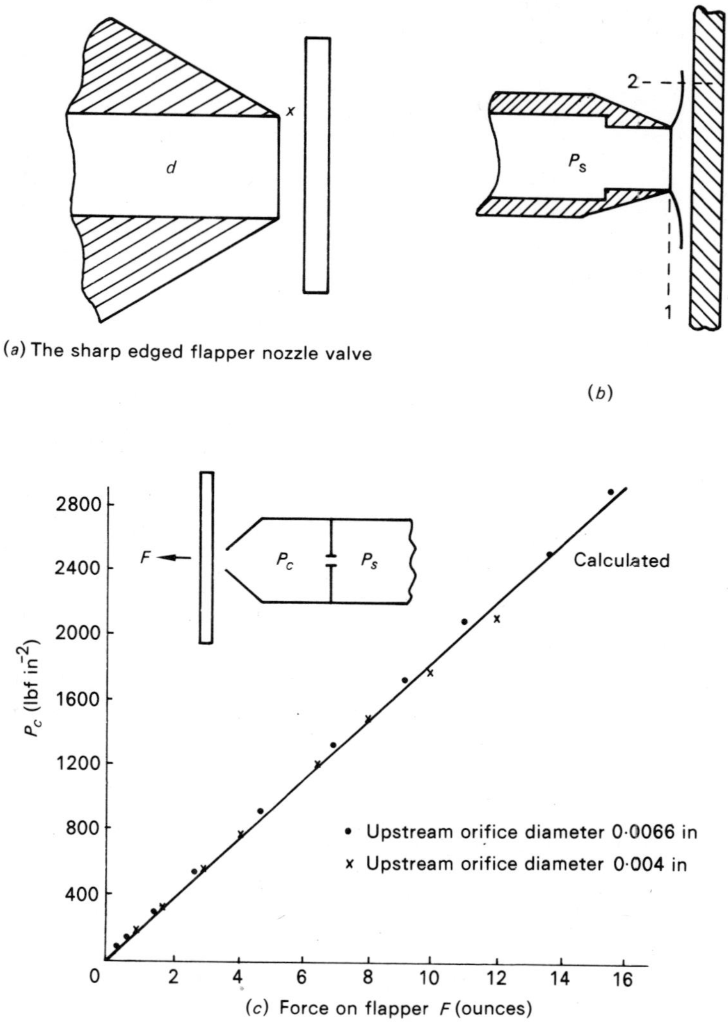

Fig. 4.12. Flow forces in the hydraulic flapper nozzle valve (*a*) the sharp-edged flapper nozzle valve (*b*) the control volume (*c*) flapper force against chamber pressure (after Feng [27])

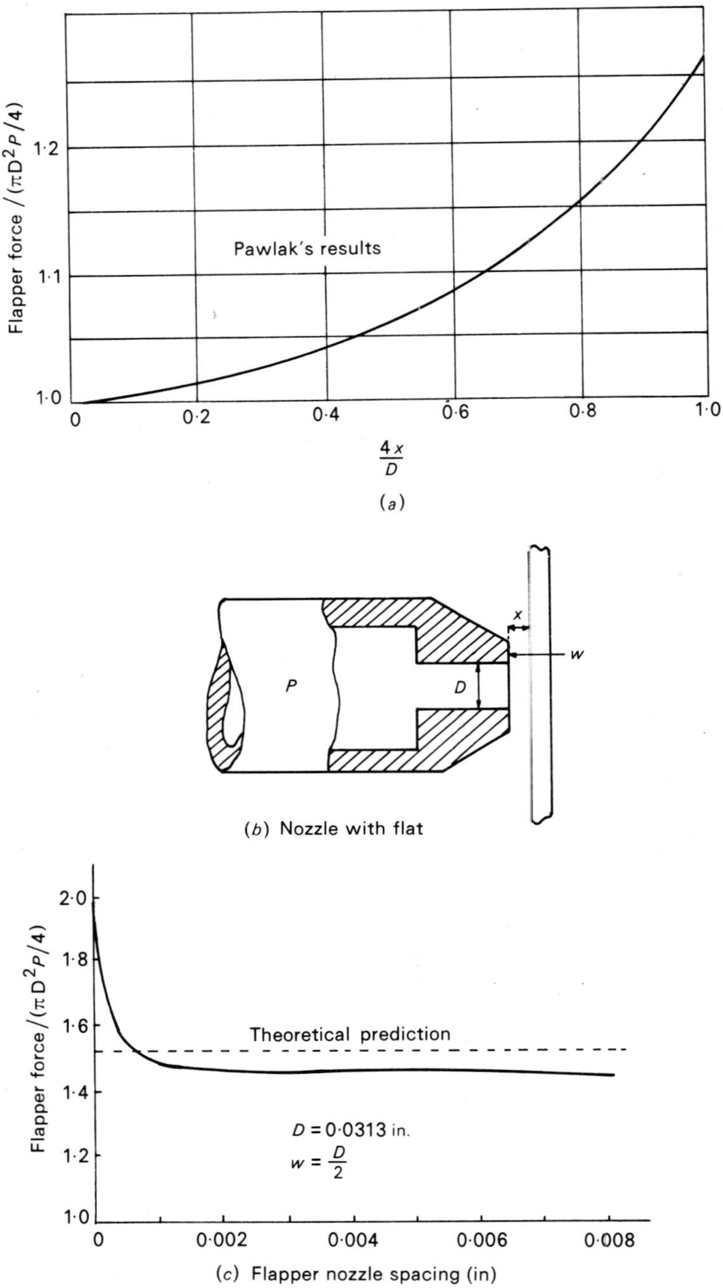

Fig. 4.13. Flow forces in the pneumatic flapper nozzle valve (*a*) theoretical flapper force against flapper lift (after Pawlak [45]) (*b*) the flat-edged nozzle (*c*) measured forces (after Pawlak [45])

where w is the width of the nozzle flat. If we write this equation as

$$F = C_f \pi r^2 \Delta P \tag{4.24}$$

the following variation of C_f with w can be presented.

Table 4.1.

w	0	$0{\cdot}5r$	$1{\cdot}0r$	$1{\cdot}5r$
C_f	1	1·54	2·16	2·87

Hence for very small flapper displacements the flat causes an increase in force (i.e. $C_f > 1$).

As the flapper displacement is increased a stage is reached when the flow separates from the orifice corner and a vena contracta forms. Turbulent flow exists and equation (4.22) applies in this case. The change from one flow condition to another is difficult to predict accurately and in practice such a flapper nozzle exhibits erratic flow and force characteristics.

Pawlak has also examined the effect of nozzle flats in pneumatic flapper nozzle valves. Figure 4.13(*b*) shows that the continuously increasing curtain area in the radial direction forms a diverging nozzle and this flow can be analyzed using the equations of isentropic flow with simple area change. As the curtain area increases in the radial direction, the Mach number and hence the pressure changes. Pawlak has determined the pressure distribution on the flapper and hence the force on the flapper. He examined two possibilities (*a*) subsonic flow, (*b*) supersonic flow in the passage. His results are

Supersonic flow

Table 4.2.

w	0	$0{\cdot}5r$	$1{\cdot}0r$	$1{\cdot}5r$
C_f	1·0	1·308	1·519	1·691

Subsonic flow

Table 4.3.

w	0	$0{\cdot}5r$	$1{\cdot}0r$	$1{\cdot}5r$
C_f	1·0	2·002	3·603	5·742

The pressures in supersonic flow are lower than those in subsonic flow and hence the flapper forces are lower for supersonic flow. Pawlak found that supersonic flow existed in all tests on his apparatus. A sample of his results is shown in Fig. 4.13(*c*). Agreement between theory and practice is very good except at very small spacings, where the measured forces are much larger than the theoretical. It would seem that at these very small openings the flow could be considered laminar and incompressible and the results in Table 4.1 could be used. This indicates that as the spacing tends to zero, the coefficient C_f should rise to 2·16.

Chapter 5

System dynamics

5.1. Introduction

In order to meet the requirements dictated by the specification, the engineer must have some means of estimating the parameters of the system, and its behaviour under dynamic conditions. This requires 'sizing' of the dimensions under steady state conditions and then formulating the dynamic equations for further assessment. The 'goodness' of the mathematical model will depend on how detailed the equations are. Having reached this stage it becomes clear that many types of system; mechanical, pneumatic, electric, etc., result in very similar equations and on this basis analogies can be drawn.

Because hydraulic and pneumatic systems operate using a fluid medium, strong analogies [46] can be drawn with electrical circuits. This allows many of the techniques used by electrical engineers to be used and often gives a clearer understanding of the problems. This approach can be developed further in the analogue computer, where active rather than passive networks can be developed. Care should be taken, however, as there are limitations on the analogies between the flow of fluid and the flow of electrical current.

5.2. Basic relationships

The types of elements to be considered have corresponding electrical analogies. For example, the behaviour of the connecting tubes, or pipes for a network of flow paths is analogous to that of an electrical network. Volume rate of flow (Q) of the working fluid is analogous to the electric current flow (I), and fluid pressure (P) is analogous to electrical voltage (V).

The working fluid may be either a liquid or a gas, and many of the elements may be non-linear, particularly in the case of pneumatic elements. Hence, direct linear analogies may often be limited to a small range.

Fluid systems can be considered as a collection of inter-connected basic elements. The differential equations which represent them are obtained by the application of the laws of networks and junctions.

If ΔP_i is the pressure difference across the *i*th element, then around a closed circuit,

$$\sum \Delta P_i = 0$$

also at any junction of Q_i is the flow,

$$\sum Q_i = 0$$

Hence it is possible to construct an electrical network analogous to a fluid circuit, starting from a knowledge of the electrical components which are analogous to the basic fluid elements.

One basic link between all physical systems is energy or work. For example, the mechanical equivalent of heat is the link between mechanical and thermal systems.

5.2.1. Fluid resistance

The effect of a fluid resistance, or restrictor, is characterised by a pressure drop in the direction of the flow. This is analogous to the drop in voltage across an electrical resistance. Care has to be taken with this analogy, since Ohms law is a linear relationship and a fluid restrictor will only exhibit linear characteristics under laminar flow conditions. The relationships are shown in Table 5.1. There are four basic types of fluid resistance commonly encountered,

Porous plug
Laminar tube flow
Turbulent pipe flow
Orifice

For small pressure drops, compared with the average pressure of the system, the characteristics of these elements can be taken as identical for liquid or gas flows. This allows compressibility effects to be ignored.

The fluid resistance R_h is measured in fluid ohms.

5.2.2 Fluid inductance

The equivalent of electrical inductance is fluid inertance which is a measure of the inertia forces developed when a fluid is accelerated in a pipe or passage. It can be determined experimentally by measuring the rate of change of flow rate in a fluid line, when a given pressure difference is slowly applied across its ends. If the change is too rapid, compressibility effects cannot be ignored. The relationship is shown in Table 5.1, where the fluid inductance L_a is measured in fluid henries.

5.2.3 Fluid capacitance

A fluid element in which potential energy is stored may be thought of as a fluid capacitor, equivalent to an electrical condenser. The capacitance can be

Table 5.1.

Fluid Element	Symbol	Equation	Electrical Analogue	Symbol	Equation
Restrictor	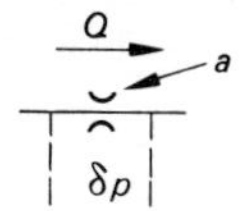	Laminar flow $\delta p = R_h Q$ where $R_h = K_a$	Resistance	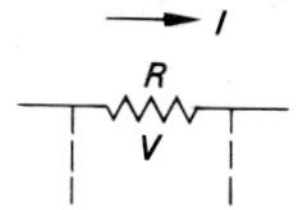	$V = RI$
		turbulent flow $\delta p = R_h Q\|Q\|$	none		none
Inductance (inertance)	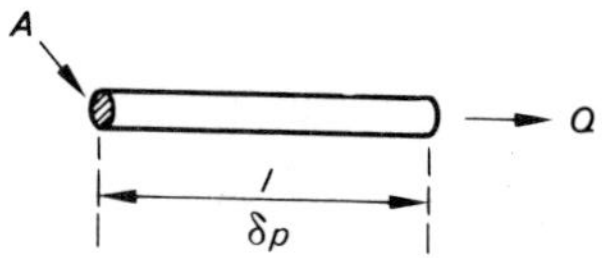 fluid of density in the pipe	$\delta p = L_h \frac{dQ}{dt}$ $L_h = \frac{\rho l}{A}$	Coil	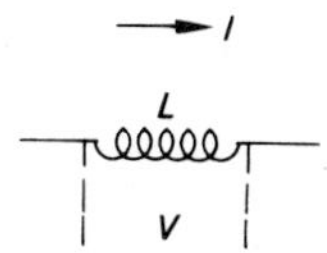	$V = L \frac{dI}{dt}$
Capacitance	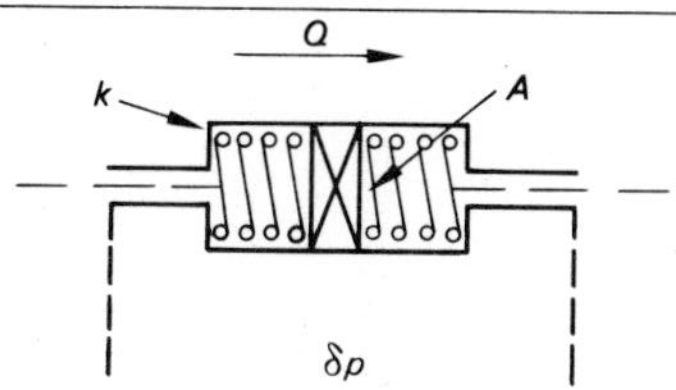	$\frac{d(\delta p)}{dt} = \frac{K}{A^2} Q$ $\delta p = \frac{1}{c_h} \int Q\,dt$ $C_h = A^2/k$	Capacitor	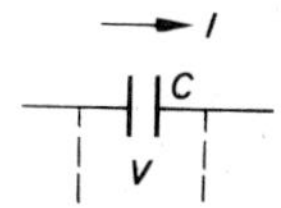	$V = \frac{I}{c} \int I\,dt$
	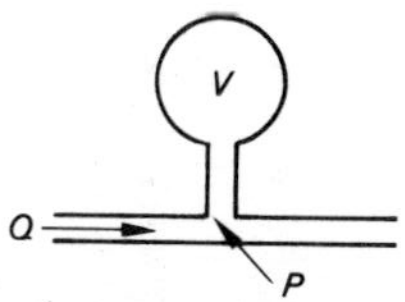 Compressibility of liquid or elasticity of a cylinder	$\frac{dP}{dt} = \frac{N}{V} Q$ or $\frac{dP}{dt} = \frac{1}{k} Q$ N = Bulk modulus k = elasticity		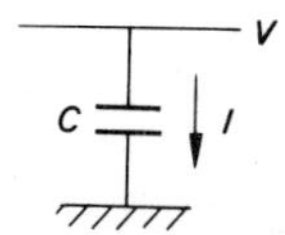	Same

determined experimentally by recording the amount of fluid that must be slowly introduced into a pipe line to produce a unit rise in pressure. Rapid introduction of the fluid could produce fluid density changes. Energy storage can take the form of compression of the liquid, or elasticity of the cylinder or piping walls. The relationships are shown in Table 5.1, where the fluid capacitance C_h is measured in fluid farads.

In general, therefore, R_h, L_h and C_h can be termed the fluid impedances of a fluid system, and hence the general relationship is

$$\text{Pressure} = \text{hydraulic impedance} \times \text{flow rate}$$
$$P = Z_h \times Q$$

Under dynamic conditions, the relationships can be treated in the same way as normal electrical relationships, in which the basic frequency forms are obtained by replacing the operator $\mathrm{d}/\mathrm{d}t$ by $j\omega$.

Example 5.1
Consider a section of pipe of length l, which offers a restriction and has inductance. Assume the line capacitance is zero and the oscillating flow is laminar, as shown in Fig. 5.1. Since the circuits are in series

$$\Delta P = (Z_1 + Z_2)Q = Z_h Q$$

Now

$$Z_h = R_h + L_h \frac{\mathrm{d}}{\mathrm{d}t}$$

or

$$Z_h(j\omega) = R_h + j\omega L_h$$

This is a complex number indicating that there is a phase difference between the pressure and the flow. Suppose that R_h is found to be 2 ohms and ωL_h to be 10 ohms for the quoted frequency condition, then

$$Z_h = \sqrt{\{(2)^2 + (10)^2\}} = 10{\cdot}2\Omega \text{ at } \omega \text{rad/s}^{-1}$$
$$\phi = \tan^{-1}\left(\frac{10}{2}\right) = 78{\cdot}5^\circ$$

Hence the flow lags the pressure by 78·5° at the flow condition of ω rad s^{-1}. Impedance is therefore a vector quantity and can be handled by conventional mathematical methods. Series and parallel circuits are also handled in the normal fashion.

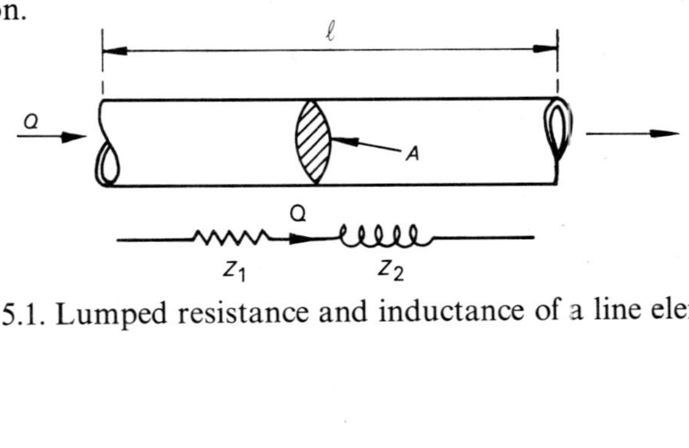

Fig. 5.1. Lumped resistance and inductance of a line element

5.3. A simple system

Consider a simple pump circuit and the electrical equivalent shown in Fig. 5.2. The mass and spring elements are referred to in the electrical diagram as inductive and capacitive elements. For the hydraulic system,

$$F = M\frac{d^2x}{dt^2} + kx = M\frac{dv}{dt} + kx$$

Dividing by the ram area A, and rearranging into the impedance form

$$P = \frac{M}{A^2}\frac{dQ}{dt} + \frac{k}{A^2}\int Q dt$$

$$= L_h\frac{dQ}{dt} + \frac{1}{C_h}\int Q dt$$

Fluid compressibility has been neglected.

This equation can now be converted into its frequency response form as,

$$\frac{P}{Q}(\omega j) = L_h\omega j - \frac{j}{C_h\omega}$$

$$= j\left[\frac{\omega^2 L_h C_h - 1}{\omega C_h}\right] \quad (5.1)$$

For the equivalent electrical circuit,

$$E = L\frac{di}{dt} + \frac{1}{C}\int i\, dt$$

showing the equivalence between the two analogous forms,

$$E \equiv P$$

$$L \equiv L_h = \frac{M}{A^2}$$

$$C \equiv C_h = \frac{A^2}{k}$$

$$i \equiv Q$$

The frequency response of the system can be determined from equation (5.1).

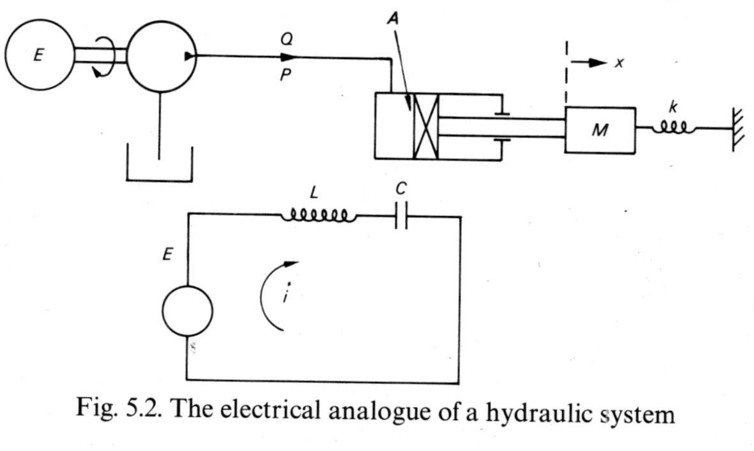

Fig. 5.2. The electrical analogue of a hydraulic system

5.4. Controlling elements

Controlling elements of fluid circuits, i.e. control valves, can be considered as analogous to the electronic valve or the transistor. The former can normally be treated as a variable resistor, where the resistance of the main current path can be varied by control of the grid bias signal. The main voltage and current distribution are changed accordingly. A hydraulic valve can perform the same function in a pipe line if its uncovered orifice is easily varied, as in the case of a spool valve.

There is an analogy between the valve restriction and the resistance of a transistor or of a thermionic valve. There is also an analogy between the valve stroke, the transistor base current and the thermionic valve grid voltage. The equivalent circuits are shown in Fig. 5.3.

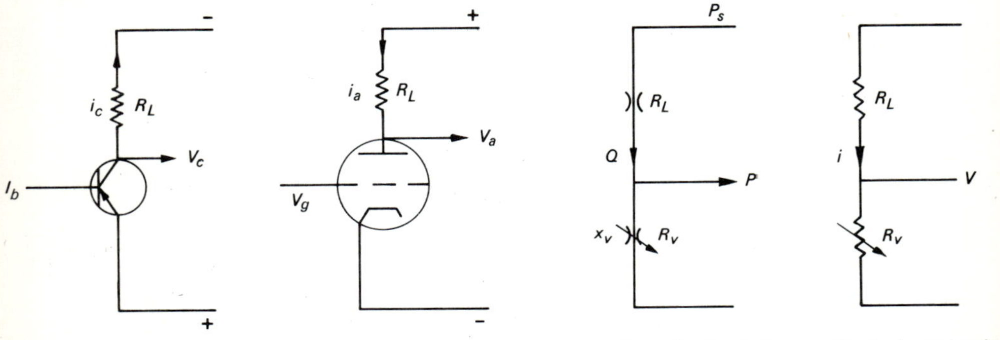

Fig. 5.3. Controlling elements and their equivalent electrical circuit

The static characteristics of each of the devices shown in Fig. 5.3 can be represented by the load line curves, where the control signals I_b, V_g and stroke x_v are equivalent parameters. These curves are shown in Fig. 5.4. The square root relationship for the hydraulic load restrictor, under turbulent flow conditions, is quite noticeable.

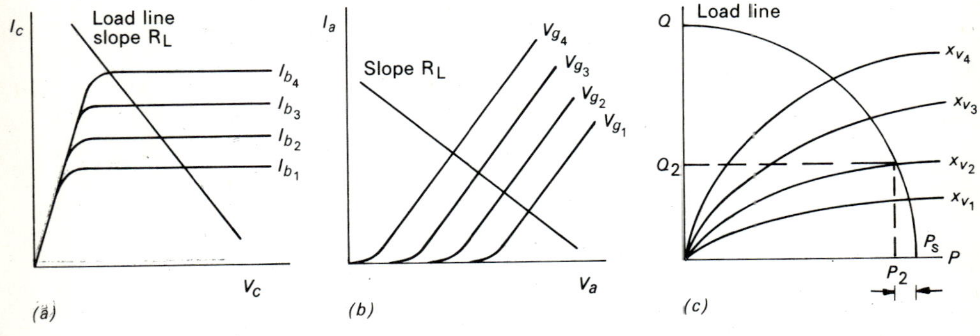

Fig. 5.4. Static characteristics of controlling elements (*a*) the transistor (*b*) the thermionic valve (*c*) the hydraulic valve

The load line establishes the working conditions for the device. For example, in Fig. 5.4(*c*) the flow restrictor R_L fixes the flow Q_2 and pressure P_2 when the valve position is at x_{v2}. One end of the load line is fixed by no flow conditions when the valve is shut. The pressure at the valve must then be the system pressure since there is no flow through R_L. As the valve opens and the fluid flows, the pressure at the 'tee' junction drops along a line determined by the load restrictor R_L. The other end of the load line is fixed by the maximum allowable valve chosen. If no load restrictor was used the valve characteristics could not be controlled within reasonable limits.

The oscillating conditions can be superimposed onto the static characteristics as shown in Fig. 5.5. In both the electronic and hydraulic systems the oscillations

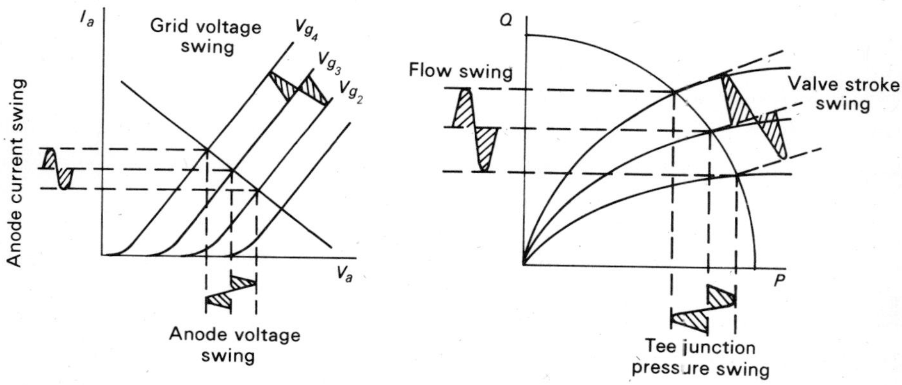

Fig. 5.5. Use of static characteristics to predict response to an oscillatory input when load is purely resistive

occur around the direct current (or flow) level. It is assumed at this stage that the frequency is low in order that inductive and inertia effects can be neglected. For dynamic work, the static characteristics can be ignored once the working conditions have been fixed.

5.5. Dynamic characteristics

The frequency of the input signals in electronic equipment can be raised quite high before inductive effects cause any significant change. This is not the case with hydraulic devices since the fluid inertia in pipe runs is significant and flow attenuation can normally be expected at comparatively low frequencies. Attempts to analyse these effects by extending the characteristics shown in Figs. 5.4 and 5.5 are normally unsatisfactory. When inertia effects are significant, each characteristic curve of both the valve and the load restrictor tend to become elliptical due to the square root effects. These ellipses are traversed once per cycle. The major and minor axes also vary with amplitude and frequency, making manipulation unwieldy.

The alternative approach, which only considers changes about the steady levels, is that of network analysis and this, once again, draws from analogy with electrical practice. The first step is to draw the equivalent circuits, in which the hydraulic header tanks (electrical H.T. supplies) are omitted. An example is shown in Fig. 5.6.

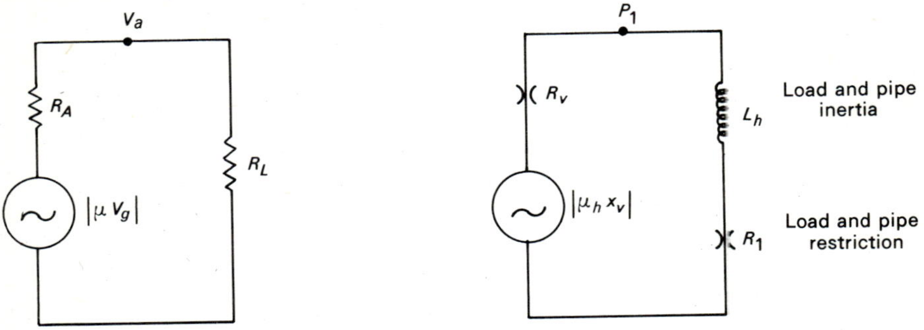

Fig. 5.6. Circuits used in dynamic analysis (*a*) electrical circuit with thermionic valve and resistive load (*b*) hydraulic circuit with valve and resistive and inductive loading

In the case of the electrical circuit, the thermionic valve is regarded as an a.c. generator producing an off-load voltage of μV_g. The current flow is determined by this voltage and $(R_A + R_L)$, where R_A and R_L are the internal and load resistances respectively. The anode voltage V_a, which is measured directly, is less than μV_g by the voltage drop across R_A. The amplification of the valve is μ.

The hydraulic case is similar in that the spool valve is regarded as a generator of alternating flow, producing an off-load pressure source of $\mu_h x_v$. The valve flow is determined by this pressure and the total reactance due to inertia effects and valve and load restrictions.

$$Q = \frac{\mu_h x_v}{(R_v + R_1) + L_h \dfrac{\mathrm{d}}{\mathrm{d}t}} \tag{5.2}$$

The valve output pressure at the 'tee' junction is less than $\mu_h x_v$ because of the drop across the valve restriction.

Pressure P_1 = (flow $Q \times$ impedance Z_h) at that section

$$= \frac{\mu_h x_v}{R(1 + T\omega j)} \times R_1(1 + T_1\omega j) \tag{5.3}$$

where

$$R = (R_v + R_1)$$
$$T = L_h/R$$
$$T_1 = L_h/R_1$$

μV_g and $\mu_h x_v$ are determined from the static characteristics, i.e. μ_h is the change in P_1 with respect to valve stroke when the output flow is constant, measured at the mean operating point x_v, as in Fig. 5.5.

The amount of capacitance in the hydraulic line is inherently small and is only of significance at high frequencies. If, however, capacitance is necessary for shaping the frequency response, then it may be increased with the use of hydraulic accumulators, although with low pressure systems these will tend to be bulky.

If by definition,

$$Q_0 = \mu_h x_v / R \text{ and } P_0 = \mu_h x_v$$

then the frequency response equations or transfer function for the system in Fig. 5·6 will be

$$\frac{Q}{Q_0}(\omega j) = \frac{1}{1 + T\omega j} \tag{5.4}$$

$$\frac{P_1}{P_0}(\omega j) = \frac{R_1(1 + T_1\omega j)}{R(1 + T\omega j)} \tag{5.5}$$

When a valve closes rapidly, pressure builds up in proportion to the hydraulic inductance and the rate of flow change,

$$P = L_h \frac{dQ}{dt}$$

This may cause a shock wave in the pipe resulting in water hammer.

5.6. Design example

For an experimental flow measurement project, [47], it was necessary to design a test rig to produce low frequency pulsating flow. The rig was required to handle

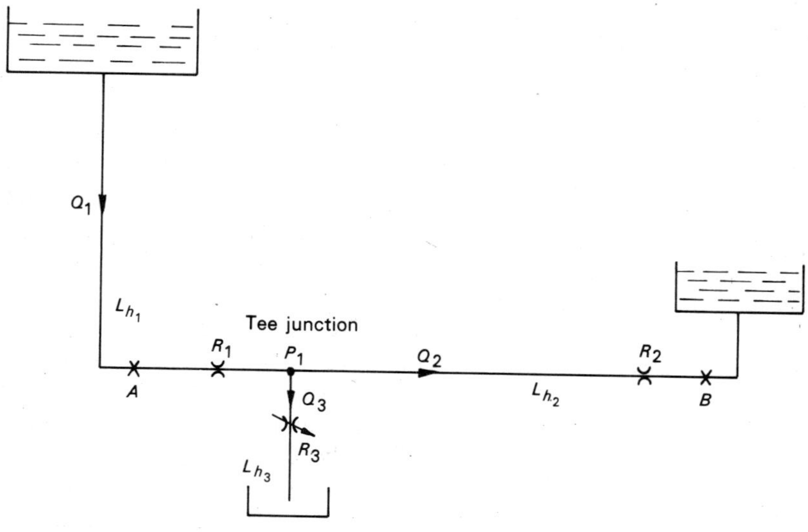

Fig. 5.7. A hydraulic test circuit

maximum steady flows up to 60 gal min^{-1}, of water under pressures up to 20 lbf in^{-2}. Under pulsating conditions, around 10 gal min^{-1} peak to peak was required to be bled off at frequencies up to 10 Hz.

In low pressure large flow systems it is not advisable to use the simple series circuits, since even at quite low frequencies, valve movements may lead to pressure peaks resulting in water hammer. Instead a bleed system is used in which the control valve R_3 is a continuously variable leak in the test pipe. The arrangement is shown in Fig. 5.7.

To make the valve fully effective two restrictors are added upstream and downstream of the variable bleed. The test pipe is supplied from a header tank which discharges into the downstream valve tapping, and also into the downstream tank giving a slight reverse head to ensure that the pipe is always full.

The static characteristics for flow in each pipe branch are shown in Fig. 5.8.

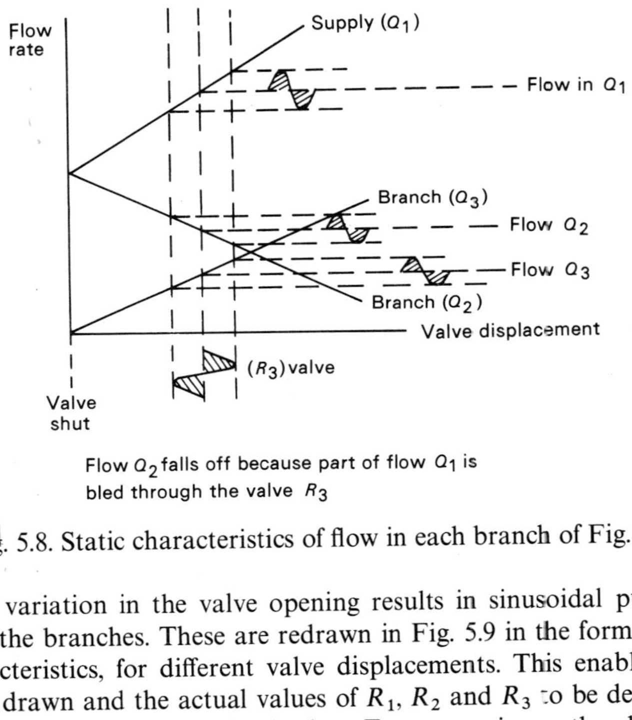

Fig. 5.8. Static characteristics of flow in each branch of Fig. 5.7

Sinusoidal variation in the valve opening results in sinusoidal pulsations in flow in all the branches. These are redrawn in Fig. 5.9 in the form of pressure flow characteristics, for different valve displacements. This enables the load lines to be drawn and the actual values of R_1, R_2 and R_3 to be determined.

In Fig. 5.9(*a*) the control valve is shut. For convenience, the characteristic for the upstream restrictor R_1 is reversed, using the maximum head at the origin. On opening main cocks *A* and *B*, the flow stabilises at a value corresponding to the crossover point. In Fig. 5.9(*b*), the downstream restrictor R_2 is inoperative by closing cock *B*. The valve is set at selected openings at which the flow stabilises at the crossover points depending on the valve opening.

In Fig. 5.9(*c*) the downstream restrictor R_2 curve is combined with the valve curves.

The vertical ordinate crosses the characteristics at the average working points,

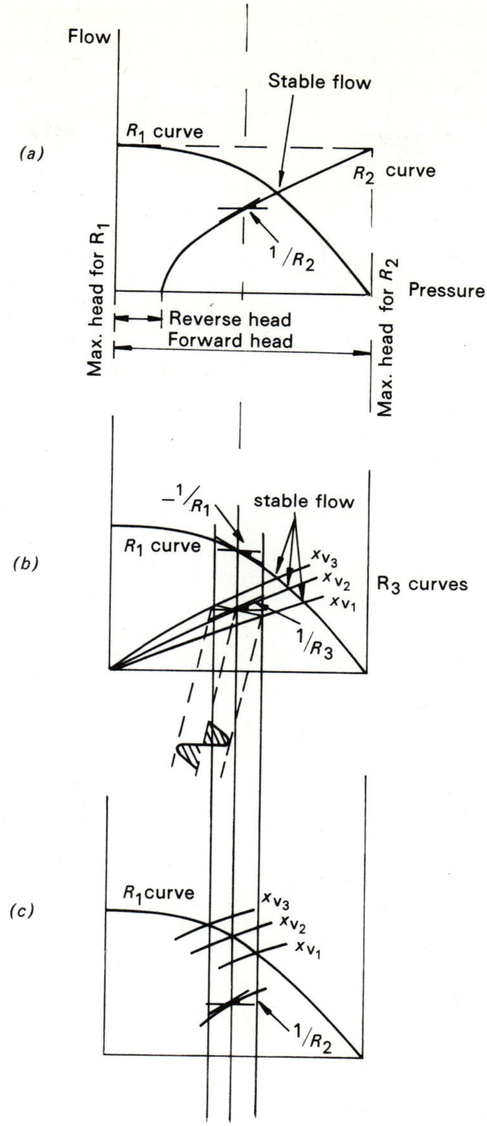

Fig. 5.9. Pressure flow characteristics for the system in Fig. 5.7 (*a*) control valve R_3 closed (*b*) valve R_2 closed (*c*) combination of control valve R_3 and valve R_2

and the slope of each characteristic at these points give the values for R_1, R_2 and R_3 to be used. As discussed earlier it is not feasible to extend this approach into the dynamic case; this is best carried out using network theory.

The approach to a dynamic analysis of the system commences by considering only perturbations about the steady flow levels, and regarding the valve as feeding an alternating flow into the two branches. The two tanks are considered as equivalent to an electrical common earth. The valve produces an alternating pressure P_0 of amplitude $\mu_h x_v$. The tee junction is the valve output joint and the pressure P_1 at this point can be measured. It is less than P_0 by the pressure drop due to the impedance of the valve.

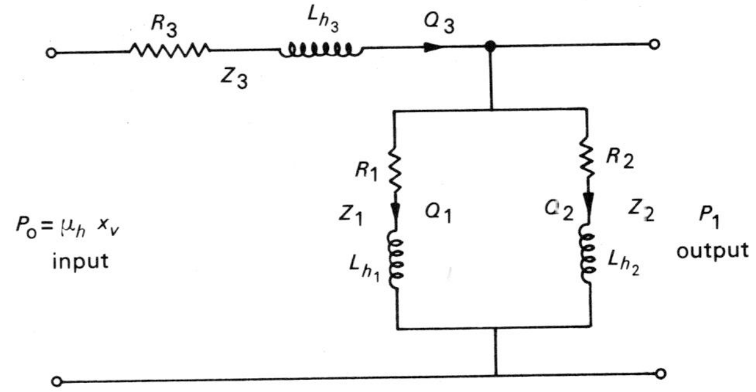

Fig. 5.10. Equivalent electrical circuit for the system in Fig. 5.7.

The equivalent circuit for Fig. 5.7 can now be drawn and is shown in Fig. 5.10. Hence,

$$Z_1 = (R_1 + j\omega L_{h1}) = R_1(1 + j\omega T_1)$$
$$Z_2 = R_2(1 + j\omega T_2)$$
$$Z_3 = R_3(1 + j\omega T_3)$$

$$Z_4 = \frac{Z_1 Z_2}{Z_1 + Z_2}$$

and therefore,

$$\frac{P_1}{P_0}(j\omega) = \frac{Z_4}{Z_3 + Z_4} \tag{5.6}$$

$$Q_2(j\omega) = \frac{P_1}{Z_2} = \frac{P_0}{Z_2} \times \frac{Z_4}{Z_3 + Z_4} \tag{5.7}$$

$$Q_3(j\omega) = \frac{P_0}{Z_3 + Z_4} \tag{5.8}$$

5.7. Fluid transmission lines

Example 5.2 shows that from a simple assessment, this particular pipe has an impedance of 10·2 ohm at the particular frequency of excitation. This sort of information is becoming increasingly more important with the advent of fluid logic devices and a.c. hydraulic systems. In the case of a pneumatic line for example, it takes a finite time for a pressure signal to traverse the length of the line, resulting in a signal delay time. Standing waves can also occur resulting in the phase lag for harmonic inputs exceeding 180°. It is therefore important that enough information be available about line dynamics to ensure adequate stability and optimum impedance matching between elements.

The lumped parameter approach discussed earlier is valid when the time required for a pressure wave to travel the length of the line is short with respect to the period of the highest frequency wave that is to be transmitted. This approach is certainly not suitable for step or impulse responses for example, since the functions contain all frequencies. In other words, single lumped parameter models are only useful for frequency response characteristics over frequency ranges considerably below the natural frequency of the line. A much more accurate and general approach to line dynamics can be achieved using a distributed parameter model.

In the case of liquid transmission lines, the rapid closing of a valve can result in a large pressure transient. The fluid at the valve is stopped suddenly and a pressure wave travels back to the 'incompressible' fluid source at velocity c, given by

$$c = \sqrt{\frac{N}{\rho}}$$

where c = velocity of sound in the fluid
N = effective bulk modulus
ρ = mass density of fluid

When the pressure wave reaches the source at the upstream end of the pipe (in l/c seconds), then the kinetic energy of the moving mass of fluid has been completely stored as potential energy in the elastic pipe and fluid. The pressure of the compressed fluid is a maximum. At this point a decompression wave forms and travels back to the valve. These waves continue to travel up and down the pipe with associated interchanges of kinetic and potential energies until friction expends all the energy.

5.8. Distributed parameter model

Consider frictionless flow in a flexible pipe of length l, as shown in Fig. 5.11. Analysis of the dynamic behaviour of the line, involves the application of the continuity and momentum equations to the elemental control volume of length dx. Now,

$$\frac{\partial}{\partial t}(\rho A)\,\mathrm{d}x = -\frac{\partial}{\partial x}(\rho V A)\,\mathrm{d}x \tag{5.9}$$

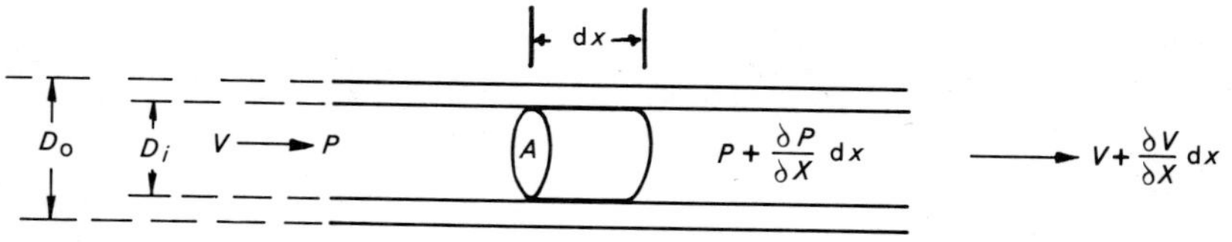

Fig. 5.11. Element of fluid in a flexible pipe

where V is the average velocity of the cross-section. Partial differential of (5.9) gives,

$$A\frac{\partial \rho}{\partial t} + \rho\frac{\partial A}{\partial t} = -VA\left[\frac{\partial \rho}{\partial x} + \rho A\frac{\partial V}{\partial x} + \rho V\frac{\partial A}{\partial x}\right] \tag{5.10}$$

For a compressible fluid

$$\mathrm{d}\rho = \left(\frac{\rho}{N}\right)\mathrm{d}P \tag{5.11}$$

where N is the bulk modulus of the fluid and hence

$$\frac{\partial \rho}{\partial t} = \left(\frac{\rho}{N}\right)\frac{\partial P}{\partial t} \text{ and } \frac{\partial \rho}{\partial x} = \left(\frac{\rho}{N}\right)\frac{\partial P}{\partial x}$$

For a thin walled elastic pipe

$$\mathrm{d}A = \frac{A}{E\left(\frac{D_o}{D_i} - 1\right)}\mathrm{d}P \tag{5.12}$$

and hence

$$\frac{\partial A}{\partial t} = \frac{A}{E(r-1)}\frac{\mathrm{d}P}{\mathrm{d}t} \text{ and } \frac{\partial A}{\partial x} = \frac{A}{E(r-1)}\frac{\partial P}{\partial x}$$

where $r = D_o/D_i$ and E = modulus of elasticity.
From equations (5.10) and (5.11)

$$-VA\frac{\partial \rho}{\partial x} - \rho A\frac{\partial V}{\partial x} - \rho V\frac{\partial A}{\partial x} = \frac{A\rho}{N}\frac{\partial P}{\partial t} + \rho\frac{\partial A}{\partial t}$$

or

$$-\frac{VA}{N}\rho\frac{\partial P}{\partial x} - \rho A\frac{\partial V}{\partial x} - \frac{\partial A}{\partial x} = \frac{A\rho}{N}\frac{\partial P}{\partial t} + \rho\frac{\partial A}{\partial t}$$

Substitution from (5.12)

$$-\frac{VA\rho}{N}\frac{\partial P}{\partial x} - \rho A\frac{\partial V}{\partial x} - \frac{\rho VA}{E(r-1)}\frac{\partial P}{\partial x} = \frac{A\rho}{N}\frac{\partial P}{\partial t} + \frac{\rho A}{E(r-1)}\frac{\partial P}{\partial t}$$

or

$$-\frac{V}{N}\frac{\partial P}{\partial x} - \frac{V}{E(r-1)}\frac{\partial P}{\partial x} - \frac{\partial V}{\partial x} = \frac{1}{N}\frac{\partial P}{\partial t} + \frac{1}{E(r-1)}\frac{\partial P}{\partial t}$$

Now

$$\frac{1}{N_e} = \frac{1}{N} + \frac{1}{E(r-1)} \tag{5.13}$$

Hence

$$-\frac{V}{N_e}\frac{\partial P}{\partial x} - \frac{\partial V}{\partial x} = \frac{1}{N_e}\frac{\partial P}{\partial t} \tag{5.14}$$

The net force acting on the element in the x direction is

$$\text{Force} = -A\left(\frac{\partial P}{\partial x}\right)\mathrm{d}x$$

Hence

$$A\frac{\partial P}{\partial x} + \frac{\partial}{\partial t}(\rho V A) + \frac{\partial}{\partial x}(\rho V^2 A) = 0 \tag{5.15}$$

and completing the partial differentiation

$$A\frac{\partial P}{\partial x} + VA\frac{\partial \rho}{\partial t} + \rho A\frac{\partial V}{\partial t} + \rho V\frac{\partial A}{\partial t} + V^2 A\frac{\partial \rho}{\partial x} + V^2\rho\frac{\partial A}{\partial x} + 2V\rho A\frac{\partial V}{\partial x} = 0$$

Substitute from equations (5.11) and (5.12):

$$\left(A + \frac{V^2\rho A}{E(r-1)} + \frac{V^2 A\rho}{N}\right)\frac{\partial P}{\partial x} + 2V\rho\frac{\partial V}{\partial x} + \rho A\frac{\partial V}{\partial t} + \left(\frac{VA\rho}{N} + \frac{VA\rho}{E(r-1)}\right)\frac{\partial P}{\partial t}$$
$$= 0$$

From (5.13) and rearranging

$$\left(1 + \frac{V^2\rho}{N_e}\right)\frac{\partial P}{\partial x} + \rho\frac{\partial V}{\partial t} + 2V\rho\frac{\partial V}{\partial x} + \frac{V\rho}{N_e}\frac{\partial P}{\partial t} = 0 \tag{5.16}$$

For most fluid power applications, V is small with respect to the velocity of sound in the fluid, and therefore for frictionless compressible flow.

Continuity equation (from (5.14))

$$-\frac{\partial Q}{\partial x} = \frac{A}{N_e}\frac{\partial P}{\partial t} \tag{5.17}$$

Momentum equation (from (5.16))

$$-\frac{\partial P}{\partial x} = \frac{\rho}{A}\frac{\partial Q}{\partial t} \tag{5.18}$$

where Q is the volumetric flow rate.

In the presence of friction forces, the momentum equation becomes

$$-\frac{\partial P}{\partial x} = \frac{\rho}{A}\frac{\partial Q}{\partial t} + K_e Q \tag{5.19}$$

where ρ and K_e are the effective values of the coefficients under pulsating conditions.

For laminar steady flow

$$K_e = \frac{128}{\pi}\frac{\mu}{D^4} \text{ per unit length}$$

For turbulent steady flow

$$K_e = \frac{8 f\mu}{\pi D^3} \text{ per unit length}$$

Equations (5.17) and (5.19) can be written in a more general form,

$$\frac{\partial E}{\partial x} = - Z(s)F \tag{5.20}$$

$$\frac{\partial F}{\partial x} = - Y(s)E \tag{5.21}$$

where E is the across variable or 'effort' and F is the through variable or 'flow'.

Since the line possesses the general properties of inertance, capacitance and resistance, the series impedance and shunt admittance are given by, fig. 5.12.

$$Z(s) = Is + R$$
$$Y(s) = Cs$$

where I = inertance per unit length of line
C = capacitance per unit length of line
R = resistance per unit length of line
s = operator d/dt

By further partial differentiation of (5.20) and substitution,

$$\frac{\partial^2 E}{\partial x^2} = - Z(s)\frac{\partial F}{\partial x} = Z(s)Y(s)E$$

or

$$E = \frac{1}{Z(s)Y(s)}\frac{\partial^2 E}{\partial x^2} \tag{5.22}$$

also

$$F = \frac{1}{Z(s)Y(s)}\frac{\partial^2 F}{\partial x^2} \tag{5.23}$$

These are the well known wave equations.

Hence, in the case of a frictionless transmission line

$$Z(s) = \frac{\rho}{A} s \text{ and } Y(s) = \frac{A}{N_e} s$$

It is convenient to define

$$\gamma = \sqrt{(Z(s)Y(s))}$$
$$= \sqrt{\left(\frac{\rho}{N_e}\right)} s$$

called the 'propagation constant'.

The 'propagation operator' is defined as

$$\Gamma = \gamma x$$

and the 'surge' or 'characteristic impedance' as

$$Z_c = \sqrt{\frac{Z(s)}{Y(s)}}$$

Readers familiar with electrical transmission line theory [48] will recognise the strong analogies between the two systems.

If equations (5.22) and (5.23) are solved subject to the boundary conditions

$$E = E_a \text{ and } F = F_a \text{ at } x = 0$$
$$E = E_b \text{ and } F = F_b \text{ at } x = L$$

the results can be conveniently expressed in terms of hyperbolic functions:

$$E_b = (\cosh \Gamma)E_a - Z_c(\sinh \Gamma)F_a$$

$$F_b = -\frac{1}{Z_c}(\sinh \Gamma)E_a + (\cosh \Gamma)F_a$$

On this basis a transmission line can be considered as a continuous distribution of series impedance Z and shunt admittance Y per unit length as shown in Fig. 5.12.

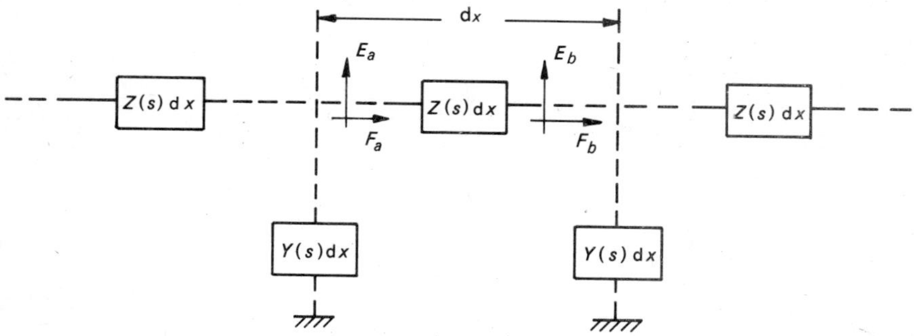

Fig. 5.12. Representation of a fluid line as a distribution of series impedances Z and shunt admittances Y

Space considerations do not permit further discussion of distributed parameter fluid systems. It is hoped that this short introduction will encourage the reader to study the appropriate literature, e.g. [159].

Chapter 6

Hydraulic servomechanisms

6.1. Introduction

The hydraulic servomechanism in its basic form is shown in Fig. 6.1(*a*). When lightly loaded it acts as an integrator, since the flow rate, controlled by the spool valve position, produces a rate of movement of the output piston rod.

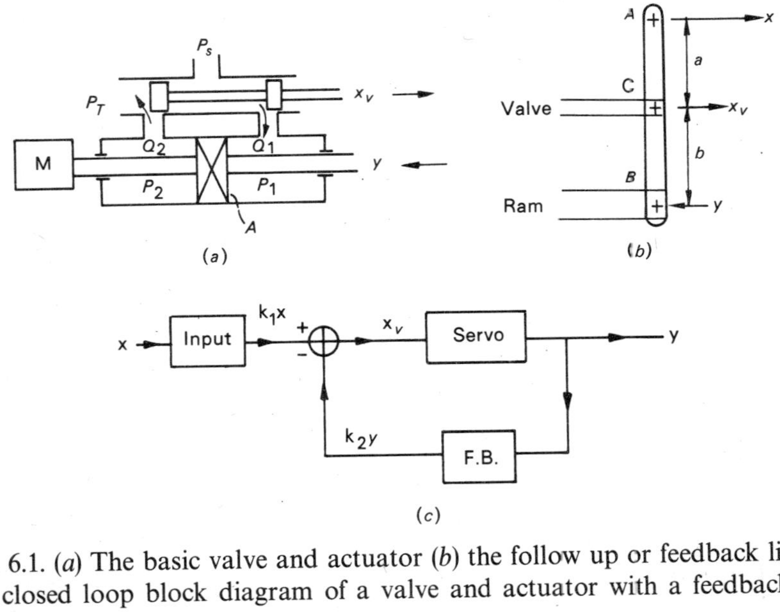

Fig. 6.1. (*a*) The basic valve and actuator (*b*) the follow up or feedback link (*c*) the closed loop block diagram of a valve and actuator with a feedback link

The flow through the valve is, for positive valve displacement

$$Q_1 = C_q a \sqrt{\frac{2}{\rho}(P_s - P_1)} \tag{6.1}$$

and

$$Q_2 = C_q a \sqrt{\frac{2}{\rho}(P_2 - P_T)} \tag{6.2}$$

Normally, the tank pressure is assumed atmospheric, and the flow rates Q_1 and Q_2 are assumed equal. The effects of compressibility are also ignored in this simple analysis.
Hence,

$$\text{mean flow rate } Q = \frac{Q_1 + Q_2}{2} = \frac{C_q a}{2}\sqrt{\frac{2}{\rho}}[\sqrt{(P_s - P_1)} + \sqrt{P_2}] \qquad (6.3)$$

but $P_s - P_1 = P_2$ from equations (6.1) and (6.2)

and therefore

$$Q = C_q a\sqrt{\frac{2}{\rho}}[\sqrt{(P_s - P_1)}] \qquad (6.4)$$

Now if

the load pressure $P_L = P_1 - P_2$

then $\qquad 2P_1 = P_L + P_s$

Hence

$$P_s - P_1 = P_s - \tfrac{1}{2}P_L - \tfrac{1}{2}P_s = \tfrac{1}{2}(P_s - P_L)$$

$$Q = C_q a\sqrt{\frac{2}{\rho}}\sqrt{\left(\frac{P_s - P_L}{2}\right)} \qquad (6.5)$$

If the load at the output is small, then the load pressure is small and $P_1 = P_2 \approx \frac{1}{2}P_S$

$$\therefore Q = A\frac{dy}{dt} = K_1 x_v\sqrt{\frac{P_s}{2}} \text{ where } K_1 x_v = C_q a\sqrt{\frac{2}{\rho}}$$

or

$$y = \frac{K_1}{A}\sqrt{\frac{P_s}{2}}\int x_v \, dt \qquad (6.6)$$

In most cases it is required to have position control of the output, and a simple way to achieve this is to introduce a follow up lever as shown in Fig. 6.1(*b*). A movement at the input x is initially pivoted about B, causing a displacement of the spool at C. The spool displacement causes flow into the ram to move in a direction to cancel the spool displacement. During this movement the lever pivot point is at A. The block diagram is shown in Fig. 6.1(*c*).

Consider similar triangles on the lever geometry.

Pivot about B: $\qquad k_1 x = \dfrac{b}{a + b}x$

Pivot about A: $\qquad k_2 y = \dfrac{a}{a + b}y$

Hence

$$x_v = k_1 x - k_2 y = \frac{b}{a+b}x - \frac{a}{a+b}y \tag{6.7}$$

but

$$x_v = \frac{A}{K\sqrt{(P_s/2)}}\frac{dy}{dt}$$

$$\frac{A}{K_1\sqrt{(P_s/2)}}\frac{dy}{dt} + \frac{a}{a+b}y = \frac{b}{a+b}x$$

$$\frac{y}{x} = \frac{b/a}{1 + \left(\dfrac{a+b}{a}\dfrac{A}{K_1\sqrt{(P_s/2)}}\right)s} \tag{6.8}$$

where $s \equiv d/dt$ operator.

It is easily seen from this transfer function, that in the steady state, $y = (b/a)x$. This is a position control. Dynamically the system is a simple lag with,

$$\text{Gain} = b/a$$

$$\text{Time constant} = \frac{a+b}{a}\frac{A}{K_1\sqrt{(P_s/2)}}$$

6.2. Precision control systems in hydraulics

Generally speaking, the hydraulic servomotor provides a fast response, high torque and short stroke characteristics which are satisfactory for many fields of application. In the aircraft industry this type of actuator is used for power operated controls, autostabilisers and autopilots. In general industry it finds favour in speed control systems for prime movers, and in the operation of process control valves.

The machine tool industry requires control systems that are able to produce high accuracy with good finish. The machine must be rigid and must be able to produce a good standard of repeatability. The introduction of automatic control techniques and hydraulic servos has brought vast improvements and even further improvement can be achieved by making the system computer controlled. These methods, however, could not be fully effective without the introduction of the special techniques developed in engineering measurements, so that accurate positioning to 0·0004 in or better can be obtained.

Many machines tools use hydraulic systems for both the basic drive mechanisms and the control of the cutting tool. In general the advantages of hydraulic components are,

Ease of accurate control of workable velocity and position.
Good stiffness characteristics.
Zero backlash.
Rapid response to change in speed or direction.
Low rate of wear.

Good stiffness characteristics are especially important in the light of recent developments in automatic machine tool control and metal cutting which has emphasised the need for rigid, backlash free, drives with an adequate response to control demands.

Less sophisticated machine tool control systems use direct input signals to the control valve from a template. For more complex metal cutting problems numerical control is commonly used.

The effects of compressibility of the oil and the contribution to stiffness by the mechanical structure associated with the simple hydraulic servomotor are often not fully appreciated, especially if the servomotor is accelerating a large mass. Furthermore, the output may be subjected to external forces, these introducing signals into the loop via the servomotor output. Some idea of the effect of these conditions on performance can be gathered from a linearised analysis of the system.

6.3. Effects of oil compressibility

A simplified model of a hydraulic servomotor can be developed, taking into account compressibility and the output load inertia. In this case the load pressure which acts across the piston in Fig. 6.1(*a*) can become large during acceleration and retardation periods. Under these conditions oil compressibility must be taken into account as this can be shown to have a destabilising effect. It will be seen that the equations describing the system are non-linear in nature and it is necessary to linearise them by considering small perturbations about a mean value. This approach was first introduced by N. F. Harpur [49].

From Fig. 6.1(*a*) it is seen that flow into the ram is expressed as,

$$Q_1 = f(x_v, P_s - P_1) \tag{6.9}$$

in other words the flow rate into the ram is a function of valve displacement and pressure drop across the control orifice,—we are not concerned with the exact relationship at this stage.

Similarly, the flow rate out of the other side of the ram is given by

$$Q_2 = f(x_v, P_s - P_T) \tag{6.10}$$

where P_T is normally zero gauge.

Consideration of the flow relationship through the ram must take into account the compressibility of the working fluid.

If the bulk modulus of the fluid is defined as:

$$N = -V\frac{dP}{dV} \tag{6.11}$$

then it can be shown, equation 1.2, that the flow rate into the driving side of the ram is:

$$Q_1 = \frac{V_1}{N_1}\frac{dP_1}{dt} + A\frac{dx_0}{dt} \text{ where } y = x_0 \tag{6.12}$$

Or in physical terms,

Q_1 = (flow rate due to fluid compressibility)
+ (flow rate due to piston movement)

By a similar argument,

$$Q_2 = -\frac{V_2}{N_2}\frac{dP_2}{dt} + A\frac{dx_0}{dt} \tag{6.13}$$

Equations (6.12) and (6.13) can now be linearised, giving

$$\Delta Q_1 = A\frac{d}{dt}\Delta x_0 + \frac{V_1}{N_1}\frac{d}{dt}\Delta P_1$$

and

$$\Delta Q_2 = A\frac{d}{dt}\Delta x_0 - \frac{V_2}{N_2}\frac{d}{dt}\Delta P_2$$

Hence the average flow through the ram is given by:

$$\frac{\Delta Q_1 + \Delta Q_2}{2} \quad \frac{1}{2}\left[\frac{V_1}{N_1}\frac{d}{dt}\Delta P_1 - \frac{V_2}{N_2}\frac{d}{dt}\Delta P_2 + 2A\frac{d}{dt}\Delta x_0\right] = \Delta Q$$

If V is the total swept volume of the ram and N the common bulk modulus, then about the centre position of the ram:

$$V_1 = V_2 = V/2$$

and hence,

$$\Delta Q = A\frac{d}{dt}\Delta x_0 + \frac{V}{4N}\frac{d}{dt}(\Delta P_1 - \Delta P_2) \tag{6.14}$$

The most interesting case from a stability point of view, is when the external load is mainly inertia and internal or external friction is ignored. This means that any system damping is due to the control valve characteristics.
The load equation is then

$$P_1 - P_2 = \frac{M}{A}\frac{d^2x_0}{dt^2}$$

or about the mean

$$\Delta P_1 - \Delta P_2 = \frac{M}{A}\frac{d^2}{dt^2}\Delta x_0 \tag{6.15}$$

The non-linear relationship associated with flow through the valve, equations (6.1) and (6.2) are linearised by applying Taylor's formula for two variables and neglecting all powers above the first.

$$\Delta Q_1 = \frac{\partial Q_1}{\partial x_v}\Delta x_v + \frac{\partial Q_1}{\partial P_1}\Delta P_1 \tag{6.16}$$

$$\Delta Q_2 = \frac{\partial Q_2}{\partial x_v}\Delta x_v + \frac{\partial Q_2}{\partial P_2}\Delta P_2 \tag{6.17}$$

where $Q_1 = Q_s + \Delta Q_1$; $x_v = x_{vs} + \Delta x_v$, etc., and the partial derivatives are evaluated at the particular steady state conditions (P_{1s}, P_{2s}, x_{vs}, Q_s). In other words the flow is made up of a contribution due to valve displacement, and a contribution due to pressure change. (see section 3.8).

These equations are now linear although their coefficients are functions of the steady state conditions at which the system is being studied. In the steady state the flows into and out of the ram are equal, hence,

$$\frac{\partial Q_1}{\partial x_v} = \frac{\partial Q_2}{\partial x_v} = C_{xv} \tag{6.18}$$

$$-\frac{\partial Q_1}{\partial P_1} = \frac{\partial Q_2}{\partial P_2} = C_p \tag{6.19}$$

These results are shown in Fig. 6.2.

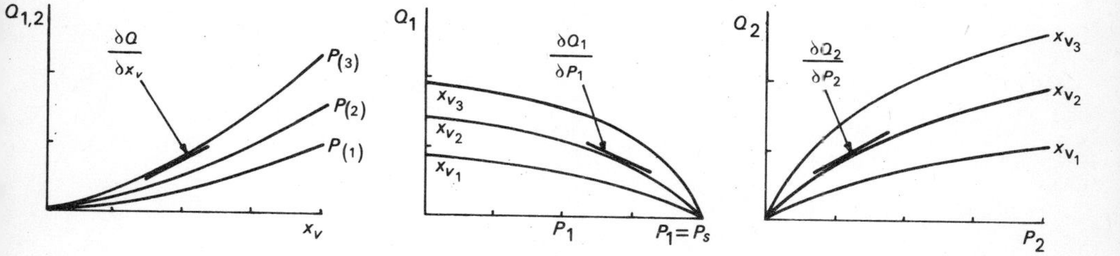

Fig. 6.2. Static characteristics of the critical centre four-way valve of Fig. 6.1.

The average flow through the valve can, therefore, be written from equations (6.16) and (6.17).

$$\Delta Q = C_{xv}\,\Delta x_v - \frac{C_p}{2}(\Delta P_1 - \Delta P_2) \tag{6.20}$$

Equating (6.14) and (6.20)

$$C_{xv}\,\Delta x_v - \frac{C_p}{2}(\Delta P_1 - \Delta P_2) = \frac{V}{4N}\frac{\mathrm{d}}{\mathrm{d}t}(\Delta P_1 - \Delta P_2) + A\frac{\mathrm{d}}{\mathrm{d}t}\Delta x_0 \tag{6.21}$$

and introducing the load term from equation (6.15)

$$C_{xv}\,\Delta x_v - \frac{C_p}{2}\frac{M}{A}\frac{\mathrm{d}^2}{\mathrm{d}t^2}\Delta x_0 = \frac{VM}{4NA}\frac{\mathrm{d}^3}{\mathrm{d}t^3}\Delta x_0 + A\frac{\mathrm{d}}{\mathrm{d}t}\Delta x_0 \tag{6.22}$$

Equation (6.22) is the forward path equation for the servomotor. If now a unity feedback to the valve is arranged, then,

$\Delta x_v = \Delta x_i - \Delta x_0$ where Δx_i is the input movement.

The student should also examine the effect of non-unity feedback, (e.g. 6.7). Combining with equation (6.22) the transfer operator for the closed loop condition is

$$\Delta x_o = \left[\frac{C_{xv}/A}{\dfrac{VM}{4NA^2} s^3 + \dfrac{C_p M}{2A^2} s^2 + s + \dfrac{C_{xv}}{A}} \right] \Delta x_i \tag{6.23}$$

The combined stiffness of the two oil columns, each of length $L/2$, is easily shown to be

$$k_o = \frac{4NA^2}{V} \tag{6.24}$$

In the steady state, equation (6.22) reduces to

$$\frac{\mathrm{d}}{\mathrm{d}t} \Delta x_o = \frac{C_{xv}}{A} \Delta x_v = \frac{1}{A} \frac{\partial Q}{\partial x_v} \Delta x_v$$

giving the open path gain relationship between the piston rate and the spool valve displacement.

The second term of (6.23) can be written

$$\frac{C_p M}{2A^2} = \frac{C_{xv}}{A} M \frac{C_p}{2AC_{xv}}$$

$$= \frac{KM}{k_v}$$

where $K = C_{xv}/A$.

$k_v = 2A(C_{xv}/C_p)$ is called the valve stiffness.

The closed loop equation (6.23) can now be rewritten in a more general form

$$\Delta x_o = \left[\frac{K}{\dfrac{M}{k_o} s^3 + \dfrac{KM}{k_v} s^2 + s + K} \right] \Delta x_i \tag{6.25}$$

The open path transfer operator from (6.23) then becomes

$$\Delta x_o = \left[\frac{K}{s\left(\dfrac{M}{k_o} s^2 + \dfrac{KM}{k_v} s + 1\right)} \right] \Delta x_v \tag{6.26}$$

The effect of compressibility is to introduce a quadratic lag term to the basic integration of equation (6.6). The system can, therefore, be subject to unstable oscillations.

Equation (6.26) can be written in the more familiar form

$$\Delta x_o = \left[\frac{K}{s\left(\dfrac{1}{\omega_n{}^2} s^2 + \dfrac{2c}{\omega_n} s + 1\right)} \right] \Delta x_v \tag{6.27}$$

where $\omega_n^2 = k_o/M = 4NA^2/MV$ and c is the damping ratio.

If there are long pipe runs between the components, the inertia of the fluid in the pipes must be considered in addition to the load inertia.

Consider a ram of area 'A' acted on by a pressure P_1 and accelerating a mass 'M' at an acceleration of $\mathrm{d}^2x_o/\mathrm{d}t^2$. The fluid is exhausted through a pipe of area 'a', with a mass 'm' of fluid in the pipe. Let the fluid in the pipe accelerate at $\mathrm{d}^2z/\mathrm{d}t^2 = (A/a)\mathrm{d}^2x_o/\mathrm{d}t^2$ and P_2 be the exhaust pressure.

Hence,

$$A(P_1 - P_2) = M\frac{\mathrm{d}^2x_o}{\mathrm{d}t^2}$$

The force applied to the fluid in the ram to accelerate the mass of fluid in the pipe will be

$$P_2a = m\frac{\mathrm{d}^2z}{\mathrm{d}t^2} = m\frac{A}{a}\frac{\mathrm{d}^2x_o}{\mathrm{d}t^2} \tag{6.28}$$

and therefore

$$P_1A = \left(M + \frac{A^2}{a^2}m\right)\frac{\mathrm{d}^2x_o}{\mathrm{d}t^2} \tag{6.29}$$

This can considerably increase the effective inertia seen by the ram.

6.4. Stability considerations

Applying the Routh–Hurwitz stability criterion, to the characteristic equation of the transfer function (6.25), gives the relationship

$$\frac{M}{k_o}K < \frac{KM}{k_v} \text{ for stability}$$

or

$$k_o > k_v$$

In other words the stiffness of the oil column must be greater than the effective valve stiffness to maintain stability.

Comparing equations (6.26) and (6.27) produces the relationship

$$\text{damping ratio } c = \frac{KM}{k_v}\frac{\omega_n}{2} = \frac{K\sqrt{(Mk_o)}}{2k_v}$$

This shows that the stiffness of the valve, k_v, must be reduced to improve stability. This is more effective than increasing the oil column stiffness. In practice the damping term will include other friction terms and the structural stiffness.

It should be noted however, that the valve stiffness, k_v, is rather complex to assess, since, as can be seen from Fig. 6.2, C_p and C_{xv} vary considerably during working conditions. Intelligent assessment of the results is therefore required.

It has been shown that the flow through the valve is,

$$Q = K_1 x_v \sqrt{(P_s - P_1)} = K_1 x_v \sqrt{P_2}$$

Differentiating this equation with respect to pressure,

$$\frac{\partial Q}{\partial P_{1,2}} = C_p = \frac{K_1 x_v}{2\sqrt{(P_s - P_1)}} = \frac{K_1 x_v}{2\sqrt{P_2}}$$

and with respect to valve displacement,

$$\frac{\partial Q}{\partial x_v} = C_{xv} = K_1 \sqrt{(P_s - P_1)} = K_1 \sqrt{P_2}$$

But $P_1 - P_2 = P_L$ and hence,

$$C_p = \frac{K_1 x_v}{\sqrt{\{2(P_s - P_L)\}}} \qquad C_{xv} = K_1 \sqrt{\left(\frac{P_s - P_L}{2}\right)}$$

Now for the valve,

$$k_v = 2A\frac{C_{xv}}{C_p} = K_1 \sqrt{\left(\frac{P_s - P_L}{2}\right)} \times \frac{\sqrt{\{2(P_s - P_L)\}}}{K_1 x_v} \times 2A$$

$$= \frac{2A(P_s - P_L)}{x_v}$$

Now if, as is usual in a servo, the valve is operating about $x_v = 0$, the valve stiffness $k_v = \infty$ and the stability criterion predicts instability. In practice the situation is not quite so drastic as there is generally some form of leakage present, and certainly other forms of damping. Underlap can be used to improve stability by ensuring that $x_v \neq 0$.

6.5. Determination of optimum response

While the use of Routh–Hurwitz stability criterion in the previous section determines the conditions for stability, there is no guide to the relationship between k_v and k_o to maintain a satisfactory transient response. This can easily be assessed as follows.

Making the substitution $s \equiv j\omega$ in equation (6.26),

$$\frac{\Delta x_o}{\Delta x_v}(j\omega) = \frac{K}{j\omega\left(1 - \frac{M}{k_o}\omega^2 + \frac{KM}{k_v}\omega j\right)}.$$

The phase angle ϕ from this equation is given by,

$$\tan\phi = \frac{\omega - (M/k_o)\omega^3}{-\omega^2 MK/k_{ov}}$$

If the system is excited at the natural frequency ω_n, then the numerator becomes,

$$\left(\omega_n - \frac{M}{k_o}\omega_n{}^3\right) = \sqrt{\frac{k_o}{M}} - \frac{M}{k_o} \times \frac{k_o}{M} \times \sqrt{\frac{k_o}{M}} = 0$$

and hence, $\phi = -180°$

Under this condition the amplitude ratio is given by,

$$\frac{\Delta x_o}{\Delta x_v} = \frac{K}{\sqrt{\left\{\left(\dfrac{\omega_n{}^2 KM}{k_v}\right)^2 + \left(\omega_n - \dfrac{M}{k_o}\omega_n{}^3\right)^2\right\}}}$$

$$= \frac{K}{\sqrt{\left[\left(\dfrac{k_o KM}{Mk_v}\right)^2 + \left\{\sqrt{\dfrac{k_o}{M}}\left(1 - \dfrac{M}{k_o}\dfrac{k_o}{M}\right)\right\}^2\right]}} = \frac{k_v}{k_o}$$

Hence, as shown in Fig. 6.3, the gain margin must be

$$\text{Gain margin} = 1 - k_v/k_o$$

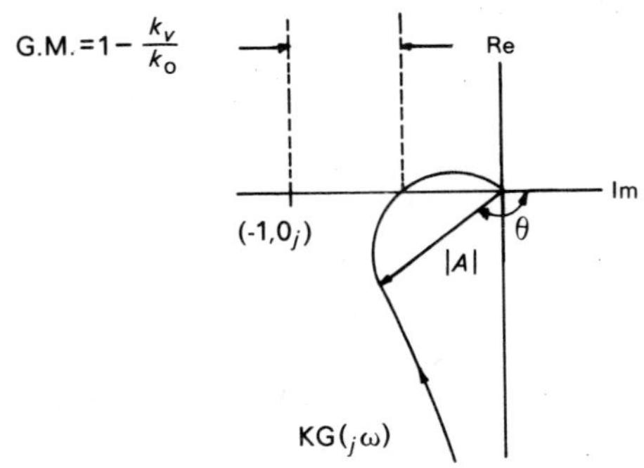

Fig. 6.3. System gain margin

This agrees with the Routh–Hurwitz condition that if $k_v = k_o$, the gain margin is zero and the system is unstable.

For a good transient response the gain margin should be around 0·75 and hence,

$$k_v = 0{\cdot}25k_o \tag{6.30}$$

Therefore, the oil column should be four times stiffer than the valve stiffness. In practice this is not always possible and some form of artificial stabilizing loop may be needed.

To determine an optimum value of the gain K, a phase margin of 45° is chosen as consistent with the gain margin. It has been shown by Nightingale [50] that for this condition,

$$K = 1{\cdot}2\sqrt{\left\{\frac{k_v}{M}\left(1 - \frac{k_v}{\sqrt{2k_o}}\right)\right\}}$$

6.6. Response characteristics

The closed loop equation, can be written

$$\frac{\Delta x_o}{\Delta x_i}(s) = \frac{K\omega_n^{\,2}}{s^3 + 2c\omega_n s^2 + \omega_n^{\,2}s + K\omega_n^{\,2}}$$

giving the characteristic equation

$$\frac{K\omega_n^{\,2}}{s(s^2 + 2c\omega_n s + \omega_n^{\,2})} = -1 \tag{6.31}$$

Three types of transient response can result depending on whether the roots are real or complex. A study of the patterns that result from parameter variation can rapidly be assessed using the root locus method. The transient response can be assessed by examining the dominant root or roots, which are positioned nearest the imaginary axis.

For $c \geqslant 1{\cdot}0$, the open loop poles of (6.31) all lie on the real axis, Fig. 6.4. The transient response, Fig. 6.5, is exponential and shows no oscillation until the critical gain $K = K_{\text{critical}}$ is reached. Up to this condition the system is sluggish and the response time is dominated by the lag term with the biggest time constant.

If now the damping ratio is reduced so that $c < 1{\cdot}0$, the open loop poles are complex and the transient response is never purely exponential. The root locus for this case is whown in Fig. 6.6. It can be seen from the transient behaviour that at lower gains (K_1 for example), the simple lag term dominates making the transient response very unsuitable, Fig. 6.7. Ideally, a servomeachanism with this type of characteristic equation should have $c > 1{\cdot}0$ and the transient response adjusted by means of the gain K as in Fig. 6.5. However, an inherent damping ratio $c > 1{\cdot}0$ is very rare in practice.

6.7. Effects of load disturbances

Since it has been indicated that rigidity is of importance in machine tool work, it is worthwhile examining the dynamic stiffness of a hydraulic servo. The stiffness of importance here is that resisting forces applied at the output with the input held fixed. It will be shown that if this applied force is oscillated harmonically, the stiffness presented to it by the servo loop varies considerably with frequency.

From equation (6.21) for a hydraulic servomotor,

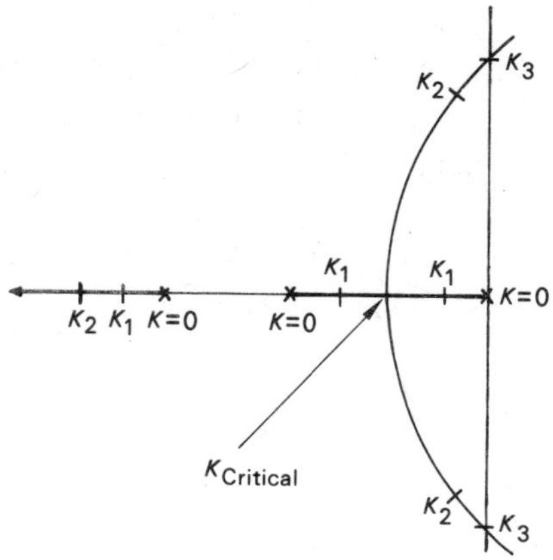

Fig. 6.4. Root locus for equation (6.31) when damping ratio $c > 1{\cdot}0$

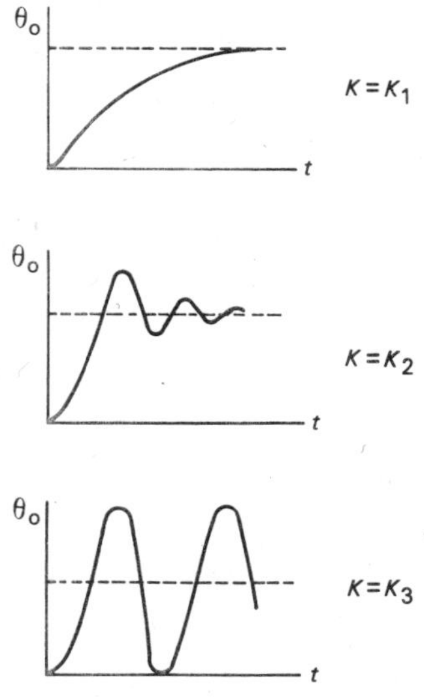

Fig. 6.5. Step response for pole locations indicated on Fig. 6.4.

$$C_x\,\Delta x_v - \frac{C_p}{2}(\Delta P_1 - \Delta P_2) = \frac{V}{4N}\frac{\mathrm{d}}{\mathrm{d}t}(\Delta P_1 - \Delta P_2) + A\frac{\mathrm{d}}{\mathrm{d}t}\Delta x_o \tag{6.32}$$

and to have closed loop control with unity feedback

$$\Delta x_v = \Delta x_i - \Delta x_o$$

If the input is held fixed at $x_i = 0$, and if the output is forced harmonically by an external force F, then since

$$\Delta F = -(\Delta P_1 - \Delta P_2)A$$

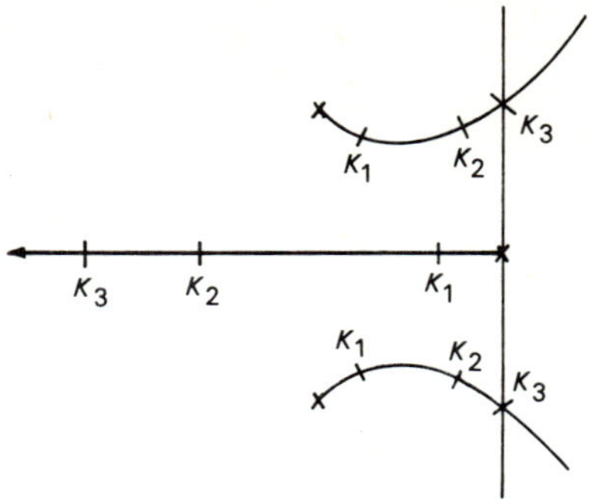

Fig. 6.6. Root locus for equation (6.31) when damping ratio $c < 1{\cdot}0$

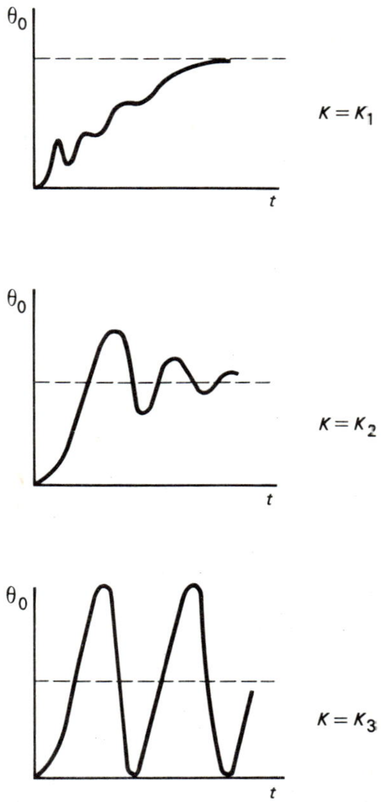

Fig. 6.7. Step response for pole locations indicated on Fig. 6.6.

equation (6.32) becomes,

$$C_{xv}\,\Delta x_o + \frac{C_p}{2}\frac{\Delta F}{A} = -\frac{V}{4NA}\frac{\mathrm{d}}{\mathrm{d}t}\Delta F + A\frac{\mathrm{d}}{\mathrm{d}t}\Delta x_o \tag{6.33}$$

Note that the sign of the force is negative since it is a force applied to the system and not one supplied by the system.

For a harmonic excitation the derivatives in (6.33) can be replaced by $j\omega$ and hence,

$$\frac{\Delta F}{\Delta x_o}(\omega j) = \frac{C_{xv} + A\omega j}{\dfrac{C_p}{2A} + \dfrac{V}{4NA}\omega j} \tag{6.34}$$

where $\Delta F/\Delta x_o(\omega j)$ is a complex number called the dynamic stiffness. Rationalise (6.34) to a single complex form,

$$\frac{\Delta F}{\Delta x_o}(\omega j) = \frac{\dfrac{C_{xv}C_p}{2A} + \dfrac{AV\omega^2}{4NA} + \left(\dfrac{C_pA\omega}{2A} - \dfrac{VC_{xv}\omega}{4NA}\right)j}{\left(\dfrac{C_p}{2A}\right)^2 + \left(\dfrac{V\omega}{4NA}\right)^2} \tag{6.35}$$

For very low frequencies of operation equation (6.35) reduces to,

$$\frac{\Delta F}{\Delta x_o} \approx 2A\frac{C_{xv}}{C_p} = k_v \tag{6.36}$$

Care must be exercised in extending this result to zero frequency since the valve might be slightly over-lapped and the resultant stiffness would be that of the oil locked in the cylinder.

For very high frequencies on the other hand, equation (6.35) reduces to

$$\frac{\Delta F}{\Delta x_o} \approx \frac{4NA^2}{V} = k_o \tag{6.37}$$

which is just the stiffness of the oil column in the ram.

Equations (6.36) and (6.37) do not mean very much unless they can be considered under some specific condition. Consider the condition for optimum transient response developed in equation (6.30) where the valve stiffness k_v must be one-quarter of the oil column stiffness in order to maintain an acceptable transient response. Hence,

$$\frac{k_v}{k_o} = 0{\cdot}25 = 2A\frac{C_{xv}}{C_p} \times \frac{V}{4NA^2}$$

therefore,

$$C_p = \frac{2C_{xv}V}{NA}$$

Substitute in equation (6.36) and at low frequencies the dynamic stiffness is,

$$\frac{\Delta F}{\Delta x_o} = 2AC_{xv} \times \frac{NA}{2C_{xv}V}$$

$$= \frac{1}{4} \times \frac{4NA^2}{V} = \frac{1}{4}k_o \tag{6.38}$$

In order to obtain the optimum transient response, the low frequency impedance needs to be one-quarter of the oil column stiffness, but to improve the low frequency stiffness would require a larger value of k_v/k_c which means a reduced stability margin. The engineer is therefore faced with a problem. Additional feedback loops may be required to maintain both transient response and rigidity.

6.8. Cavitation in the hydraulic actuator

The analysis discussed so far assumes that the inflow to the ram and the outflow are identical. This is not always the case. For example, with large inertia loading conditions, high transient pressures may be developed in the cylinder as the load is decelerated. On the inflow side of the ram, the closing valve is not able to supply enough fluid to keep that side of the ram full and cavities start to form. The load is, in fact, overrunning. Hence the outflow side of the ram piston has a high peak pressure and the inflow side pressure has been reduced to the vapour pressure of the oil. The presence of the large peaking pressures may give rise to fatigue failure and the presence of cavities in the fluid may result in implosion damage. Pressure peaks of up to five times the supply pressure have been recorded experimentally.

The presence of valve overlap [51] which is sometimes purposely employed to reduce leakage across the valve, or may be present due to manufacturing tolerances, can increase the pressure peaks considerably. On the other hand, the compressibility of the fluid helps to reduce the magnitude of the peaks. It can be shown that the worst condition occurs when the valve is within the lap region and the load is decelerating. The fluid is then trapped in the cylinder by the valve lap and absorbs the kinetic energy of the moving mass.

The conditions that give rise to cavitation can be estimated by developing the linear equations for the actuator [52].

Equation (6.22) can be written

$$\Delta x_v = a\frac{\mathrm{d}^3}{\mathrm{d}t^3}\Delta x_o + b\frac{\mathrm{d}^2}{\mathrm{d}t^2}\Delta x_o + c\frac{\mathrm{d}}{\mathrm{d}t}\Delta x_o \tag{6.39}$$

This third-order differential equation relates the actuator and valve displacements. However, it is required to find the conditions for cavitation and to do this, the relationship between actuator pressure and valve movement is required. Since the inflow and outflow are assumed equal up to cavitation

$$Q_1 = Q_2$$

Hence, from equations (6.16) and (6.17)

$$\Delta Q_1 = C_{xv}\Delta x_v - C_p\Delta P_1$$
$$\Delta Q_2 = C_{xv}\Delta x_v + C_p\Delta P_2$$

then, from Fig. 6.2

$$\Delta P_1 + \Delta P_2 = 0 \tag{6.40}$$

and hence from (6.15)

$$\Delta P_1 = \frac{M}{2A}\frac{\mathrm{d}^2}{\mathrm{d}t^2}\Delta x_o \tag{6.41}$$

Combining equations (6.41) and (6.39) and introducing the operator s,

$$\frac{2A}{M}\frac{\Delta P_1}{\Delta x_v} = \frac{s}{as^2 + bs + c} \tag{6.42}$$

The natural frequency of the load and oil column system is

$$\omega_n{}^2 = \frac{c}{a} = \frac{4NA^2}{MV}$$

If the valve is oscillated at frequency ω and amplitude $\Delta\hat{x}_v$, it can be shown that the amplitude of the pressure waveform from equation (6.42) will be,

$$\Delta\hat{P}_1 \frac{2A^2}{MK_1\omega_n\sqrt{(P_s - P_1)}} = \frac{\omega^*\Delta\hat{x}_v}{\sqrt{\{(1-\omega^{*2})^2 + 4c^2\omega^{*2}\}}} \tag{6.43}$$

where $\omega^* = \omega/\omega_n$
and the damping ratio

$$c = \frac{MC_p\omega_n}{4A^2} = \frac{MK\omega_n x_{vr}}{8A^2\sqrt{(P_s - P_1)}} \tag{6.44}$$

Let $\Delta\hat{P}_1{}^* = \Delta\hat{P}_1/P_s$, then equation (6.42) reduces to

$$\Delta\hat{P}^* = \alpha\sqrt{\left\{\frac{1}{2}\left(1 - \frac{P_{1r}}{P_s}\right)\right\}}\frac{\omega^*}{\sqrt{\{(1-\omega^{*2})^2 + 4c^2\omega^{*2}\}}} \tag{6.45}$$

where the load factor

$$\alpha = \frac{MK_1\omega_n\Delta\hat{x}_v\sqrt{(P_s/2)}}{P_sA^2}$$

and subscript r refers to the steady state condition.
Considering the particular case $x_{vr} = 0$, i.e. the valve oscillating about the centre position,

$$P_{1r} = [P_s/2]$$

From equation (6.44), $c = 0$

then

$$\Delta\hat{P}_1^* = \frac{\alpha\omega^*}{2\,|\,1 - \omega^{*2}\,|} \tag{6.46}$$

When the valve is oscillating at the natural frequency; equation (6.46) indicates that the actuator pressure amplitude is infinite. At both zero and infinite excitation frequency the pressure $\Delta\hat{P}_1$ is zero. Cavitation arises when $\Delta\hat{P}_1$ exceeds $\frac{1}{2}P_s$. For this case equation (6.40) shows that ΔP_2 will be less than $-\frac{1}{2}P_s$. The total pressure on the exhaust side of the actuator is

$$P_{2r} + \Delta P_2 = \tfrac{1}{2}P_s - \tfrac{1}{2}P_s = 0$$

The critical value of α for cavitation occurrence is found from (6.46), letting $\Delta P_1^* = 0{\cdot}5$:

$$\alpha_c = \frac{|\,1 - \omega^{*2}\,|}{\omega^*} \tag{6.47}$$

These results are shown in Fig. 6.8.

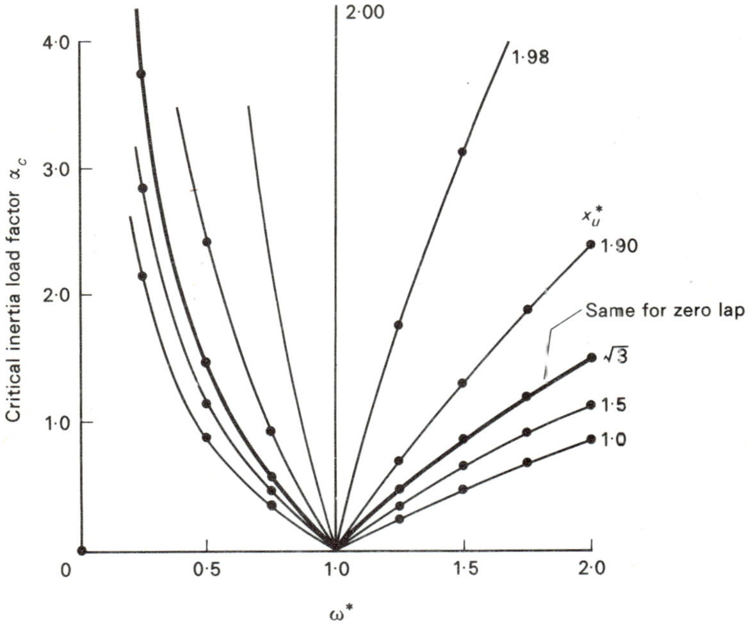

Fig. 6.8. Critical load factor α_c, above which cavitation occurs

It is clear from the results that it is difficult to avoid cavitation in the region of the natural frequency. This result is confirmed in the more detailed analysis in chapter 7, where it is shown that damping tends to improve the situation.

If the fluid is considered incompressible, the bulk modulus N is infinite and hence α must be redefined since the natural frequency is infinite.

$$\alpha_i = \alpha\omega^* = \frac{MK_1\Delta\hat{x}\sqrt{(P_s/2)}\omega}{P_sA^2}$$

and equation (6.47) becomes

$$\frac{\alpha_{ic}}{\omega^*} = \frac{1 - \omega^{*2}}{\omega^*}$$

$$\text{or} \quad \alpha_{ic} = 1{\cdot}0 \text{ since } \omega^* = 0$$

6.9. Effect on underlap on cavitation

For the underlapped valve, shown in Fig. 6.9, operating within the lap, x_v is less than the underlap x_u,

$$Q_1 = K_1(x_v + x_u)\sqrt{(P_s - P_1)} - K_1(x_u - x_v)\sqrt{P_1} \tag{6.48}$$

For small oscillations about valve central position, $x_{vr} = 0$, $P_{1r} = \frac{1}{2}P_s$ giving

$$Q_1 = C_{xv}\Delta x_v + C_p\Delta P_1$$

where

$$C_{xv} = 2K_1\sqrt{(P_s/2)}$$

$$C_p = \frac{K_1x_u}{\sqrt{(P_s/2)}}$$

From these values, the pressure amplitude can be found as before, giving,

$$\Delta\hat{P}_1^* = \frac{\alpha\omega^*}{\sqrt{\{(1 - \omega^{*2})^2 + 4c^2\omega^{*2}\}}} \tag{6.49}$$

It is easily shown that the damping ratio is,

$$C = \frac{MC_p\omega_n}{4A^2} = \frac{\alpha}{2}x_u^*$$

where the non-dimensional underlap $x_u^* = \hat{x}_u/\Delta\hat{x}_v$. For operation within the underlap $x_u^* > 1$ and equation (6.49) becomes,

$$\Delta\hat{P}_1^* = \frac{\alpha\omega^*}{\sqrt{\{(1 - \omega^{*2})^2 + \alpha^2x_u^{*2}\omega^{*2}\}}} \tag{6.50}$$

At first glance one would expect this equation to reduce to equation (6.46) when the underlap x_u^* is reduced to zero. This is not the case since it should be borne in mind that for the case considered, x_u^* must exceed unity.

The pressure amplitudes of equations (6.46) and (6.50) are compared in Fig. 6.9. Underlap is effective in reducing the pressure peaks in the region of the natural frequency. When P_1^* exceeds unity, P_2^* becomes negative and cavitation occurs.

The critical load factor α_c can also be calculated. If cavitation occurs when $\Delta\hat{P}_1^* = 0{\cdot}5$, then

$$\alpha_c = \frac{0{\cdot}5\sqrt{\{(1-\omega^{*2})^2 + \alpha_c^{\,2} x_u^{*2}\omega^{*2}\}}}{\omega^*} \tag{6.51}$$

These results are also shown in Fig. 6.8. If $x_u^* > 2$ cavitation does not occur at any frequency. It is interesting to note that the boundary for $x_u^* = \sqrt{3}$ coincides with that for zero lap. Some typical experimental results are shown in Fig. 6.10(*a*) and (*b*).

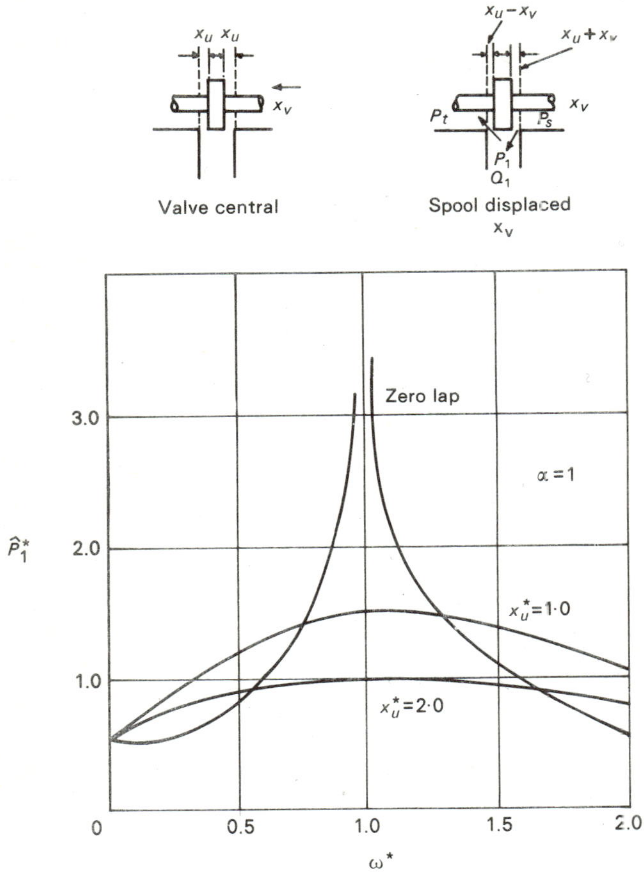

Fig. 6.9. Pressure peaks in the hydraulic actuator, with and without underlap x_u

6.10. Graphical analysis of the closed loop step response

In the development so far, great reliance has been placed on linearising the system equations, but even this is not very satisfactory. For example, in order to solve the equation (6.23) values of C_{xv} and C_p are required. There is in fact no single value here, and the engineer is faced with having to use his experience in selecting suitable values. A first step towards a fuller understanding of the non-linearities involved is based on a graphical method first proposed by

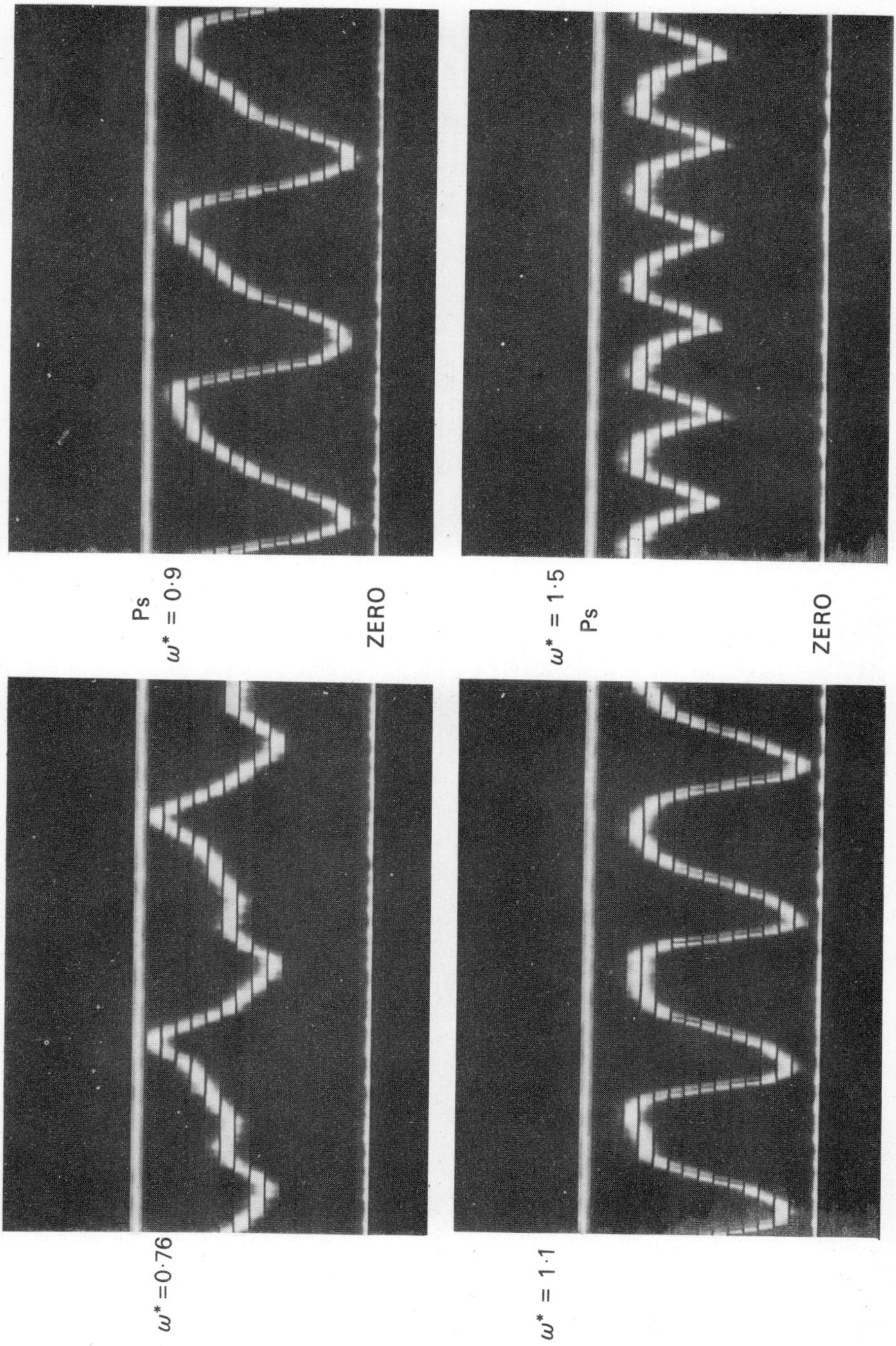

Fig. 6.10. (*a*) Measured pressure waveforms in a hydraulic cylinder driven by a critically lapped valve

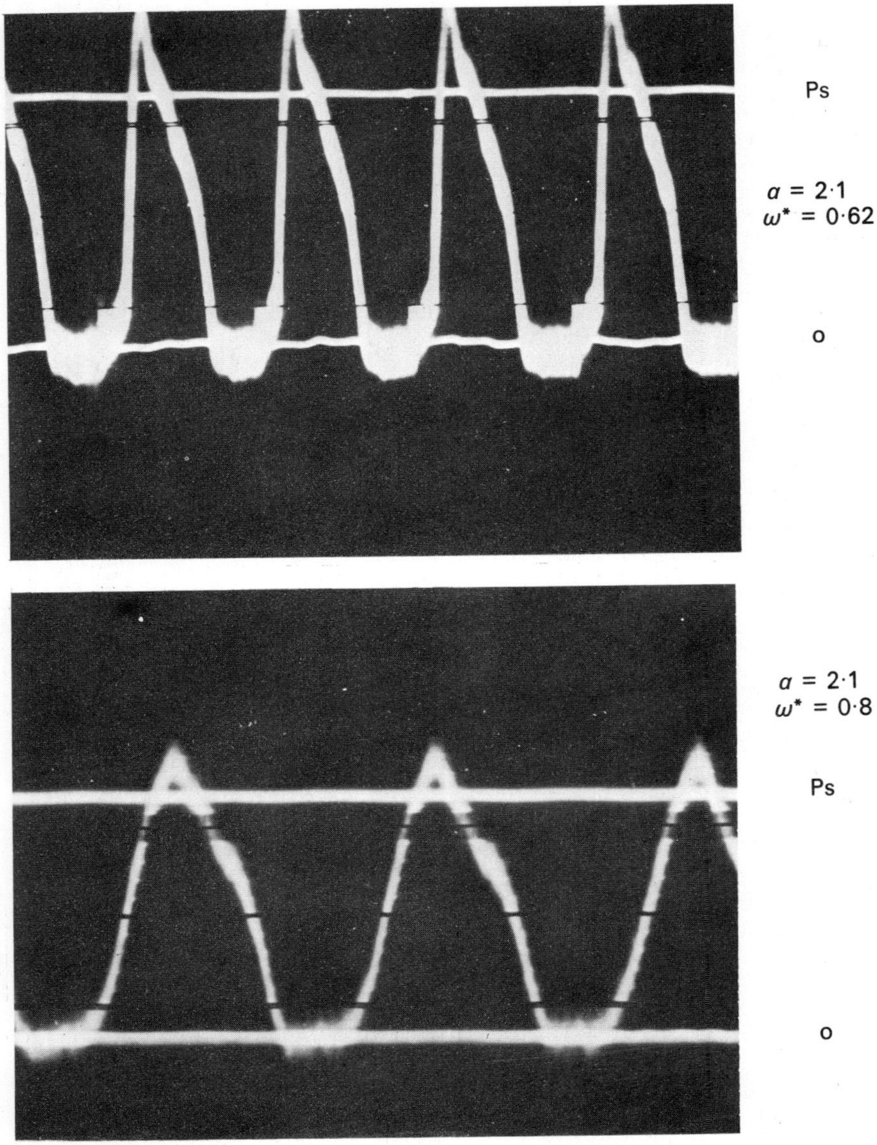

Fig. 6.10 (*b*) Pressure waveforms showing cavitation, valve overlap = 0·002 in

Cunningham [53] and developed in [54]. The method allows the effects of inertia, spring, viscous friction and coulomb friction loading to be examined in the closed loop configuration. The incompressible case is considered only.

The equations for flow through the valve, see Fig. 6.1, are equations (6.1) and (6.2), with the constants lumped into one term K_1, giving

$$Q_1 = K_1 x_v \sqrt{(P_s - P_1)} \tag{6.52}$$

$$Q_2 = K_1 x_v \sqrt{P_2} \tag{6.53}$$

In assuming an incompressible fluid, the natural frequency must be very large. In such a case the volume of oil in the actuator must be small or the actuator oil column very stiff.

The load pressure P_L is defined as,

$$P_L = P_1 - P_2 \tag{6.54}$$

The valve flow rate is related to the piston displacement by

$$Q = A\frac{dx_o}{dt} = Au \tag{6.55}$$

Equations (6.52), (6.54) and (6.55) are combined to give

$$u = \frac{K_1 x_v}{A\sqrt{2}}\sqrt{(P_s - P_L)} \tag{6.56}$$

This can be non-dimensionalised by letting $u^* = u/\hat{u}$, $p = P/P_s$, $x_v{}^* = x_v/\hat{x}_v$. The reference velocity $\hat{u}$ is,

$$\hat{u} = \frac{K_1 \hat{x}_v}{A}\sqrt{(P_s/2)} \tag{6.57}$$

Equation (6.56) now becomes,

$$u^* = x_v{}^*\sqrt{(1 - p_L)} \tag{6.58}$$

From the mechanical feedback linkage shown in Fig. 6.1(*b*) and equation (6.7),

$$x_v = \frac{b}{a+b}x - \frac{a}{a+b}x_o \tag{6.59}$$

The reference valve movement is,

$$\hat{x}_v = \hat{x}\left(\frac{b}{a+b}\right) \tag{6.60}$$

where x is the magnitude of the step input. Equation (6.59) can now be non-dimensionalised, giving

$$x_v{}^* = 1 - \beta x_o{}^* \tag{6.61}$$

where $x_o{}^* = x_o/\hat{x}_o$ and $\beta = (\hat{x}_o/\hat{x})\,(a/b)$ is the feedback factor. The reference output movement is $\hat{x}_o = \hat{u}\hat{t}$. The reference time $\hat{t}$ is defined in the load equation.

The load equation is,

$$P_L A = M\frac{du}{dt} + Cu + F \operatorname{sgn} u + kx_o \tag{6.62}$$

where C is a viscous friction constant
F is the magnitude of the coulomb friction force
k is the spring stiffness

Equation (6.62) can be non-dimensionalised to

$$p_L = \frac{du^*}{d\tau} + \gamma u^* + \eta \operatorname{sgn} u^* + \Psi x_o^* \tag{6.63}$$

where $\gamma = C\hat{u}/P_sA$ the viscous friction factor
$\eta = F/P_sA$ the coulomb friction factor
$\Psi = k\hat{t}\hat{u}/P_sA$ the spring load factor
$\hat{t} = M\hat{u}/P_sA$ the reference time
$\tau = t/\hat{t}$

The relationship between output displacement and velocity can be written

$$x_o^* = \int_o^\tau u^* \, d\tau$$

6.11. Method of solution

Consider the simple case of inertia loading only, the system equations are,

$$x^* = 1 - \beta x_o^* \tag{6.65}$$

$$u^* = x_v^* \sqrt{(1 - p_L)} \tag{6.66}$$

$$p_L = du^*/d\tau \text{ (inertia load only)} \tag{6.67}$$

$$x_o^* = \int_o^\tau u^* \, d\tau \tag{6.68}$$

The method of solution is shown in Fig. 6.11. In Fig. 6.11(*a*) the flow equation is plotted with the load pressure as the independent variable and the valve opening as parameter. Equation (6.67) can be approximated by

$$p_L = \frac{\Delta u^*}{\Delta \tau} \tag{6.69}$$

where Δ indicates a small change in the variable. This is more conveniently written as,

$$\frac{\Delta u^*}{p_L} = \Delta \tau \tag{6.70}$$

where p_L is assumed constant during the interval $\Delta\tau$.
Equation (6.68) can be approximated by,

$$\frac{\Delta x_o^*}{u^*} = \Delta \tau \tag{6.71}$$

where u^* is assumed constant during the interval $\Delta\tau$ and equal to the value at the end of the interval. The solution is as follows:

Initially the system is at rest and the valve is closed, $x_v^* = 0$. A step input is applied and the valve opening $x_v{}^*$ (the value of x_v^* at $\tau = 0$) becomes unity.

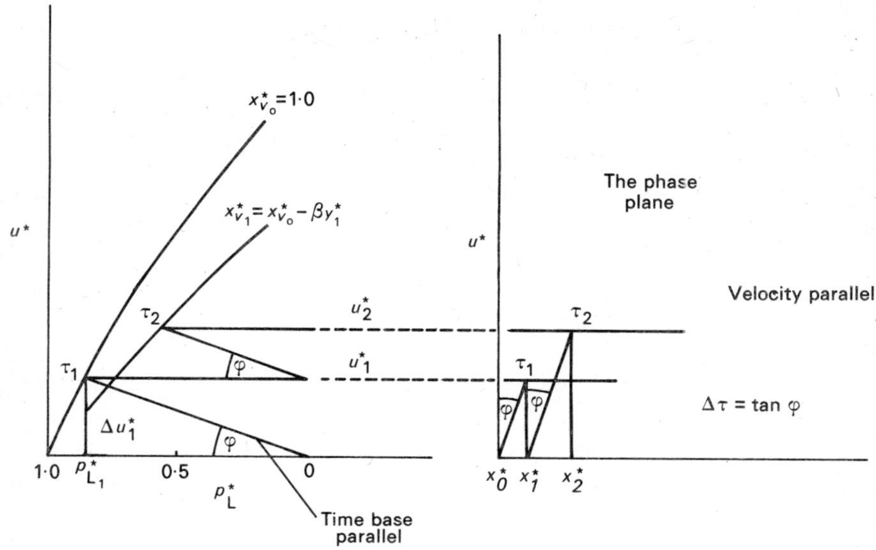

(*a*) Determination of load velocity and pressure

(*b*) Determination of load position and feedback

Fig. 6.11. Basis of method of determining closed loop response

The valve is assumed to remain in the position during the interval $\Delta\tau$. During the interval the change in velocity is given by equation (6.70) and this assumes p_L to remain constant during the interval. However, p_L does not remain constant and a representative value of p_L during the interval must be chosen. For the purpose of this method the most convenient assumption is that the load pressure during the interval is constant and equal to the load pressure p_{L1} at the end of the interval. This is shown in Fig. 6.10(*a*). The angle ϕ is $\tan^{-1} \Delta\tau$ and the geometry shows that

$$\frac{\Delta u^*}{p_{L1}} = \Delta\tau = \tan\phi \tag{6.72}$$

A more accurate solution can be obtained if the representative load pressure is taken to be the mean load pressure during the interval. This is referred to as method II.

Hence the velocity change can be determined from the point of intersection of the line Oτ (referred to as a time base parallel) and the flow line. The change in output position can also be determined by graphical integration and this is illustrated in Fig. 6.11(*b*). Initially the output displacement is zero, $x_o^* = 0$. The change in displacement after the interval $\Delta\tau$ is determined by the intersection of the time base parallel in the phase plane (u^*, x_c^*). The velocity is assumed constant during the interval. Knowing x_o *, the output displacement at the end of the first interval, the valve displacement $(1 - \beta x_1{}^*)$ during the second interval can be calculated and a new flow line drawn. The above procedure can then be used to find the conditions at τ_2, etc.

As an example, consider unity feedback $\beta = 1$. A convenient time interval is $\Delta\tau = 0{\cdot}06$. The graphical construction is shown in Fig. 6.12. Note that in the (u^*, p_L) plane it is not necessary to draw complete flow lines. It is sufficient to show a small segment of a flow line in the region where the time base parallel and the flow line are expected to intersect. In Fig. 6.12 the values of valve displacement are shown on the flow lines. From this figure the changes of velocity, load pressure and valve displacement with time can be determined. The output displacement as a function of time can be found from the (u^*, x_o^*) phase plane, Fig. 6.13. During the initial period, Fig. 6.12 shows that the load pressure (and the velocity) is decreasing and hence the assumption that the load pressure is constant during an interval results in an under-estimation of the velocity. This can be improved by the use of method II.

The results shown in Fig. 6.13, were determined in roughly one hour with the help of a set square and ruler. Method II requires a slightly longer time.

The output displacement is plotted against time in Fig. 6.14, which shows the effect of varying the feedback factor β. As β increases the speed of response improves. However, the quicker response is the result of large load pressures and ultimately, as β is increased, cavitation occurs. When $\beta = 2{\cdot}25$ the trajectory in the (u^*, p_L) plane just crosses the line $p_L = -1$, indicating the occurrence of cavitation. In Fig. 6.12 the minimum load pressure is $-0{\cdot}56$ and at this point the pressure on the tank side of the actuator is $0{\cdot}78\ P_s$ and on the pump side $0{\cdot}22\ P_s$.

When cavitation exists the flow equation must be changed. In this condition $p_2 > 1{\cdot}0$ and p_1 is zero. Hence for $p_L < -1$ the flow equation becomes

$$u^* = x_v^* \sqrt{(-2p_L)}$$

The method can also be used to examine the effects of other types of load. For instance, when viscous friction is present the load equation is

$$p_L = \frac{du^*}{d\tau} + \gamma u^* \tag{6.73}$$

In incremental form this gives

$$(p_L)_{\tau+\Delta\tau} = \frac{\Delta u^*}{\Delta\tau} + \gamma[(u^*)_\tau + \Delta u^*]$$

or

$$\Delta u^* = \frac{(p_L)_{\tau+\Delta\tau} - \gamma(u^*)_\tau}{\gamma + 1/\Delta\tau}$$

The construction is illustrated in Fig. 6.15. In this case the time base parallels are inclined at an angle θ to the horizontal, where $\theta = \tan^{-1}\Delta\tau/(1 + \gamma\Delta\tau)$. The base line for construction of the time base parallels is a straight line inclined at an angle $\tan^{-1}\gamma$ to the vertical. Examination of Fig. 6.15 shows that, as

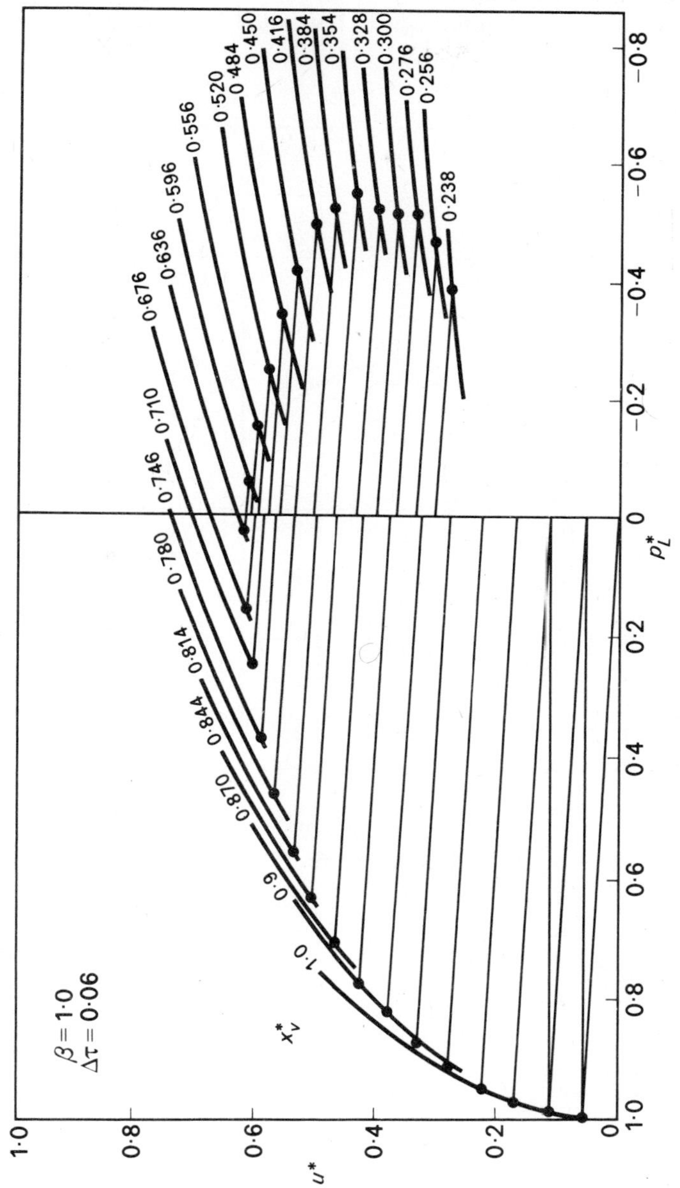

Fig. 6.12. Graphical construction with $\beta = 1{\cdot}0$ and $\Delta\tau = 0{\cdot}06$

expected, viscous friction reduces the maximum velocity during the transient. When coulomb friction is also present it is easily shown that the base line for construction of the time base parallels is moved to the left a distance δ.

This graphical technique can be extended [55] to include compressibility effects.

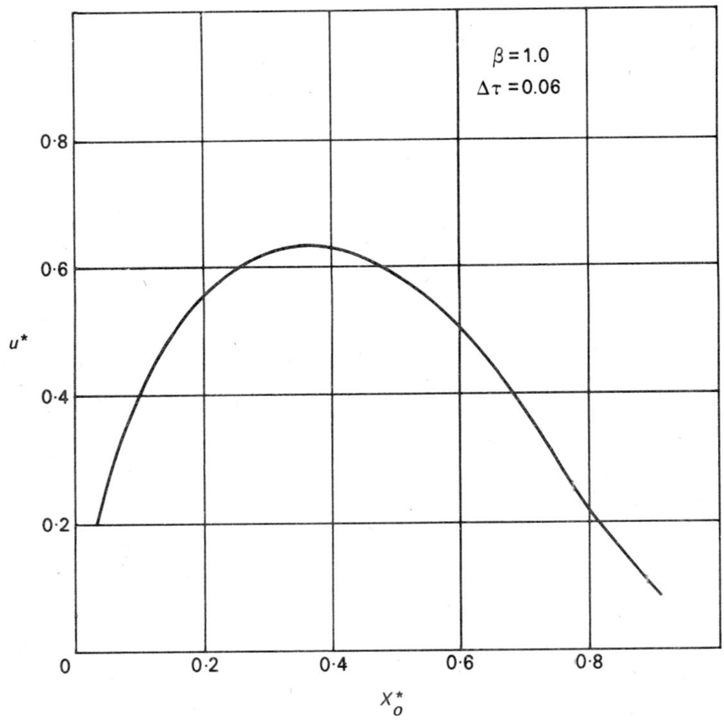

Fig. 6.13. Phase plane plot derived from Fig. 6.12.

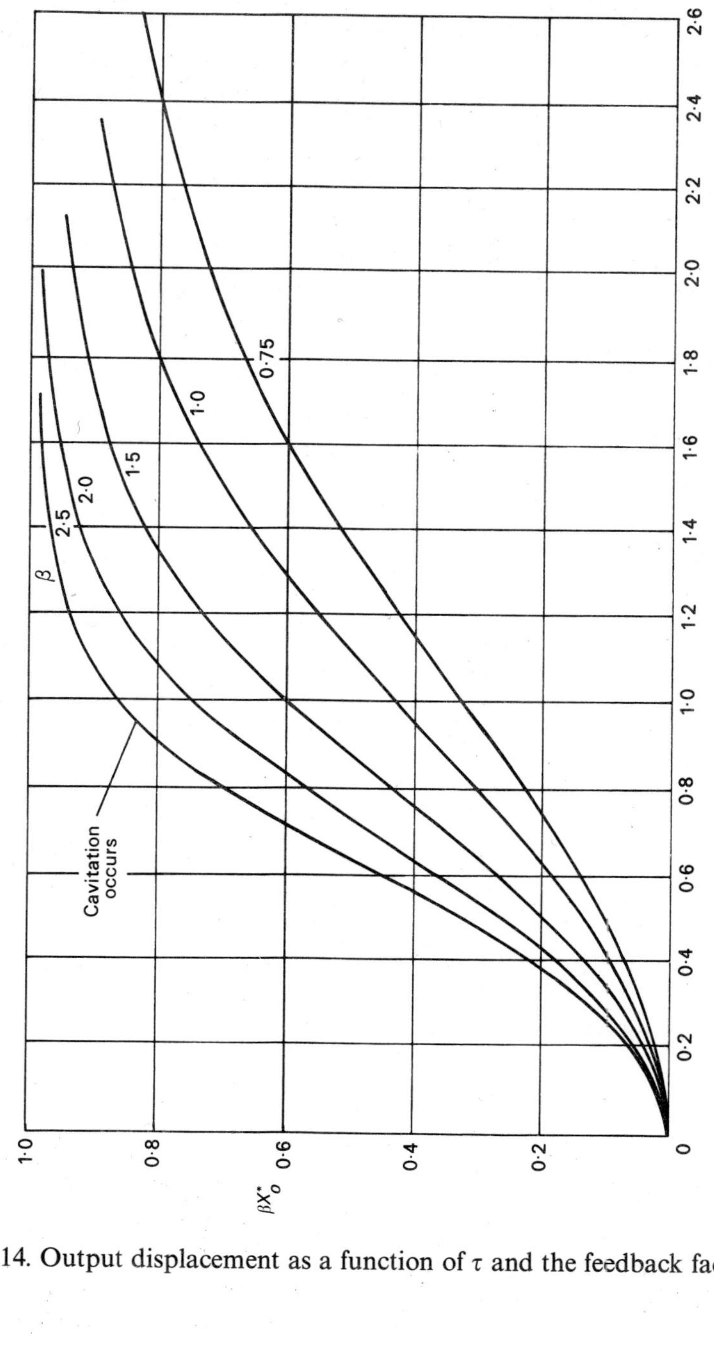

Fig. 6.14. Output displacement as a function of τ and the feedback factor β

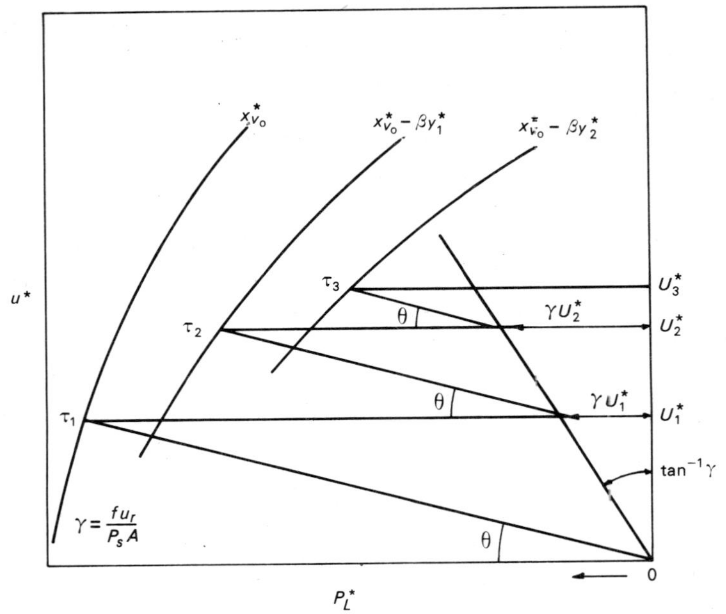

Fig. 6.15. A means of dealing with inertia and viscous friction loads

Chapter 7

Electrohydraulic servomechanisms and general aspects of analysis

7.1. Introduction

The increasing demands from modern control systems necessitate greater flexibility between controlling elements. This has resulted in the combining of electrical signalling with fluid power actuators. The need to change from electrical to hydraulic control at some point in the loop has led to the development of an electric torque motor to actuate the hydraulic servo valve. The torque motor receives its inputs from an electronic or magnetic amplifier. These components are discussed in more detail in chapter 10.

A block diagram of a typical servo loop is shown in Fig. 7.1. It will be shown in chapter 10, that the characteristics of the torque motor is a function of two variables,

$$T_o = K_g i - K_s \theta_r \tag{7.1}$$

where K_g = torque motor gain
i = current
K_s = torque motor stiffness
θ_r = torque motor shaft deflection.

This characteristic complicates the problem, since the valve movement is determined by the interaction of the torque motor force F_R which tries to open

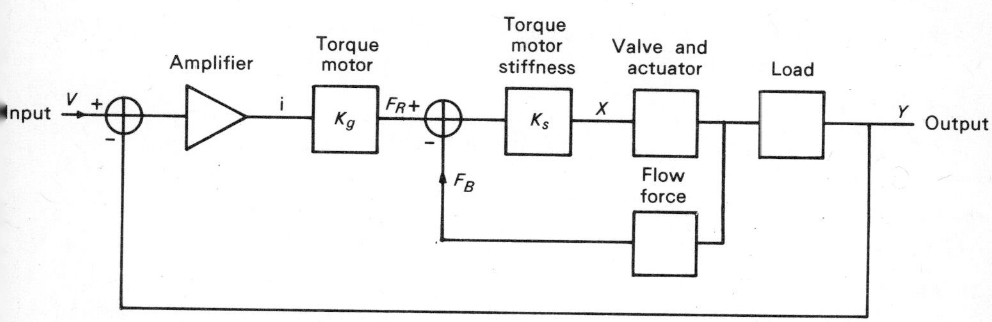

Fig. 7.1. Block diagram of an electrohydraulic servo

the valve, and the flow force which tries to close it. This interaction will be discussed in more detail later. On the other hand, the servo amplifier can deal with a variety of signals, such as those generated by electrical compensation networks.

7.2. Complete system equations

The analytical development in chapter 6 was basically a linearised approach. In this chapter a more detailed analysis is considered and solutions are obtained using an analogue computer simulation.

Considering the actuator arrangement shown in Fig. 7.2 and assuming the valve to have zero lap, the flow rates into and out of the actuator are:

$$\begin{aligned} &\text{For } x_v > 0 \quad & Q_1 &= \operatorname{sgn}(P_s - P_1)K_1 x_v\sqrt{(|P_s - P_1|)} \\ & & Q_2 &= K_1 x_v \sqrt{P_2} \\ &\text{For } x_v < 0 \quad & Q_1 &= K x_v \sqrt{P_1} \\ & & Q_2 &= \operatorname{sgn}(P_s - P_2)K_1 x_v\sqrt{(|P_s - P_2|)} \end{aligned} \tag{7.2}$$

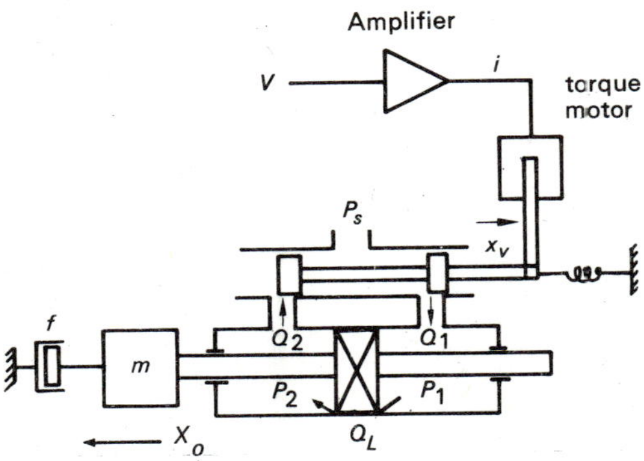

Fig. 7.2. Basic open loop system

Note that on a full simulation it is necessary to consider negative spool movements and the possibility of reverse flow [56, 57, 58, 59]. This was not essential in the earlier analysis. These equations are made non-dimensional by putting $x_v^* = x_v/\hat{x}_v$, $p = P/P_s$, $Q^* = Q/\hat{Q}$ with $\hat{Q} = A\hat{u}$ and $\hat{u} = \{K_1\hat{x}_v\sqrt{(P_s/2)}\}/A$. Hence

$$\begin{aligned} &x_v^* > 0 \quad & Q_1^* &= \sqrt{2}x_v^* \operatorname{sgn}(1 - p_1)\sqrt{(|1 - p_1|)} \\ & & Q_2^* &= \sqrt{2}x_v^* \sqrt{p_2} \\ &x_v^* < 0 \quad & Q_1^* &= \sqrt{2}x_v^* \sqrt{p_1} \\ & & Q_2^* &= \sqrt{2}x_v^* \operatorname{sgn}(1 - p_2)\sqrt{(|1 - p_1|)} \end{aligned} \tag{7.3}$$

The continuity equations (6.12) and (6.13) can also be made non-dimensional. In addition to the above reference quantities non-dimensional time is defined as $\tau = \omega t$ where ω is the driving frequency. Equation (6.12) becomes:

$$Q_1{}^* = u^* + \frac{V}{2N}\omega P_s \frac{\mathrm{d}p_1}{\mathrm{d}\tau}\frac{1}{K\hat{x}_v\sqrt{(P_s/2)}} \tag{7.4}$$

or

$$Q_1{}^* = u^* + \frac{2\omega^* \mathrm{d}p_1}{\alpha \ \ \mathrm{d}\tau} \tag{7.5}$$

where the inertia load factor α is

$$\alpha = \frac{\text{maximum inertia load at the natural frequency}}{\text{stalled load}}$$

or

$$\alpha = \frac{M\omega_n K\hat{x}_v\sqrt{(P_s \mid 2)}}{P_s A^2}$$

and $\omega^* = \omega/\omega_n$ where ω_n is the mass and oil column natural frequency $\sqrt{(4NA^2/MV)}$

By a similar argument equation (6.13) reduces to

$$Q_2{}^* = u^* - \frac{2\omega^*}{\alpha}\frac{\mathrm{d}p_2}{\mathrm{d}\tau} \tag{7.6}$$

If the load consists of inertia M and viscous friction f then the load equation is

$$P_l = P_1 - P_2 = \frac{M}{A}\frac{\mathrm{d}^2x_o}{\mathrm{d}t^2} + \frac{f}{A}\frac{\mathrm{d}x_o}{\mathrm{d}t} \tag{7.7}$$

or

$$p_l = p_1 - p_2 = \alpha\omega^* \frac{\mathrm{d}u^*}{\mathrm{d}\tau} + \gamma u^* \tag{7.8}$$

where the viscous load factor γ is

$$\gamma = \frac{\text{max. viscous force}}{\text{stall force}} = \frac{f K_1 \hat{x}_v}{P_s A^2}\sqrt{(P_s/2)} \tag{7.9}$$

7.3. Simple case

If cavitation is not allowed to occur there is no reversal of flow and a simplified equation results [59]. Since under this condition the flow into the ram approximately equals the flow out of the ram at all times, equations (7.3) can be combined to give,

for $x_v{}^* > 0$ $\qquad Q_L{}^* = \sqrt{2}x_v{}^*\sqrt{(1 - p_1)}$ $\qquad$ (7.10)
and for $x_v{}^* < 0$ $\qquad Q_L{}^* = \sqrt{2}x_v{}^*\sqrt{p_1}$

where $Q_L{}^*$ is the average load flow through the valve.

From equations (7.5) and (7.6), the flow into the ram is defined by,

$$Q_L{}^* = u^* + \frac{1}{\alpha}\omega^* \frac{dp_l}{d\tau} \tag{7.11}$$

Since inflow and outflow are equal

$$p_1 + p_2 = 1{\cdot}0$$

also

$$p_L = p_1 - p_2$$

Hence,

$$p_1 = \tfrac{1}{2}(1 + p_L) \tag{7.12}$$

Hence from equations (7.8), (7.10), (7.11) and (7.12),

$$x_v{}^* \sqrt{\left\{1 - \left(\alpha\omega^* \frac{du^*}{d\tau} + \gamma u^*\right)\frac{x_v{}^*}{|x_v{}^*|}\right\}} = u^* + \frac{\gamma}{\alpha}\omega^* \frac{du^*}{d\tau} + (\omega^*)^2 \frac{d^2u^*}{d\tau^2} \tag{7.13}$$

The input is sinusoidal $\qquad x_v{}^* = \sin\tau \qquad$ (7.14)

It has been established earlier that the flow force in the negative x direction is given by (see chapter 4)

$$\text{for } x_v{}^* > 0 \quad F_\beta = \sqrt{(2\rho)}\,[Q_1\sqrt{(P_s - P_1)} + Q_2\sqrt{P_2}]\cos\theta \tag{7.15}$$
$$\text{for } x_v{}^* < 0 \quad F_\beta = -\sqrt{(2\rho)}[Q_1\sqrt{P_1} + Q_2\sqrt{(P_s - P_2)}]\cos\theta$$

It can be seen from Fig. 7.1 that this force opposes the torque motor output so that the valve displacement will be determined by,

$$x_v = K_s(F_R - F_\beta)$$

Since the inflow and outflow are equal, equation (7.15) can be written,

$$x_v{}^* > 0 \quad F_\beta = 2\sqrt{(2\rho)}[Q_1\sqrt{(P_s - P_1)}]\cos\theta$$
$$x_v{}^* < 0 \quad F_\beta = -2\sqrt{(2\rho)}[Q_1\sqrt{P_1}]\cos\theta \tag{7.16}$$

If the reference flow force $F_\beta = 2\sqrt{\rho}\cos\theta\, K_1\hat{x}_v P_s/\sqrt{2}$ then equation (7.16) becomes,

$$F_\beta{}^* = x_v{}^*\left[1 - \left(\alpha\frac{du^*}{d\tau} + \gamma u^*\right)\frac{x_v{}^*}{|x_v{}^*|}\right] \tag{7.17}$$

The valve displacement for a sinusoidal electrical input signal will be,

$$x_v = K_s(K_g\hat{V}\sin\omega t - F_\beta)$$

assuming for simplicity that K_g includes the amplifier gain.
Hence,

$$x_v{}^* = \sin\tau - \delta F_\beta{}^* \tag{7.18}$$

where the flow force factor δ is

$$\delta = \frac{2\sqrt{\rho}\cos\theta\, K\hat{x}_v P_s}{\sqrt{2K_g\hat{V}}}$$

The system equations to be solved are then

$$x_v^*\sqrt{\left\{1 - \left(\alpha\omega^*\frac{du^*}{d\tau} + \gamma u^*\right)\frac{x_v^*}{|x_v^*|}\right\}} = u^* + \frac{\gamma\omega^*}{\alpha}\frac{du^*}{d\tau} + \omega^{*2}\frac{d^2u^*}{d\tau^2} \tag{7.19}$$

$$F_\beta^* = x_v^*\left[1 - \left(\alpha\frac{du^*}{d\tau} + \gamma u^*\right)\frac{x_v^*}{|x_v^*|}\right] \tag{7.20}$$

$$x_v^* = \sin\tau - \delta F_\beta^* \tag{7.21}$$

The circuit for the solution of these equations is shown in Fig. 7.3.

7.4. Results from analogue computer study

Although this analysis does not allow results to be recorded when the actuator is cavitating, it is possible to determine the cavitation boundaries. This is achieved by increasing the load factor α until p_L becomes equal to unity. Results from this examination [56, 58, 59, 60, 65] are shown in Fig. 7.4 and should be compared with those shown in Fig. 6.8. Also included in this figure is the effect of viscous friction on the cavitation boundary. It is seen that a little friction is quite effective particularly at the natural frequency. (Flow force is neglected at this stage).

A similar study of the effects of leakage gives results very similar to those of Fig. 7.4, except that the coefficients are 50 per cent lower in magnitude. The sensitivity of the cavitation boundary to small changes in leakage factor is to be expected, since the use of an orifice bleed is one of the established methods of reducing cylinder pressure peaks.

The leakage factor ϕ is defined as follows. Since the leakage past the ram can be assumed laminar,

$$Q_l = l(P_1 - P_2)$$

and therefore equations (6.12) and (6.13) became

$$Q_1 = A\frac{dx_o}{dt} + \frac{V}{2N}\frac{dP_1}{dt} + l(P_1 - P_2) \tag{7.22}$$

$$Q_2 = A\frac{dx_o}{dt} - \frac{V}{2N}\frac{dP_2}{dt} + l(P_1 - P_2) \tag{7.23}$$

Using the same process of non-dimensionalising

$$Q_1^* = u^* + \frac{2}{\alpha}\omega^*\frac{dp_1}{d\tau} + \frac{Ml\omega_n}{A^2}\frac{P_sA^2}{MK_1x_v\hat{\omega}_n\sqrt{(P_s/2)}}(p_1 - p_2)$$

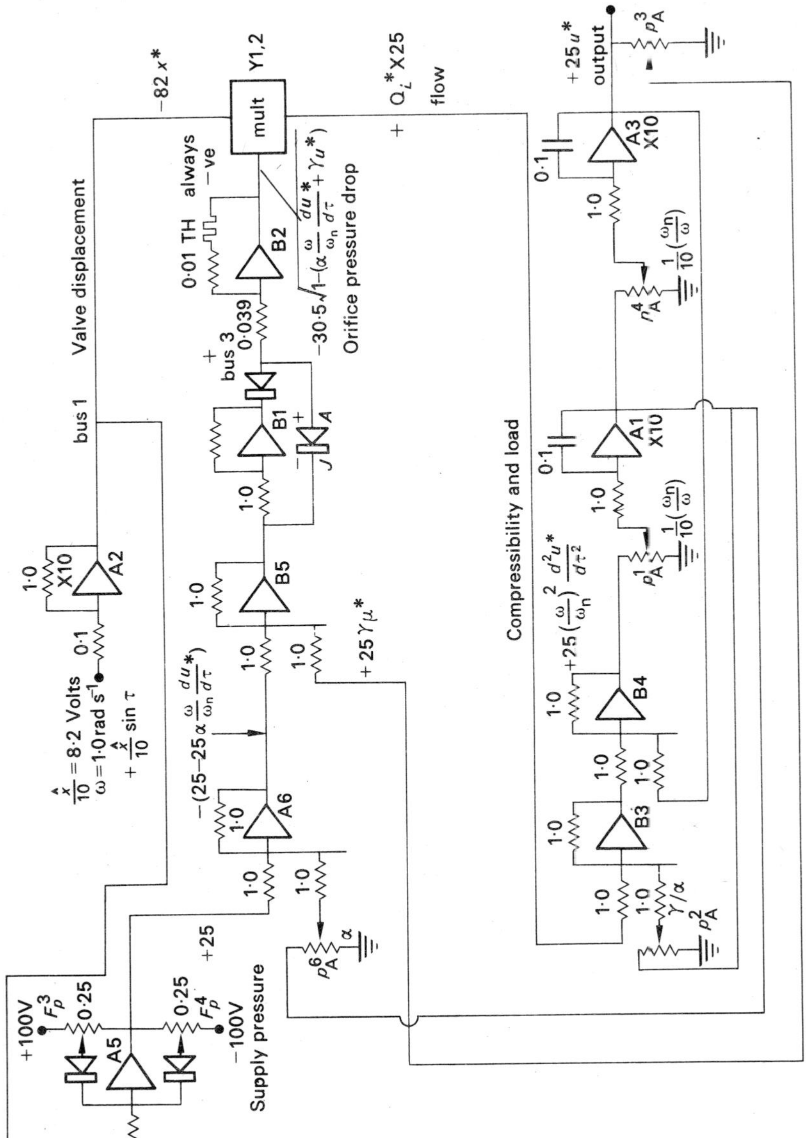

Fig. 7.3. Analogue computer circuit for non cavitating open loop system

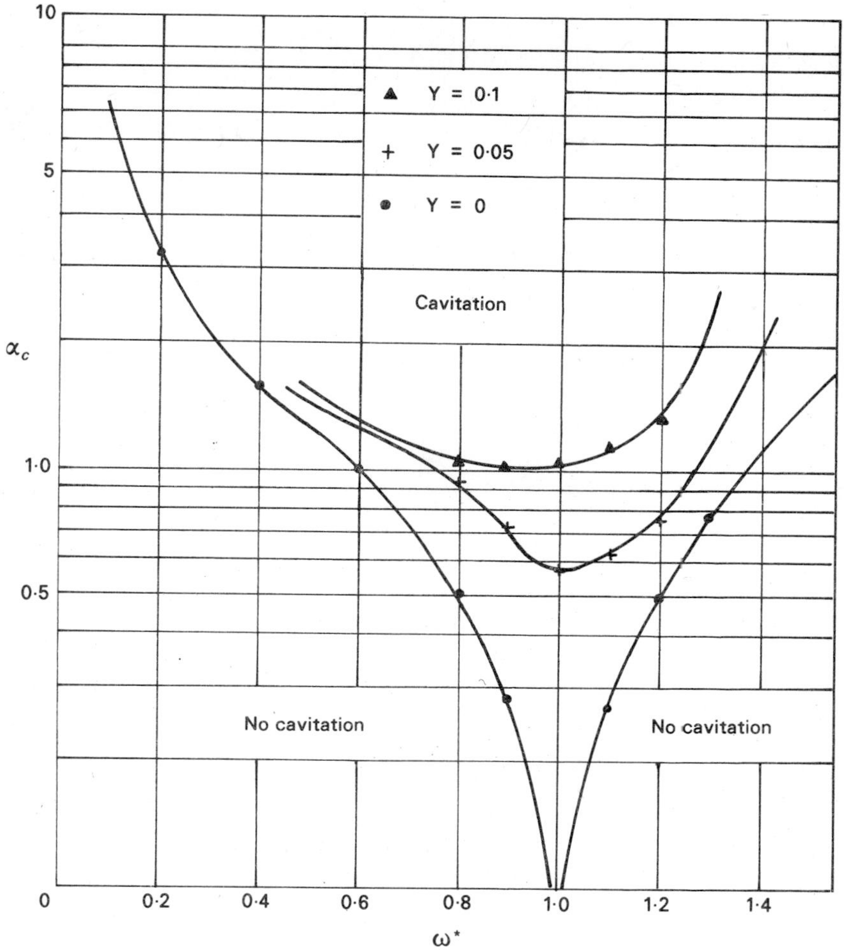

Fig. 7.4. Effect of viscous friction γ on the cavitation boundary

Define the leakage factor as

$$\phi = \frac{Ml\omega_n}{A^2}$$

Then (7.22) and (7.23) become,

$$Q_1{}^* = 2\left[\frac{u^*}{2} + \frac{1}{\alpha}\omega^*\frac{\mathrm{d}p_1}{\mathrm{d}\tau} + \frac{\phi}{2\alpha}(p_1 - p_1)\right]$$

$$Q_2{}^* = 2\left[\frac{u^*}{2} - \frac{1}{\alpha}\omega^*\frac{\mathrm{d}p_2}{\mathrm{d}\tau} + \frac{\phi}{2\alpha}(p_1 - p_2)\right]$$

and therefore equation (7.19) becomes

$$x_v^* \sqrt{\left\{1 - \left(\alpha\omega^* \frac{du^*}{d\tau} + \gamma u^*\right)\frac{x_v^*}{|x_v^*|}\right\}}$$
$$= \left(1 + \gamma\frac{\phi}{\alpha}\right)u^* + \left(\phi + \frac{\gamma}{\alpha}\right)\omega^*\frac{du^*}{d\tau} + (\omega^*)^2\frac{d^2u^*}{d\tau^2} \tag{7.24}$$

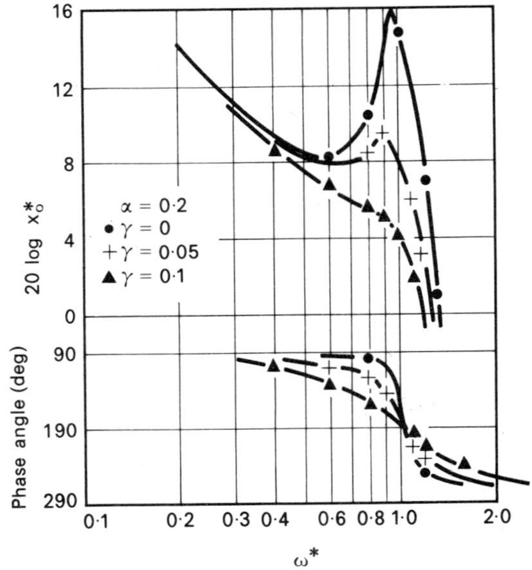

Fig. 7.5. Effect of viscous friction on open loop response

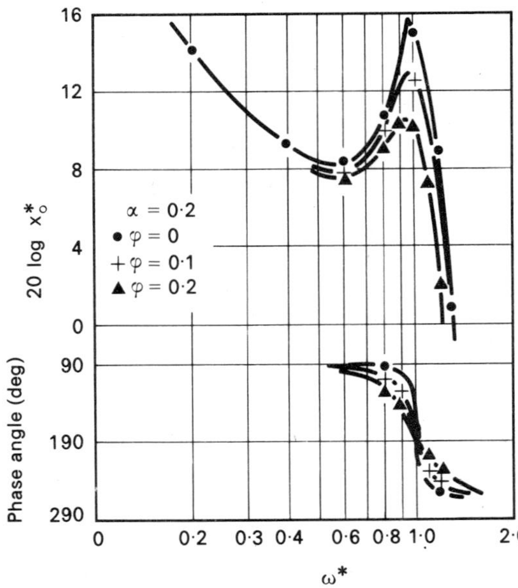

Fig. 7.6. Effect of leakage on open loop response

The effect of viscous damping and leakage on the open loop frequency responses are shown in Figs. 7.5 and 7.6. Flow forces are assumed zero. It should be noted that since the characteristic equation of the system is third order, velocity feedback and viscous loading do not have the same effect. Viscous loading like leakage, provides a form of acceleration feedback.

7.5. Practical considerations

In order to give a real meaning to the non-dimensionalised coefficients quoted, the relevant parameter values for a small 5hp servomotor are quoted. When working at a supply pressure of about 2000 lbf in^{-2}, maximum valve displacement is about 0·010 in and driving a mass of about 1000 lb the main non-dimensional parameters turn out to be: $\alpha \approx 1{\cdot}0$, $\gamma \approx 1{\cdot}0$, $\phi \approx 0{\cdot}05$.

With regards to the simulation shown in Fig. 7.3, it is convenient to carry out a frequency response by altering the ratio ω^*, i.e. changing the position of the natural frequency, rather than altering the input frequency. In following this method it is necessary to correct all other terms in the equations which contain ω_n. For example,

$$\alpha_\omega = \frac{\alpha \text{ at the natural frequency}}{\omega^*}$$

For convenience, the driving frequency ω is fixed at 0·16 Hz (1 rad s^{-1}), while the input amplitude $\hat{x}_v/10$ is fixed at 8·2 V. (This odd value arises because of the method used to derive the square root.)

7.6. Effects of flow forces

The presence of flow forces has two main effects on the dynamic behaviour of the servo loop [58, 60, 61, 62],

the open loop static gain of the servo is reduced.

the stability of the servo is affected.

It was shown earlier that the flow force opposes the torque motor driving force, hence affecting the valve position. Equation (7.17), clearly shows that the flow force varies with the load pressure in the actuator. These effects will be fed back directly to the control valve. The torque motor is in fact working into a frequency and amplitude dependent load. As the forcing frequency approaches the natural frequency determined by the output mass load and the oil column stiffness, the valve flow decreases. At the natural frequency the inertia load is simply bouncing on the oil column and the sum of the displacement flow and compressibility flow is very small. This is demonstrated in the results shown in Fig. 7.7(*a*). The explanation of why the flow decreases can easily be seen from the graphical analysis shown in Fig. 7.7(*b*).

Since there is very little flow through the valve the flow force is small and the valve displacement torque motor current relationship is high, making the forward loop gain high in this region. In other words the gain of the servo is a function of frequency, being largest in the region of the natural frequency. This is the worst possible condition with regards to loop stability. Results demonstrating the effect of flow forces are shown in Fig. 7.8.

It is seen that although the overall gain level is reduced due to the first effect mentioned above, the resonant amplitude is greatly increased in the region of the natural frequency.

In Fig. 7.9, the value of the flow force coefficient required to induce instability in the open loop condition is defined as δ_c. This condition is modified depending on the amount of viscous damping (γ) present. With the smaller inertia loads, viscous damping has a pronounced effect. From these results it is apparent that the two most important factors to be considered are the type of loading and the position of the natural frequency of the oil column and inertia relative to the operational bandwidth. If the natural frequency is well beyond this bandwidth, a reduction in open loop gain is the main effect of the flow force. In this case power is being wasted in driving an extra load on the valve; it may therefore be advantageous, especially in aircraft [63, 64] and missile applications, to reduce the flow force. The results show that, if the natural frequency lies within the operational bandwith, it may be necessary, from considerations of stability, to reduce the flow force. If the load on the servo is mainly of a viscous or of a spring nature, the stability of the system is not so drastically affected

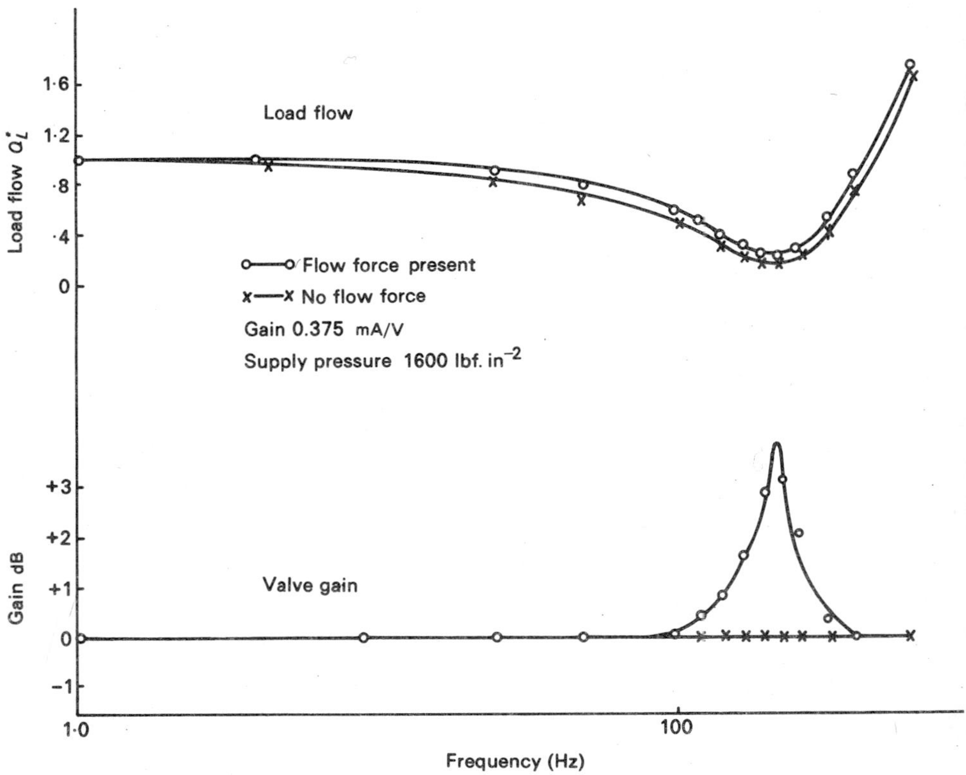

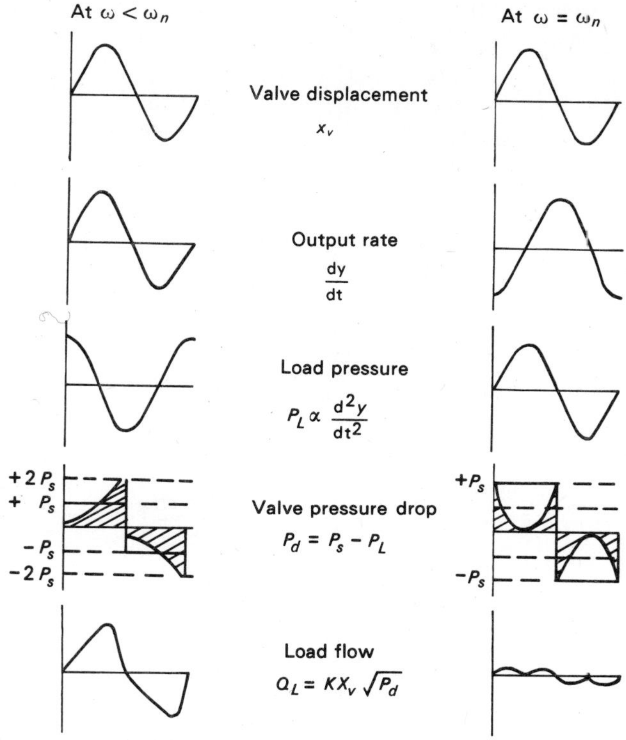

Fig. 7.7. (*a*) Experimental results showing behaviour of load flow Q_e and valve gain (*b*) Behaviour of load flow under dynamic conditions

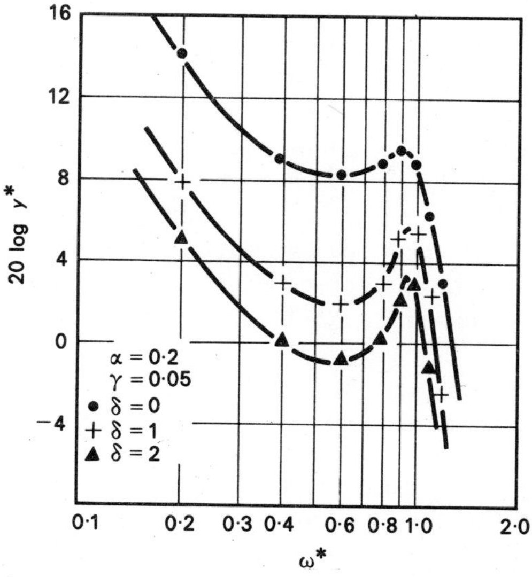

Fig. 7.8. Effect of flow force on open loop response

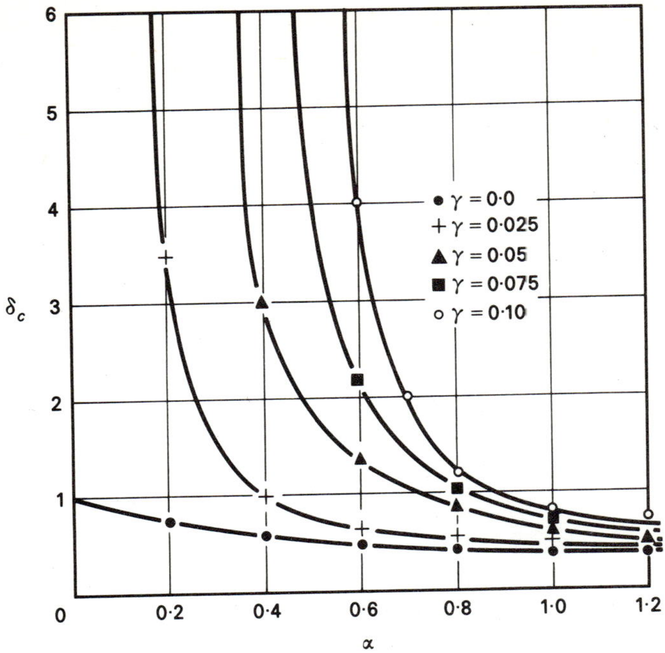

Fig. 7.9. The critical flow force factor δ_c

as when inertia load is present. In fact with a spring load, a case which has not been considered in this analysis, the Bernoulli force can be used to advantage. There are several methods of reducing flow force effects.

The flow forces may be completely removed. This can be achieved by shaping the spool and sleeve of the control valve as elaborated by Lee and Blackburn [38], and Eynon [62]. This has the disadvantage of being an expensive operation, especially if a high degree of flow force compensation is desired. It has the advantage, however, that the torque motor power required to drive the control valve can be kept within reasonable limits, and this, as has been mentioned before, may be of importance in missile applications.

The valve reaction force may be swamped by a spring load. This may be achieved by increasing the torque motor stiffness or by fitting a valve-centralising spring. By such means, the compliance K_s and hence the flow force factor δ are reduced.

Another method of swamping the reaction force is to use a more powerful torque motor. This increases the relay amplifier gain K and hence reduces the flow factor δ. This method may be limited by difficulty in obtaining good response rates from the larger torque motors.

The destabilising effect of the flow force can be reduced by the introduction of a viscous load, perhaps in the form of a dashpot or by the use of an orifice The presence of a leak tends to reduce the positional accuracy of the system but the increase in gain thus made possible can more than compensate for this. Such damping effects can of course be obtained by using electrical networks and transducers. For instance, acceleration feedback can be obtained by using transducers to measure the load pressure.

If flow forces are completely eliminated, the electrohydraulic gain, the flow per unit error voltage, is high. This feature is undesirable because most of the non-linearities are in this part of the loop and it may mean that the electrical gain, relay driving voltage per unit error in load displacement, must be reduced. This can cause loss of accuracy in a positional system. In general, it is preferable to keep the electrohydraulic gain as low as possible because the electrical gain is much more flexible. Hence one of the best methods of improving system performance is to remove some of the frequency-dependent flow force by partial shaping and to reduce the electrohydraulic gain by fitting a spring.

It should be borne in mind that, even if flow force effects are eliminated, an undesirable resonance peak can still occur when α is small, and the introduction of artificial damping may be necessary.

7.7. More exact simulation

When cavitation occurs it can no longer be assumed that the inflow and outflow are equal. The symmetry has been destroyed and the flow through each valve port must be treated separately. The equations to be solved are, for positive spool movements

Inflow

$$\sin \tau\sqrt{\{\tfrac{1}{2}1 - p_1\}} = \frac{u^*}{2} + \frac{1}{\alpha}\omega^*\frac{dp_1}{d\tau} + \frac{\phi}{2\alpha}(p_1 - p_2) \tag{7.25}$$

Outflow

$$\sin \tau\sqrt{\tfrac{1}{2}p_2} = \frac{u^*}{2} - \frac{1}{\alpha}\omega^*\frac{dp_2}{d\tau} + \frac{\phi}{2\alpha}(p_1 - p_2) \tag{7.26}$$

Load

$$p_1 - p_2 = \alpha\omega^*\frac{du^*}{d\tau} + \gamma u^* + \eta\frac{u^*}{|u^*|} \tag{7.27}$$

where η is the seal friction coefficient given by $\eta = \text{friction}/P_sA$. The equations for negative x_v also follow from (7.3). The analogue circuit is shown in Fig. 7.10.

It is necessary to obtain a clear physical picture of the mechanism of cavitation before any simplifying assumptions are made. If, as in Fig. 7.11, the cylinder pressure falls to zero at some point in the stroke but recovers before the valve crosses centre, no cavitation bubble will form. The side of the jack on which the zero pressure exists is connected to exhaust and so no outflow, except that due to piston movement, takes place. This side remains full of oil and so no cavitation bubble forms. The attainment of zero pressure will cause the oil to vaporise as shown by the experimental evidence in [140]. When the pressure on the

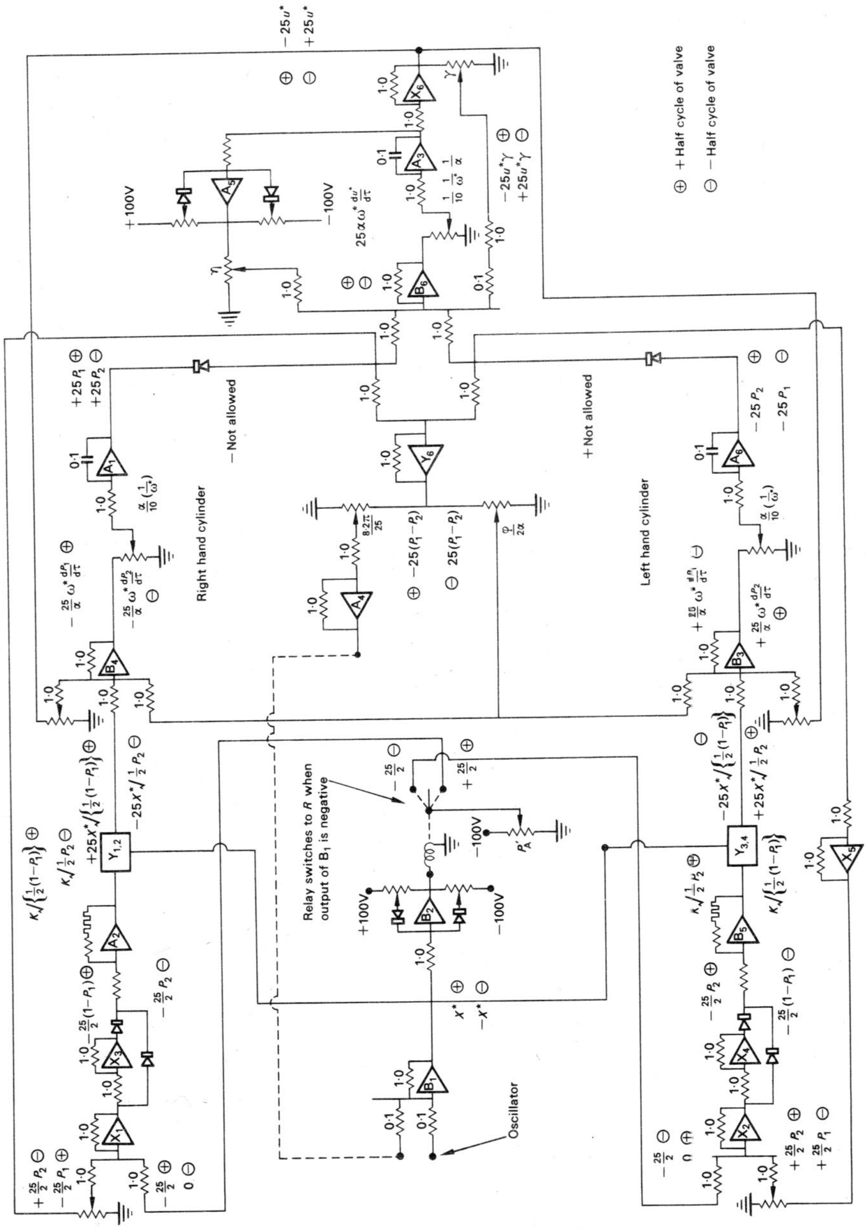

Fig. 7.10. Analogue computer circuit which takes cavitation into account

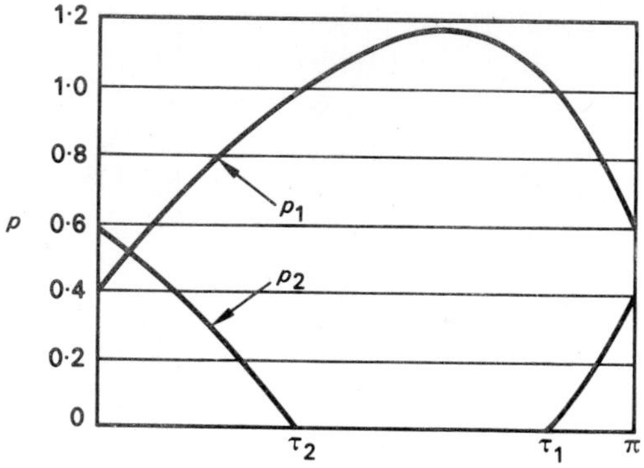

Fig. 7.11. Pressure waveform illustrating cavitation

high-pressure side drops below system pressure, the low pressure side begins to increase and continuity of flow is maintained. Inflow and outflow are equal and $p_1 + p_2 = 1$. However, during the time that the pressure on the low-pressure side is zero (and the pressure on the high-pressure side exceeds P_s) equations (7.25) and (7.27) describe the dynamics.

A cavitation bubble can form if the pressure on one side of the jack is zero when the valve crosses centre. In the case illustrated in Fig. 7.12, the pressure

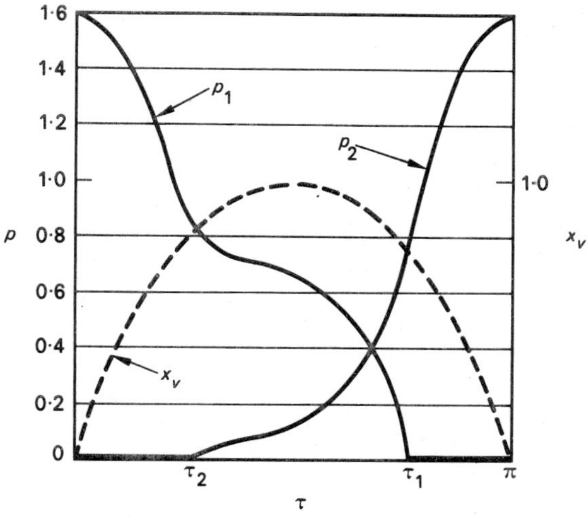

Fig. 7.12. Example of cavitation at reversal

on that side of the jack connected to the supply falls to zero and does not recover before the valve crosses centre. During the interval τ_1 to π the pressure drop across the inlet port is P_s and this does not provide sufficient flow to keep the inlet side full. The piston breaks away from the fluid and a bubble forms. When the valve crosses centre that side of the jack containing the bubble is now connected to the exhaust. The pressure in this side cannot rise until the bubble has been repaired at time τ_2. Hence from time 0 to τ_2 equations (7.25) and (7.27) are used, from time τ_2 to τ_1 equations (7.25), (7.26) and (7.27) are used, and from τ_1 to π equations (7.26) and (7.27) are used.

To simplify the analysis it is assumed that the bubble fills instantaneously and so p_2 begins to rise as soon as p_1 falls below unity. This is in effect assuming equal inflow and outflow when neither pressure is zero. Thus when $p_1 > 1$, equations (7.25) and (7.27) are used, when $p_2 > 1$ equations (7.26) and (7.27) are used, and when p_1 and p_2 are both less than one, equations (7.25), (7.26) and (7.27) are used.

The assumption that the bubble fills instantaneously simplifies the analysis considerably. With reference to Fig. 7.13, cavitation bubbles exist in the region marked 'cavitation at reversal'. It is seen that the values of inertia load factor required to produce this condition are relatively large and it is unlikely that

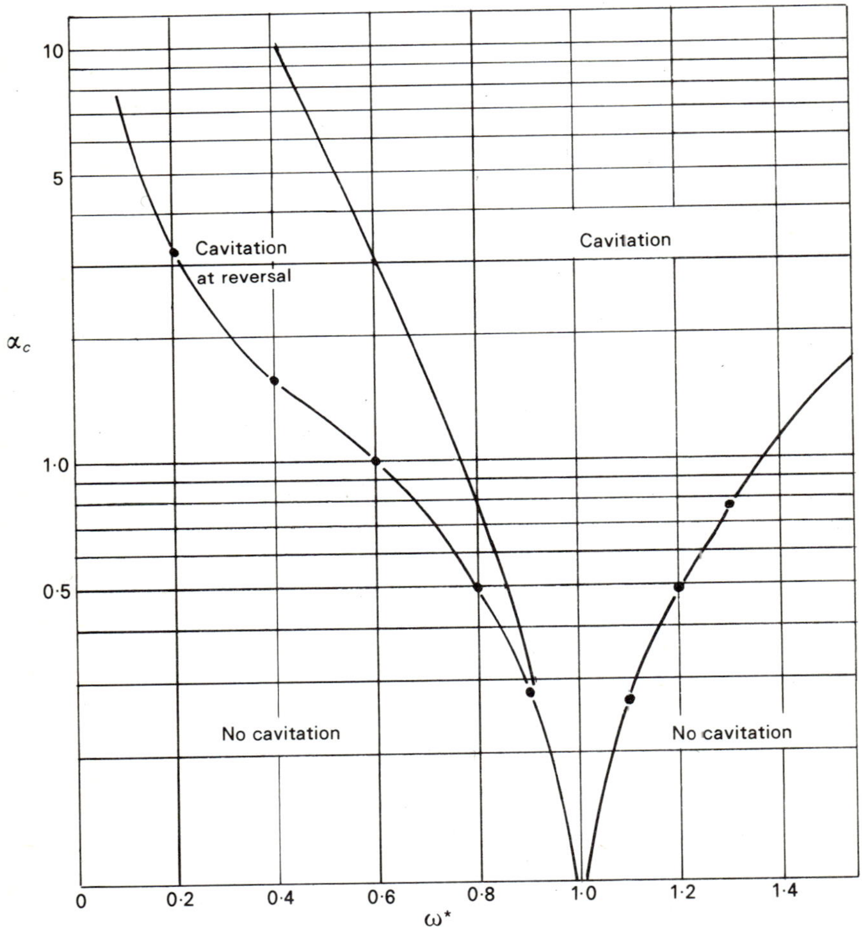

Fig. 7.13. Cavitation boundaries

such values would be achieved in the closed loop condition, except perhaps near the natural frequency. Calculations have shown that the time required to fill these cavities is relatively small and so the assumption of instantaneous filling is not unrealistic.

These conditions are used to investigate the effect of large inertia loads (Fig. 7.14). Increasing α has a strong stabilising effect owing to the large valve flows present. It is interesting to note that if Bernoulli forces are non-existent, the increasing α is stabilising as shown. If flow forces are present, however, valve flow no longer stabilises but de-stabilises, and large α is destabilising.

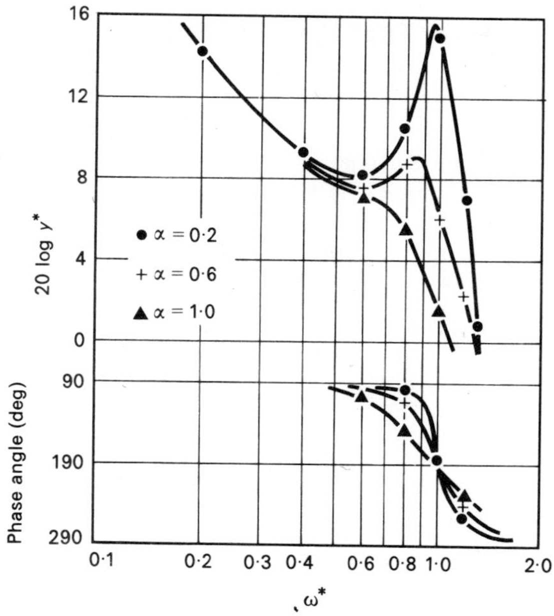

Fig. 7.14. Effect of inertia load factor on open loop response

Chapter 8

The small signal response of the hydraulic servo and means of improving performance

8.1. Introduction

The analysis of a linear servo system, excluding leakage affects, indicates that there are no 'threshold' conditions. The system responds to the smallest input

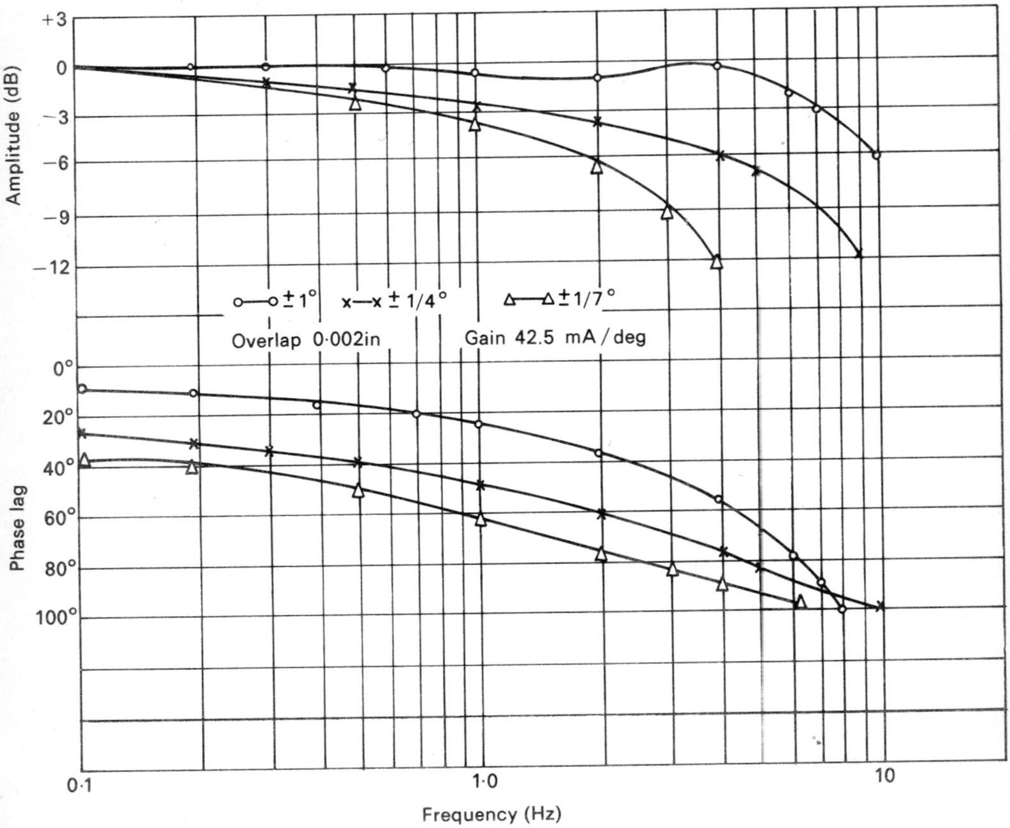

Fig. 8.1. Experimental results showing variation of servo response with input amplitude

signals and disturbances, whilst the time taken for similar movements is constant irrespective of the overall amplitude scale of movement.

In practice however, a hydraulic servo has non-linearities which detract from this ideal, and experimental results show that as the input amplitude is reduced, the output waveform is not in exact proportion to the input signal. An example of this is shown in Fig. 8.1, which is an experimental result. If the input is reduced still further, eventually a stage is reached when there is no response at the output for a finite disturbance at the input. This condition is termed the threshold region. As the input amplitude is increased, the system operates more definitely and approaches the behaviour exhibited by the linear servo. As the input is increased still further, a stage is ultimately reached where saturation occurs. This will be the result of one or more of the system components reaching the limit of their output.

Electrohydraulic servo systems are often used for applications which require a fast response to cater for an emergency or extreme condition. However it is very probable that only a small fraction of their life will be spent under these conditions, and a good deal of this time will be used during the test and development period. Much the greater part of the servo's life will be spent responding to small signals, and in fact much of this time will be spent around the threshold region.

8.2. Characteristics of the threshold region

The threshold region may be considered as the dead zone of a servomechanism, and investigations [66] reveal that there are two main characteristics

Static dead zone, defined as the maximum signal to which the system fails to respond after being quiescent for a finite period of time.

Dynamic dead zone, defined as the maximum signal to which the system fails to respond, whilst the input has some form of movement.

Some experimental results are shown in Figs. 8.2 and 8.3.

Static dead zone appears to be due mainly to friction, either of the piston seals or of the hydraulic valve, and develops with time as lubricant is squeezed from between rubbing surfaces. The pressure which elastomeric seals exert on the cylinder walls ultimately exudes the lubricant, with a consequent increase in the coefficient of friction, which reduces again after the first movement causes relubrication; dead zone from this cause is of more significance if there is a bleed orifice, or if underlapped spool valves are used.

Hydraulic spool valves are normally 'pressure balanced' to avoid large operating forces, but even so quite large frictional forces can develop, of the order of 0·5 lbf with a miniature valve, if it is quiescent for a length of time. This is because side forces are present on the valve owing to unbalanced pressure gradients around the periphery of the spool. These side forces ultimately squeeze out the fluid in the clearance between the spool and sleeve, and a high coefficient of friction again results. Inadequate filtration will also contribute greatly to increased spool valve friction.

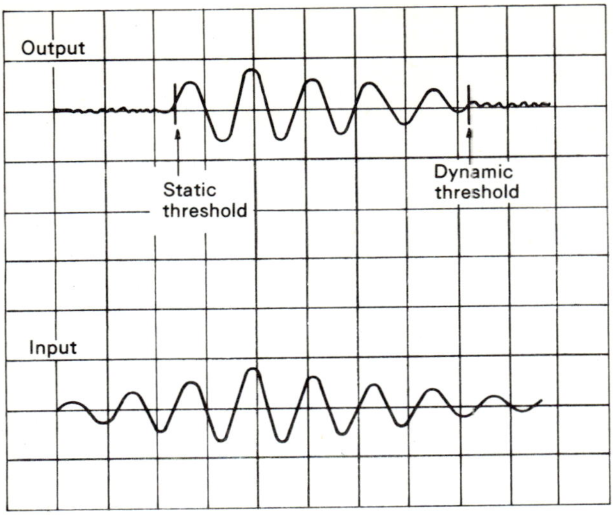

Fig. 8.2. Threshold characteristics

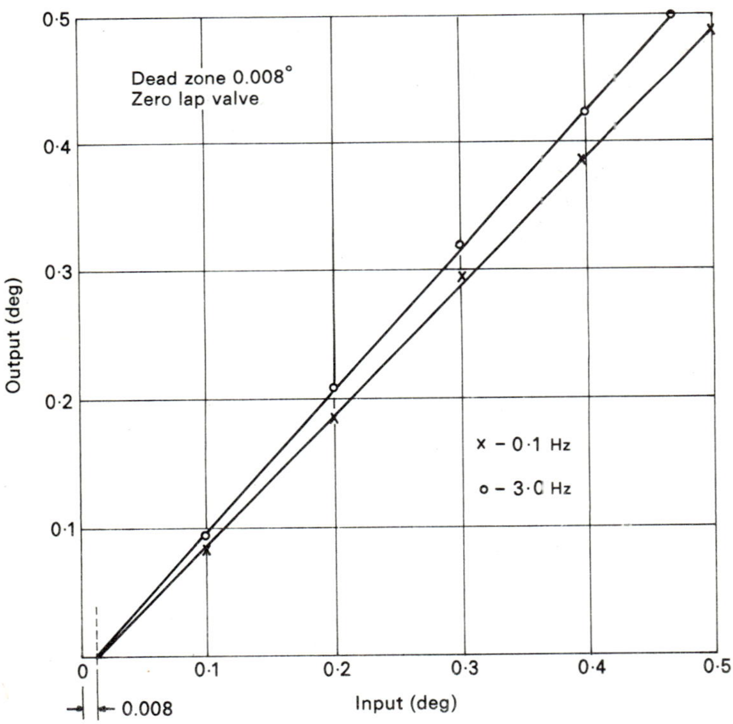

Fig. 8.3. Dynamic threshold results—servo output v input

It is fortunate that the causes of static dead zone are not inherently additive, that is, almost all the error signal necessary to break the valve friction becomes available to break the seal friction when the valve has moved. As static dead zone is mainly a function of friction it is obviously a very erratic phenomenon, as attempts at measurement soon reveal.

Dynamic dead zone is usually only about 30 per cent of the static dead zone, and is the combined effect of any valve overlap and backlash, friction, hysteresis and resolution of the various components in the system. Friction in this case will be of a more viscous nature. Continuous efforts are made to reduce the dynamic dead zone, yet in the last analysis a certain amount of dead zone is essential. With round ported overlapped and semi-overlapped valves, analytical treatment concludes that the servomechanism is unstable with the valve in the neutral position. The fact that these servomechanisms are stable is due to the friction in the system. Though the servomechanism is nominally in the dead zone it may often be observed to drift slowly and erratically with zero input signal within the dead zone until a position is found at which it can stay long enough to establish a static dead zone condition.

Predictions of the magnitude of dead zone are difficult, because the ultimate gain of the system may not be known, and it is impractical to control the equivalent dead zone of the components, except to ensure that it does not exceed a certain limit. Thus, unless very extensive testing of components has been done, any calculation will present the worst case, the final measured results possibly being up to three or four times better.

8.3. Methods of improving performance

Having accepted the fact that the threshold condition is a real problem in practice, it is now necessary to investigate a method of improving the situation. This must fall into two general approaches.

In addition to good basic design, improve the design and manufacturing tolerances of the components.

Alter the system arrangement, possibly with additional loops.

To achieve satisfactory performance by the first approach is not always the most economic approach, since development should be aimed at widening tolerances in order to reduce manufacturing costs. Hence the second approach would seem to offer the more economic method. An example of this is shown in Fig. 8.4, where the improvement achieved using small amplitude high frequency excitation of the control valve is evident. This may not always be an acceptable solution.

Consider the system shown in Fig. 8.5. In this case the threshold $[T]$ is in the forward loop path. The transfer function will be

$$\theta_o = [K_A][T][KG(s)]\theta_e$$

and for the feedback loop,

$$\theta_e = \theta_i - \theta_o' = \theta_i - K_f\theta_o$$

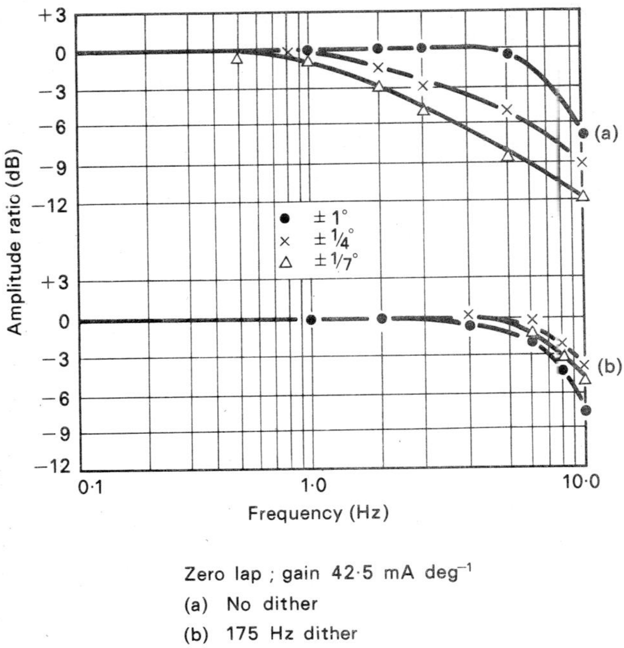

Fig. 8.4. Effect of dither on amplitude response. Zero lap, gain 42·5 mA/degree (*a*) no dither (*b*) 175 Hz dither—for different input amplitudes

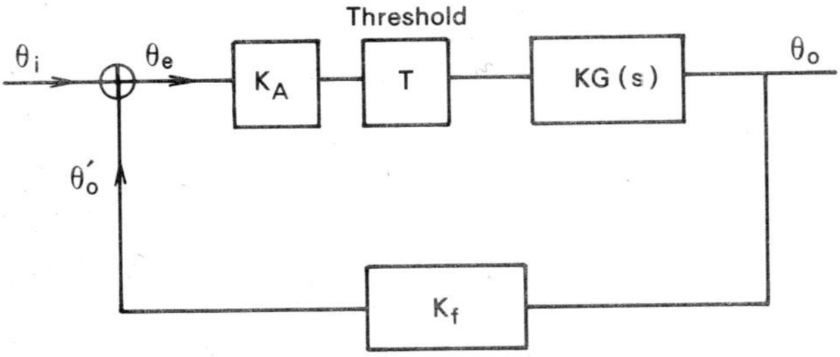

Fig. 8.5. Closed loop with threshold included

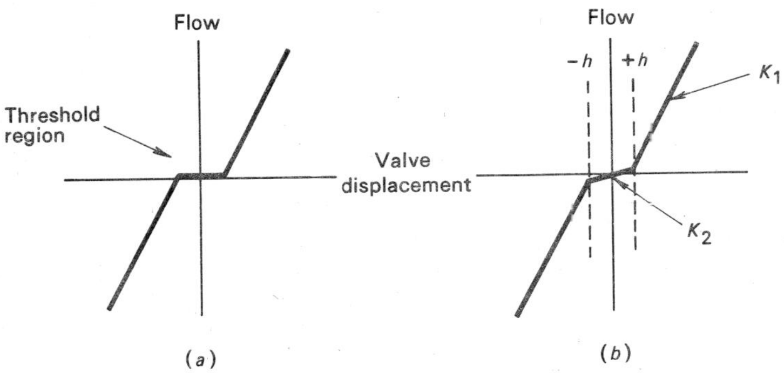

Fig. 8.6. Approximations to the threshold characteristics

These two relationships combine to give the closed loop transfer function

$$\frac{\theta_o}{\theta_i} = \frac{[K_A][T][KG(s)]}{1 + [K_f][K_A][T][KG(s)]}$$

It has been a well known practice to reduce the threshold by increasing the forward loop gain $[K_A]$ so that $[T]$ makes a smaller contribution to the overall gain.

If the characteristic for $[T]$ were a dead zone, for example valve overlap, as shown in Fig. 8.6, the characteristic could be approximated to two gain conditions of slopes K_1 and K_2. The threshold could then be described by

$$\begin{aligned} |T| &= [K_1] \text{ for } |\theta_e| \geqslant h \\ &= [K_2] \text{ for } |\theta_e| \leqslant h \end{aligned}$$

By increasing the amplifier gain K_A, an attempt is made to make the effect of K_2 small, but in doing so K_1 is also affected, so that the system gain is greatly increased outside the limits of $\pm h$. This will result in an alteration of the system stability and performance, and hence some form of additional shaping loop would be required to return the system to its original performance. A phase lag network, for example, would help to attenuate the increased gain at the higher frequencies. Clearly, this is a sledge-hammer approach. Ideally a gain increase in the threshold region only is required, and this to contribute no phase lag. This may be possible with non-linear networks.

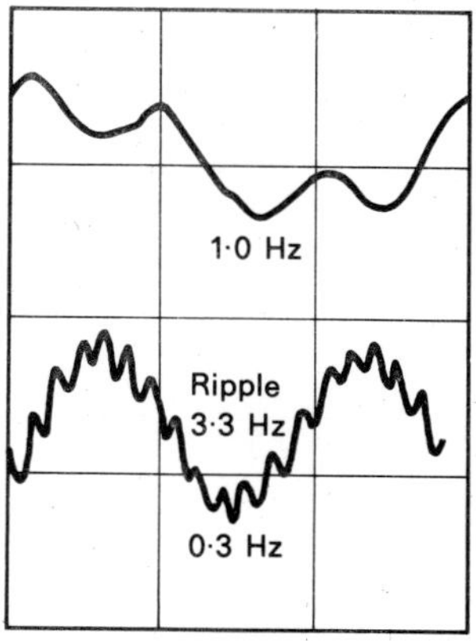

Fig. 8.7. Output waveform distortion, showing effect of inertia load bouncing on the oil column

Some practical results of the inter-relationship between attempts to improve threshold and the resultant effects on performance are shown in Table 8.1. The type of output waveform distortion experienced is shown in Fig. 8.7, where the effect of the inertia load bouncing on the oil column is clearly seen.

Table 8.1

Run	Threshold (deg)	Wave-form	Gain $mA\ \mathrm{deg}^{-1}$	Lap condi-tion (in)	Bleed dia-meter (in)	Velo-city feed-back	Accelera-tion feed-back
1	0·085	Poor	25·6	Overlap 0·001	—	—	—
2	0·29	Much better	25·6	Overlap 0·001	0·0135	—	—
3	0·12	Better	50·0	Overlap 0·001	0·0135	Yes	—
4	0·088	Worse	25·8	Overlap 0·001	—	Yes	—
5	0·044	Worse	57·5	Overlap 0·001	0·006	Yes	—
6	0·05	Much better	25·0	Overlap 0·001	—	—	Yes
7	0·008	Much better	42·5 (175 Hz dither)	Zero	—	—	Yes

8.4. Computer study of small amplitude response

The flow through the control ports of the valve has been described by

$$Q = Ka\sqrt{(P_s - P_L)}$$

where a is the uncovered orifice area. This equation is only approximately true since the discharge coefficient will vary at small openings and the orifice area–displacement curve will be non-linear. Some comparisons are drawn in Fig. 8.8.

Using the equations developed in section 7.2, and assuming that cavitation is unlikely to occur under small amplitude response conditions, then the open loop condition is,

$$a^* \sqrt{\left\{1 - \left(\alpha\,\omega^* \frac{\mathrm{d}u^*}{\mathrm{d}\tau} + \gamma u^* + \eta \frac{u^*}{|u^*|}\right)\right\}}$$
$$= \omega^{*2} \frac{\mathrm{d}^2 u^*}{\mathrm{d}\tau^2} + \left(\frac{\gamma}{\alpha} + \phi\right)\omega^* \frac{\mathrm{d}u^*}{\mathrm{d}\tau} + \left(1 + \frac{\phi\gamma}{\alpha}\right)u^* + \frac{\phi}{\alpha}\eta \frac{u^*}{|u^*|} \qquad (8.1)$$

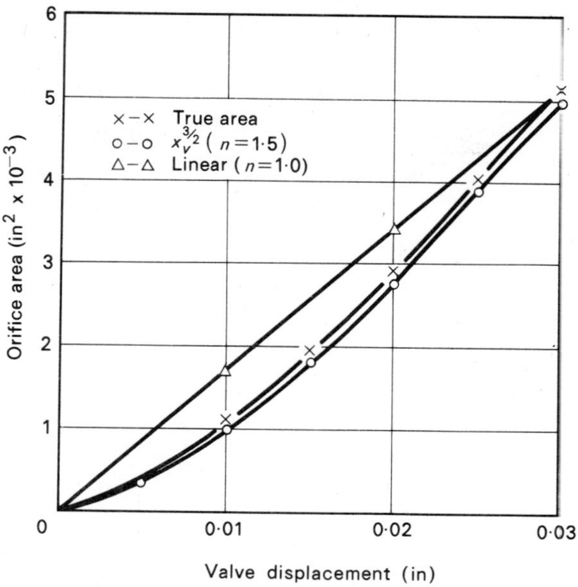

Fig. 8.8. Comparison between true and simulated orifice openings

When non-dimensional area is used, the load coefficient term is defined slightly differently,

$$\alpha = \frac{MK\hat{a}\sqrt{(P_s/2)}\omega_n}{P_s A^2}$$

where $a^* = a/\hat{a}$.

It is seen from Fig. 8.8 that the most realistic area characteristic is given by,

$$a = K_2 x_v^n$$

where $n = \frac{3}{2}$.

$n = 1$ is the usual linear relationship.

For a sinusoidal input, let

$$x_v = Tx_v \sin \tau$$

where T is some factor by which the input is altered and describes the input amplitude selected.

Hence,

$$a^* = \frac{K_2 x_v^n}{\hat{a}} = \frac{K_2(T\hat{x}_v \sin \tau)^n}{K_2 \hat{x}_v^n} \tag{8.2}$$

$$= (T \sin \tau)^n$$

It can be seen [67] from considerations of simple relationships that the minimum requirement to achieve a threshold condition is that a leakage path must be present except in the case of valve overlap. Without a leakage path any opening of the valve will allow full system pressure to act on the ram and hence overcome any resistance. However a real system will have a leakage path present. Before the ram can move, flow must take place across the leakage path to establish a pressure sufficient to overcome the load.

The theoretical absolute threshold value can be obtained from equation (8.1) if it is assumed that all output motion ceases, giving u^*, $du^*/d\tau$ and $d^2u^*/d\tau^2$ all zero. Hence,

$$T = \left(\frac{\phi\eta}{\alpha\sqrt{(1-\eta)}}\right)^{1/n} \tag{8.3}$$

From this equation it is seen that only leakage and seal friction affect the final steady state condition. These results are shown in Figs. 8.9 and 8.10. This shows

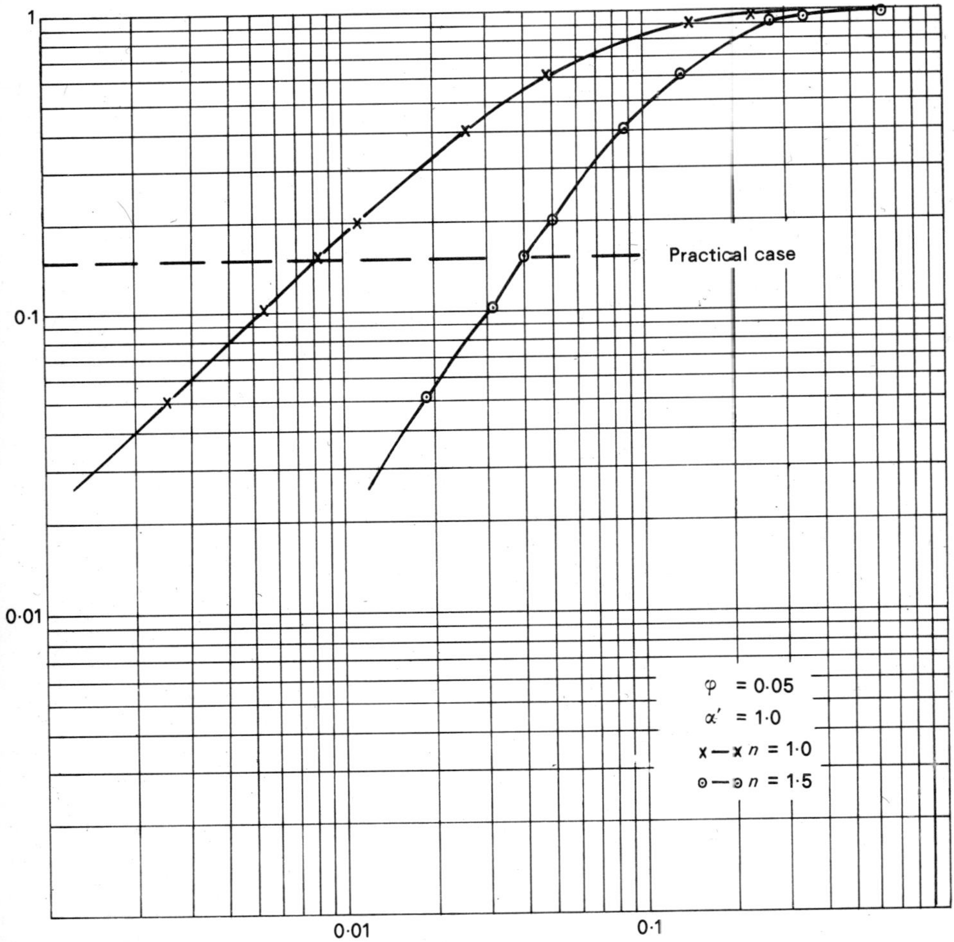

Fig. 8.9. Theoretical variation of absolute threshold with seal friction

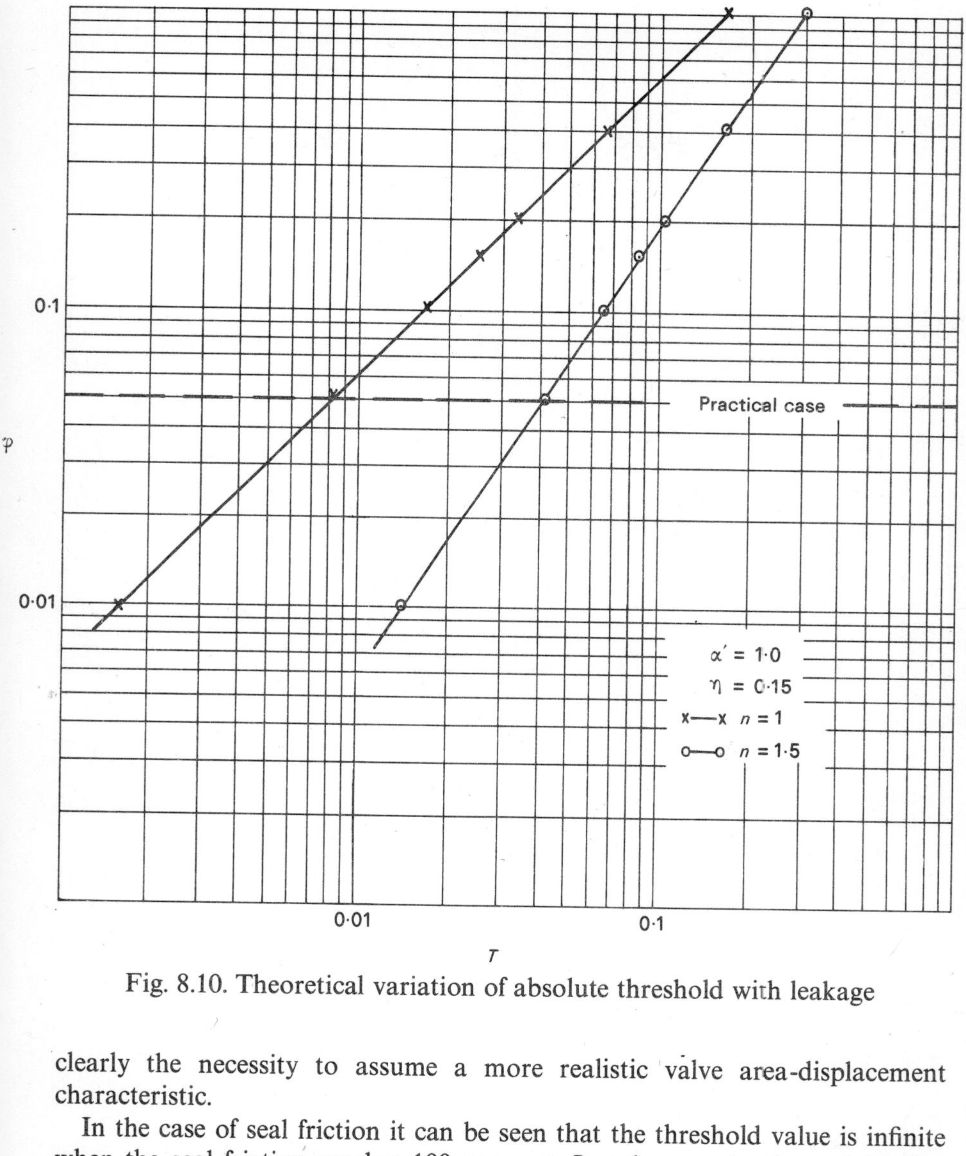

Fig. 8.10. Theoretical variation of absolute threshold with leakage

clearly the necessity to assume a more realistic valve area-displacement characteristic.

In the case of seal friction it can be seen that the threshold value is infinite when the seal friction reaches 100 per cent. In other words, the seal friction force becomes equal to available pressure force on the ram.

The computer result shows that at the smaller amplitudes ($T < 0{\cdot}0173$, 1·7%) the system becomes oscillatory, as shown in Fig. 8.11. A stage is reached when a small input will cause the system to oscillate at its natural frequency. It seems likely that at these small amplitudes, conditions can be such that the inherent damping contribution from the servo valve is not able to keep the system stable. This phenomena has also been found by Royle [68].

Results showing the effects of seal friction, viscous damping and leakage are

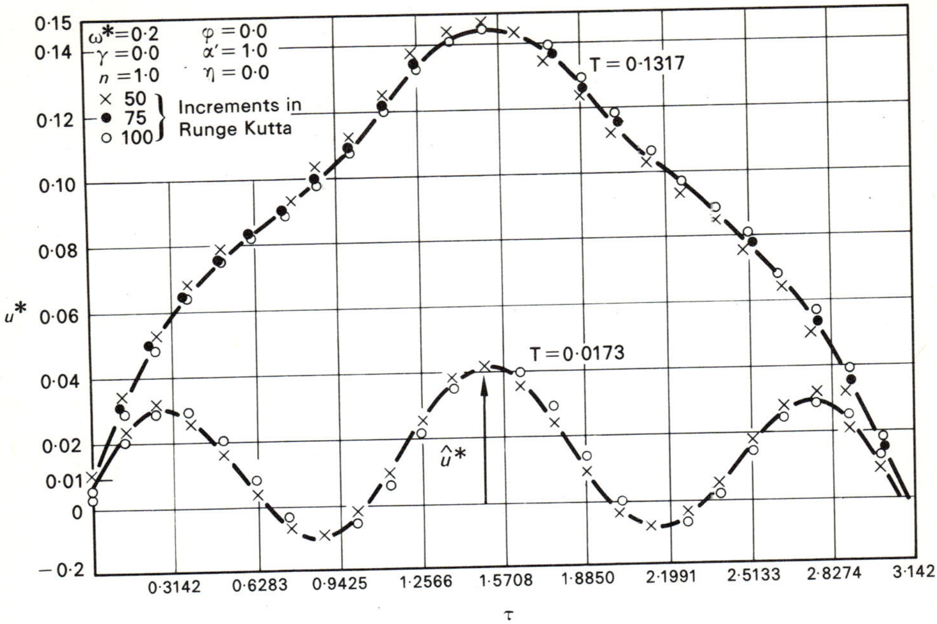

Fig. 8.11. Effect of computational increment on computer results

shown in Figs. 8.12 to 8.17. The results are based on $\omega^* = 0{\cdot}2$ and $\alpha = 1{\cdot}0$.

Viscous damping has the least effect in deviating the small amplitude response from the ideal linear condition, while leakage appears to give the worst results. The effects of seal friction lie somewhere between the two. When considered in relation to the other aspects of performance, there seems to be no straightforward method of achieving a small threshold. The optimum system can only be achieved by compromise with respect to the other aspects of performance.

8.5. Further methods of improving performances

8.5.1 Loop gain

The loop gain of an electrohydraulic servo can be conveniently divided into two groups,

The electrical gain, which includes the servo amplifier and position feedback characteristics.

The hydraulic gain, which includes the torque motor, valve and actuator characteristics.

If the servomotor has a high hydraulic gain, and an excess velocity capability over that necessary to achieve the frequency response bandwidth, then the electrical gain must be reduced in order to maintain the stability margin. The amplifier and torque motor are therefore less effective. The servo valve and actuator should be matched so that the saturation velocity is just sufficient to

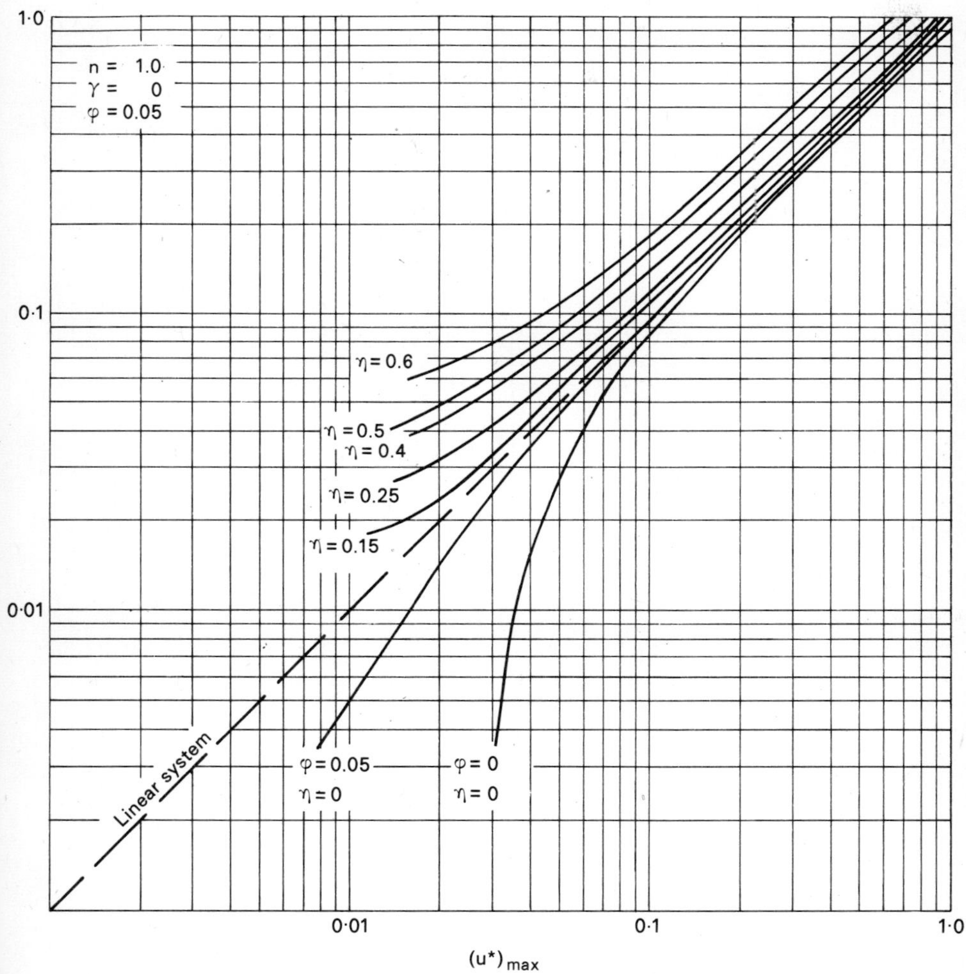

Fig. 8.12. Effect of seal friction on the response. $n = 1{\cdot}0$

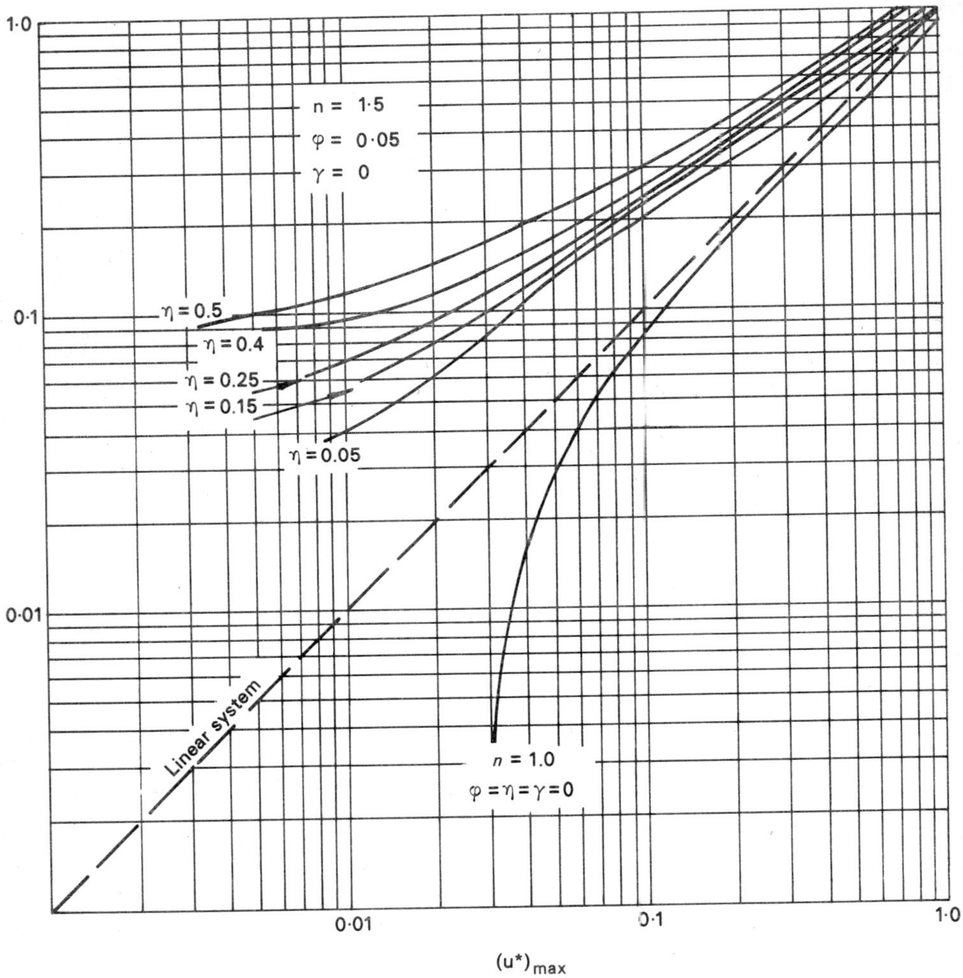

Fig. 8.13. Effect of seal friction on the response. $n = 1{\cdot}5$

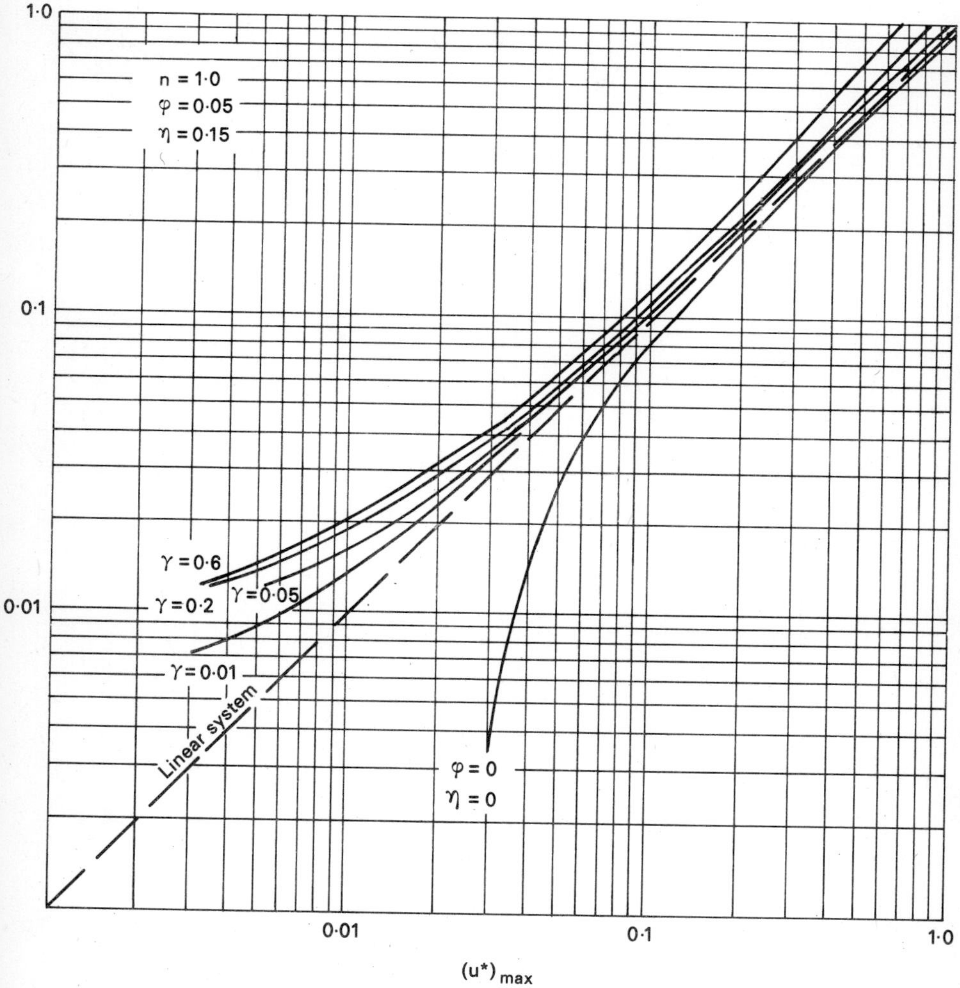

Fig. 8.14. Effect of viscous damping on the response. $n = 1{\cdot}0$

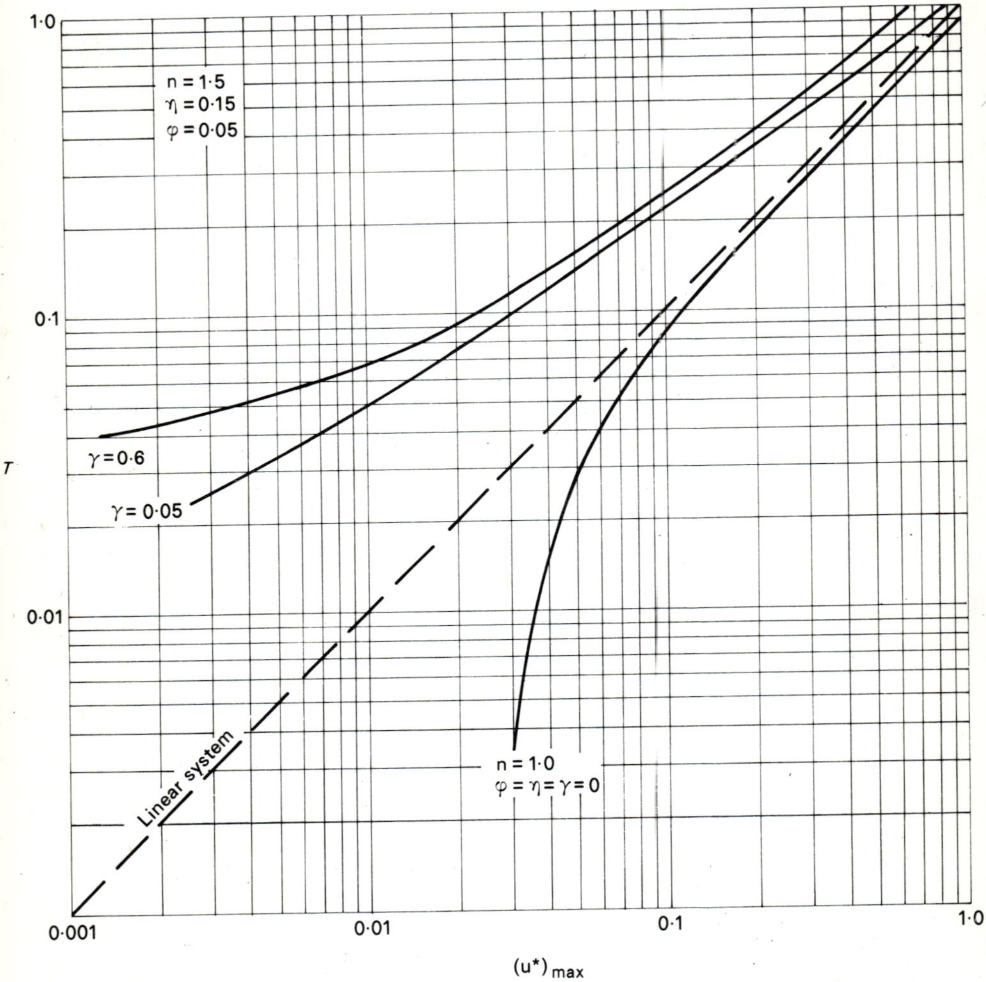

Fig. 8.15. Effect of viscous damping on the response. $n = 1{\cdot}5$

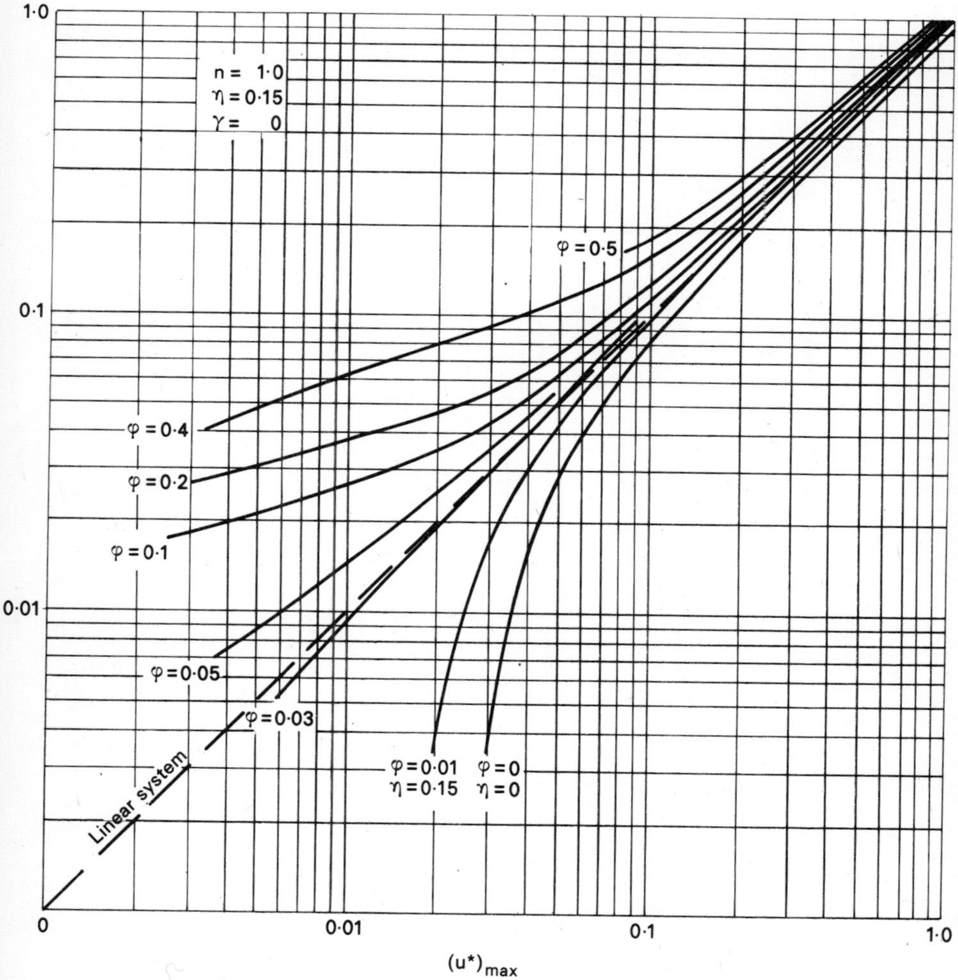

Fig. 8.16. Effect of leakage on the response. $n = 1{\cdot}0$

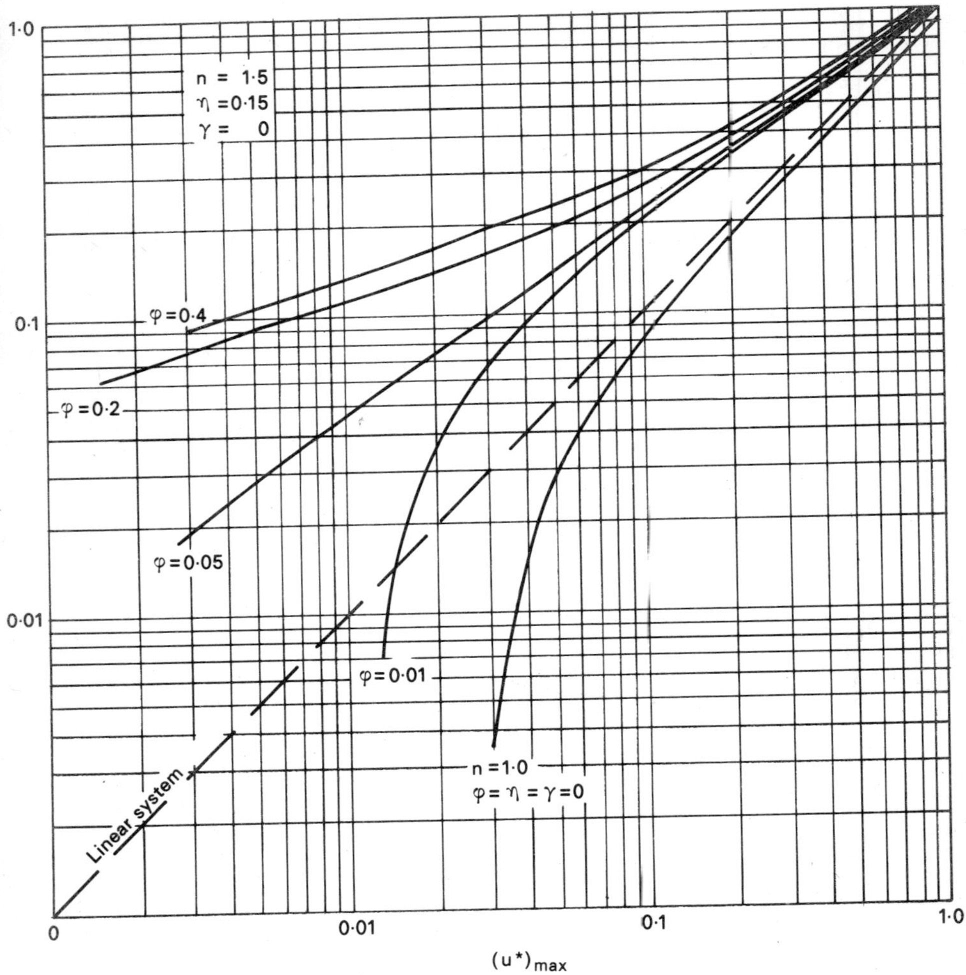

Fig. 8.17. Effect of leakage on the response. $n = 1{\cdot}5$

meet the response requirements. The occurrences of amplifier, torque motor and velocity saturations should be as near as possible. The loop gain can then be optimised in the electronics sections, which are more linear and more flexible.

8.5.2 Orifice bleeds and valve underlap

The instability and harmonic distortion which are due to the dynamic behaviour of the servo and its structure can be damped by an orifice bleed across the actuator or by underlapping the valve. Harpur [49] shows that the former is more efficient.

The orifice within the actuator is different from simple viscous damping in that the leakage is a function of the pressure difference across the cylinder, and therefore output force. This device approximates more to an acceleration feedback, but it is also operative when there is an output load at steady velocity. There is an optimum orifice size for maximum sensitivity, though this may not damp the harmonics sufficiently.

The difficulties with this method are:

It exaggerates the effects of piston seal friction; a dead zone is caused because the differential pressure necessary to overcome friction requires a flow through the orifice, with a corresponding control valve displacement and error signal.

For the same reason an error is inherent with orifice bleed stabilisation if the servo is required to hold an output load.

The pressure drop—flow characteristic is a square law, so that the damping virtually disappears at low velocities. This limits the gain of the system if a secondary instability is to be avoided, and this condition can be aggravated by the equivalent form of backlash, particularly seal slip-stick effects.

The orifice bleed significantly reduces the bandwidth of the system.

However, the orifice is a cheap and easy feature to incorporate, and is sufficient in many cases. It has an advantage in that it combats cylinder pressure peaks, and indeed, capillary type bleeds are sometimes introduced with this as the primary objective.

8.5.3 Pressure feedback

It is generally necessary to introduce an acceleration term, or an approximation to this. This may, of course, be for reasons other than threshold, for example, harmonic distortion due to peaking pressures. The general principle is to feedback the cylinder differential pressure. The use of transducers and their associated electronics is obvious, but not generally favoured because of complexity and cost. The hydraulic two-stage valve suggests an hydraulic solution, as the cylinder pressures are easily ducted to actuate the second stage spool. Simple pressure feedback of this type can result in reduced output system stiffness. Dynamic pressure feedback has been developed [69] to overcome this. This method is effective but requires careful matching.

8.5.4 Electrical networks

Standard textbooks give details of many transducer circuit passive networks which have their use in particular circumstances. The following are briefly described because of their usefulness in a more general sense.

Velocity feedback The usefulness of this form of damping is marred because its effect on the natural frequency harmonic of the load and actuator is limited. A signal which is strong enough to reduce the harmonics will give too much lag at lower frequencies. However, if another feature is introduced to deal with the harmonic, e.g. an orifice bleed, then velocity feedback will enable the gain to be increased and a decrease in dead zone will be achieved.

Acceleration feedback There is no doubt as to the effectiveness of an acceleration term in stabilising an hydraulic actuator. Accelerometers have been developed to provide a wide range of performance with high accuracy, but costly. Simple feedbacks can be achieved using transient velocity feedback and shaping networks in conjunction with the servo amplifier. These are shown in Fig. 8.18 but must be 'tuned' to the load characteristics. This method

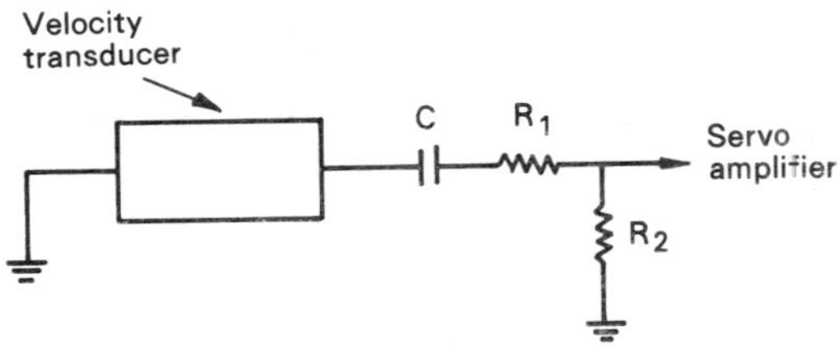

(a)

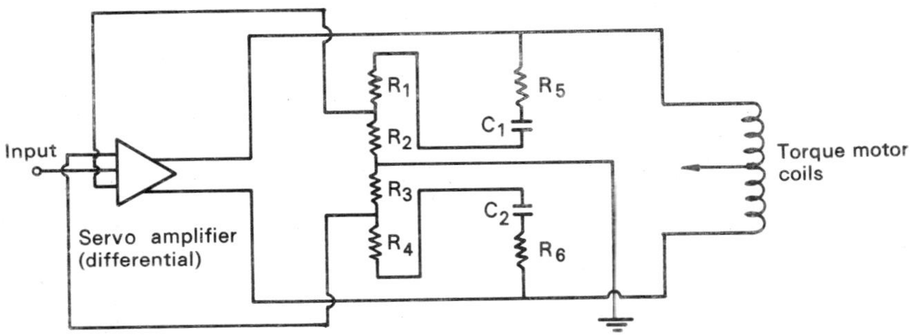

(b)

Fig. 8.18. Acceleration feedback (*a*) using velocity transducer and shaping network (*b*) shaping between the amplifier and torque motor

has the advantage of being less sensitive to structural noise which might be picked up by the accelerometer. Feedback shaping can be used to great effect, especially in closing the loop around the valve, torque motor and amplifier. This gives much greater control of the valve position which is effectively equivalent to output velocity.

8.6. Two-stage electrohydraulic valve

When large fluid flows are required, in general, it is necessary to use two stages in the control valve arrangement. Basically this is necessary in order to maintain a satisfactory response at high flow rates. Loss of response in single stage valves at large flow rates, is due to high spool inertia and larger flow forces, resulting in the need for a larger torque motor. Feedback loops are often provided at each stage, making them more flexible.

A typical arrangement in conjunction with a ram, is shown in diagrammatic form in Fig. 8.19. It consists of a pilot spool valve and main spool, having an

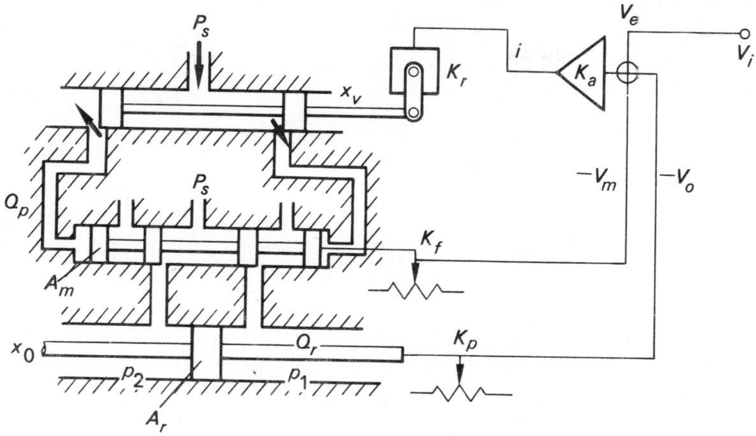

Fig. 8.19. Two-stage valve and actuator with feedback

electrical feedback path to the servo amplifier. The output from the servo amplifier actuates the torque motor attached to the pilot spool. Supply pressure is directed to both spool valves, so that the supply pressure times the main spool area produces the force to move the main spool This is clearly more effective for dealing with large forces. The output flow from the main spool is used to drive the ram piston, which is also provided with a feedback loop. Simplified transfer operators can be developed for the system shown in Fig. 8.19 if the load is assumed zero.

For the pilot spool

$$Q_p = K_1 x_v \sqrt{(P_s/2)} \tag{8.4}$$

For the main spool

$$Q_r = K_1' x_m \sqrt{(P_s/2)} \tag{8.5}$$

An interaction occurs between the two stages which can be described by,

$$Q_p = A_m \frac{dx_m}{dt} \tag{8.6}$$

From equations (8.4) and (8.5)

$$x_v = \frac{Q_p}{K_1\sqrt{(P_s/2)}} = \frac{A_m}{K_1\sqrt{(P_s/2)}}\frac{dx_m}{dt} \tag{8.7}$$

and from (8.5)

$$x_m = \frac{Q_r}{K_1'\sqrt{(P_s/2)}} \tag{8.8}$$

Differentiate equation (8.8):

$$\frac{dx_m}{dt} = \frac{1}{K'\sqrt{(P_s/2)}}\frac{dQ_r}{dt} \tag{8.9}$$

Substitute (8.9) in (8.7)

$$x_v = \frac{2A_m}{P_s K_1 K_1'}\frac{dQ_r}{dt}$$

Hence,

$$Q_r = \frac{K_1 K_1' P_s}{2A_m}\int x_v\, dt = A_r\frac{dx_o}{dt} \tag{8.10}$$

The output acceleration d^2x_o/dt^2 is then proportional to the input x_v. This is an unsatisfactory form of control giving rise to stability problems.

In order to improve the control of the system, feedback from the main spool to the servo amplifier is required. This is represented by the voltage V_m, a function of the main spool displacement.

Now,

$$x_v = V_e K_a K_r \tag{8.11}$$

and

$$V_m = K_f x_m \tag{8.12}$$

where K_f is the gain of the spool transducer.

Now if the output position feedback is ignored for the moment

$$V_e = V_i - V_m \tag{8.13}$$

Equating (8.7) and (8.11)

$$K_a K_r V_e = \frac{A_m}{K_1\sqrt{(P_s/2)}}\frac{dx_m}{dt} \tag{8.14}$$

From (8.13) and (8.12) with (8.14)

$$V_i = V_m = \frac{A_m}{K_a K_r K_1\sqrt{(P_s/2)}}\frac{dx_m}{dt}$$

$$= V_i - K_f x_m$$

The transfer function is, therefore,

$$\frac{x_m}{V_i}(s) = \frac{1}{\left\{\dfrac{A_m}{K_1 K_a K_r\sqrt{(P_s/2)}}s + K_f\right\}}$$

$$= \frac{1}{K_f(1 + T_c s)} \tag{8.15}$$

where $T_c = A_m/K_1K_aK_fK_r\sqrt{(P_s/2)}$
Bur from equation (8.8) substituted in (8.15)

$$Q_r(s) = \left[\frac{K_1'\sqrt{(P_s/2)}}{K_f(1 + T_c s)}\right] V_i \tag{8.16}$$

The output flow to the ram now reaches a steady state rate exponentially, as shown in Fig. 8.20, for a step input change in V_i, while the ram piston moves at

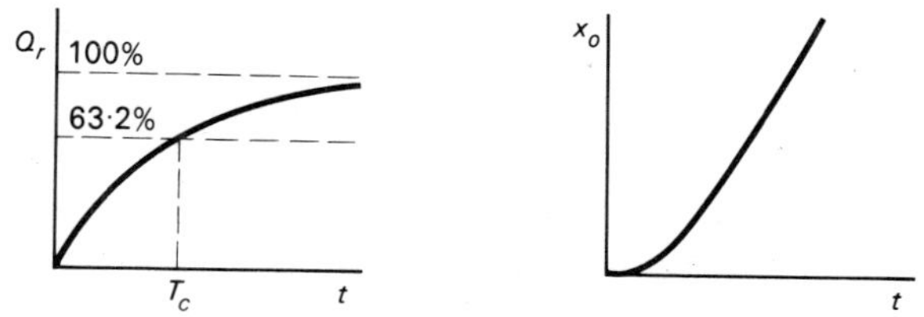

Fig. 8.20. Effect of feedback from main spool valve

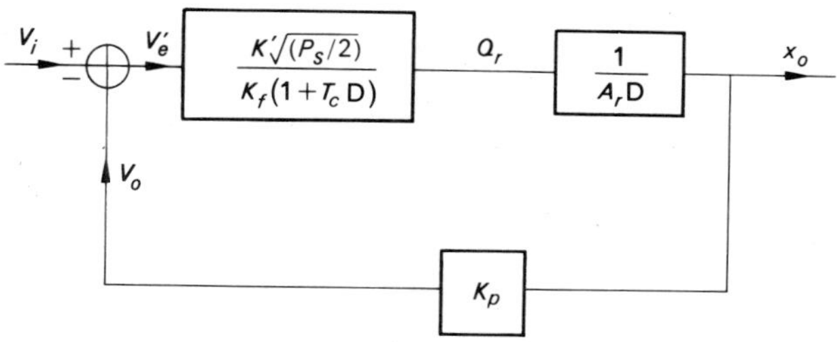

Fig. 8.21. Block diagram showing position feedback from the actuator

constant velocity, instead of constant acceleration, as was indicated by equation (8.10).

In order to maintain position control of the ram, a voltage V_o is fed back from the ram shaft.

The new arrangement is shown in the block diagram of Fig. 8.21. From equation (8.16) in relation to Fig. 8.6 where V_i becomes in fact V_e, due to the introduction of an additional feedback loop,

$$Q_r = \left[\frac{K_1'\sqrt{(P_s/2)}}{K_f(1 + T_c s)}\right] V_e = A_r \frac{dx_o}{dt} \tag{8.17}$$

But

$$\begin{aligned} V_e' &= V_i - V_o \\ &= V_i - K_p x_o \end{aligned} \tag{8.18}$$

Combining (8.17) and (8.18)

$$A_r \frac{dx_o}{dt} = \left[\frac{K_1'\sqrt{(P_s/2)}}{K_f(1 + T_c s)}\right] (V_i - K_p x_o)$$

therefore,

$$K_1'\sqrt{(P_s/2)}V_i = K_f A_r \frac{dx_o}{dt} + K_f A_r T_c \frac{d^2 x_o}{dt^2} + K_1' K_p \sqrt{(P_s/2)} x_o$$

or

$$K_1'\sqrt{(P_s/2)}V_i(s) = (K_f A_r T_c s^2 + K_f A_r s + K_1' K_p \sqrt{(P_s/2)}x_o(s)$$

The transfer function is, therefore,

$$\frac{x_o}{V_i}(s) = \frac{K_1'\sqrt{(P_s/2)}}{K_f A_r T_c s^2 + K_f A_r s + K_1' K_p \sqrt{(P_s/2)}}$$

$$= \frac{(1/K_p)}{\dfrac{K_f A_r T_c}{K_1' K_p \sqrt{(P_s/2)}} s^2 + \dfrac{K_f A_r}{K_1' K_p \sqrt{(P_s/2)}} s + 1} \qquad (8.19)$$

There is now a direct relationship between x_o and V_i in the steady state given by,

$$x_o = \frac{1}{K_p} V_i$$

and hence position control of the ram is achieved. Notice that the transfer function (8.19) is a quadratic, indicating that the system can exhibit continuous oscillations if,

$$\frac{K_f A_r}{K_1' K_p \sqrt{P_s}} \leqslant 0$$

the frequency of the resulting oscillations being given by,

$$f_n = \frac{1}{2\pi} \sqrt{\left\{ \frac{K_1' K_p \sqrt{(P_s/2)}}{K_f A_r T_c} \right\}} \text{ Hz}$$

If the ram were moving an inertia type load then the quadratic term due to compressibility would need to be included in equation (8.19), raising the order to four and introducing a possible maximum lag of 360°. Care must be taken, therefore, when combining commercially bought control valves with existing plant, to ensure that stability is maintained. A proper analytical approach is therefore needed, and not the all too often approach of 'suck it and see'. When optimising the loops in such systems one should allow sufficient margin for wear and deterioration of the system; for example, the friction in the system is bound to reduce with wear. It is therefore good practice to provide adjustment of loop gains to help maintain performance.

Chapter 9

Pneumatic servomechanisms

9.1. Introduction

The analysis of the pneumatic servo has received relatively little attention in comparison to the hydraulic servo. This is partially due to the fact that the equations for compressible flow are more difficult to handle and to the fact that this type of system is not so popular. The reason for this will become apparent as comparisons are made with its hydraulic counterpart.

The fact that pneumatic servos use a very compressible fluid accounts for the two main differences in performance. If, for example, air is introduced into a cylinder when the valve is opened, then sufficient gas must flow into the cylinder volume to increase the pressure and build up enough force to overcome the spring pre-load. In hydraulic systems a very small valve displacement will cause immediate pressure rise in the cylinder. Hence hydraulics have a rapid response while pneumatics exhibit a time delay.

Because the fluid is compressible, the servo will exhibit a lack of stiffness, especially to external load disturbances. In order to achieve the 'stiff' characteristics of hydraulics, these systems must either be large enough to absorb any expected load variation with relatively small change in pressure, or be provided with a large servo valve capable of rapidly adding or removing gas.

On the other hand, a compressible fluid can be used to store energy as in the case of the hydro-pneumatic accumulator discussed in chapter 1.

The efficiency of pneumatic systems is much lower than hydraulic systems. Two important factors causing loss are,

Cooling of the fluid which dissipates energy in the form of heat.

Use of positive displacement actuators causes loss of energy by expansion.

In addition it is difficult to achieve satisfactory sealing and several horsepower may be used in making up leakage. This puts a limit on system pressure. The poor lubrication properties of gases, compared to oil, also lead to energy losses. Table 9.1 highlights the main differences from the fluid mechanic viewpoint. One big advantage of the gas actuator is its ability to work above 800°F, where both electrical and hydraulic systems would fail.

The basic equations of pneumatics are non-linear and tend to be more complex than those of incompressible flow systems. This may explain why this field of analysis has received little attention. One of the first analyses was carried out by Shearer [70], who studied an open loop system consisting of a pneumatic servo motor driving an inertia load acted on by an external force. It is this approach that will be developed in this chapter.

Table 9.1

	Orifice flow	Capillary flow
Gas	Flow $\propto$ upstream pressure (choked)	Flow $\propto$ (upstream pressure)^2
Hydraulics	Flow $\propto$ $\sqrt{\text{(pressure drop)}}$	Flow $\propto$ pressure drop

9.2. Pneumatic servo

The pneumatic servo configuration is very similar to that considered in the chapter dealing with hydraulic systems. Most of the earlier pneumatic control systems were used in the process control industries, where low pressure air 150 lbf in^{-2} was easily obtainable and gave a sufficiently fast response. Nowadays owing to the constantly increasing temperature requirements for missile and aircraft components, a renewed interest has developed for high speed pneumatic control devices using pressures in the region of 2500 lbf in^{-2}.

In the system considered, a four-way valve is used to control a double acting cylinder, The valve can be actuated by a torque motor, although in the electro-pneumatic mode a flapper nozzle or flapper nozzle four-way valve is often used.

A typical arrangement is shown in Fig. 9.1, and for this Shearer makes the following assumptions to simplify the mathematical development.

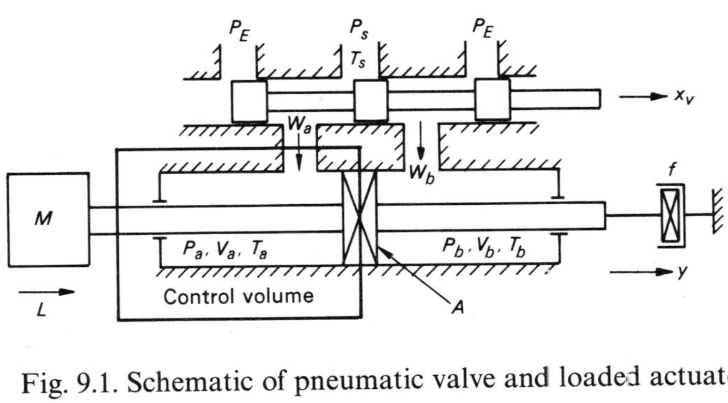

Fig. 9.1. Schematic of pneumatic valve and loaded actuator

1. Supply pressure P_s and temperature T_s are constant.
2. No heat transfer between the fluid and its environment.
3. The working fluid obeys the perfect gas law.
4. The temperature of the fluid flowing between the valve and the ram is equal to the supply temperature.
5. The piston moves only small distances from the central position.
6. The pressures developed each side of the ram vary only by small amounts from an initial steady value P_i.
7. Control valve constants k_{1a} and k_{1b}; and k_{2a} and k_{2b} are equal, and do not vary with valve displacement.
8. External load force L is small enough to make assumption (6) valid.

The valve characteristics have been developed fully in chapter 3 and can be treated as effectively two three-way valves operating with identical pressure, flow and opening conditions.

From assumption (7)

$$k_{1a} = k_{1b} \text{ and } k_{2a} = k_{2b}$$

where

$$k_{1a} = \frac{\partial W_a}{\partial x_v} \text{ and } k_{2a} = \frac{\partial W_a}{\partial P_a}$$

The mass flow rate into each chamber may be expressed,

$$W_a = f(x_v, P_a)$$
$$W_b = f(-x_v, P_b)$$

These equations can be linearised by applying Taylor's formula for two variables and neglecting all powers above the first. Small perturbations are considered about an initial steady state value.

$$\Delta W_a = \frac{\partial W_a}{\partial x_v}(\Delta x_v) + \frac{\partial W_a}{\partial P_a}(\Delta P_a) \tag{9.1}$$

$$\Delta W_b = \frac{\partial W_b}{\partial x_v}(\Delta x_v) + \frac{\partial W_b}{\partial P_b}(\Delta P_b) \tag{9.2}$$

These relationships can be expressed graphically as shown in Fig. 9.2. Equations (9.1) and (9.2) are in a similar form to those for the hydraulic control valve in chapter 6, i.e.

$$\Delta W_a = k_{1a}(\Delta x_v) + k_{2a}(\Delta P_a) \tag{9.3}$$
$$\Delta W_b = k_{1b}(-\Delta x_v) + k_{2b}(\Delta P_b) \tag{9.4}$$

The difference in mass flow rates is therefore

$$\Delta W_a - \Delta W_b = 2k_{1a}(\Delta x_v) + k_{2a}(\Delta P_a - \Delta P_b) \tag{9.5}$$

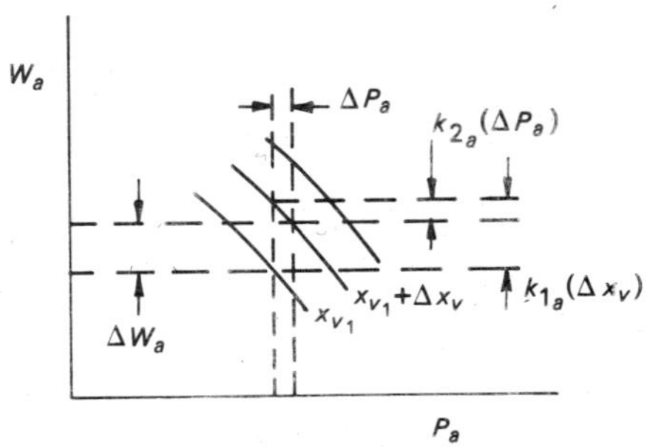

Fig. 9.2. Graphical interpretation of equation (9.2)

9.3. Ram equations

Considering the left-hand side of the ram, Fig. 9.1, the thermodynamic system is enclosed in the box, or control volume.

Applying the continuity equation,

$$W_a = \frac{d}{dt}(\rho_a V_a) \tag{9.6}$$

Since the fluid medium is assumed to be a perfect gas

$$\rho_a = \frac{P_a}{RT_a}$$

$$C_p = \frac{\gamma R}{\gamma - 1}$$

Substitution in equation (9.6), gives

$$W_a = \frac{1}{R}\frac{d}{dt}\left(\frac{P_a V_a}{T_a}\right) \tag{9.7}$$

The conservation of energy states that the energy inflow minus the energy outflow equals the rate at which energy is stored inside the control volume; this can be expressed,

$$\sum W_{in} h_{in}' - \sum W_{out} h_{out}' + \left(\frac{dQ_h}{dt} - \frac{d\Phi}{dt}\right)' = \frac{dE}{dt} \tag{9·8}$$

where Q_h is the heat flow to the control volume
Φ is the work done by the system
h' is the total energy per unit mass of fluid

$$h' = u + \frac{P}{\rho} + \frac{V^2}{2} + gz = h + \frac{V^2}{2} + gz$$

h is the enthalpy of the fluid
u is the internal energy
E is the total internal energy inside the control volume.
z is the elevation.

The system control volume gives

$$C_p W_a T_a - P_a \frac{dV_a}{dt} + \frac{dQh}{dt} = \frac{d}{dt}(C_v, \rho_a, V_a, T_a) \tag{9.9}$$

since for a perfect gas, flowing between points 1 and 2

$$C_p(T_1 - T_2) = \frac{V_2^2 - V_1^2}{2}$$

If the ram is perfectly insulated (adiabatic conditions) and again assuming a perfect gas, (9.9) can be rewritten

$$W_a T_a - \frac{P_a}{C_p}\frac{dV_a}{dt} = \frac{1}{\gamma R}\frac{d}{dt}(P_a, V_a) \tag{9.10}$$

since,

$$\gamma = \frac{C_p}{C_v} \text{ and } \rho = \frac{P}{RT}$$

By a similar argument,

$$W_b T_b - \frac{P_b}{C_p}\frac{dV_b}{dt} = \frac{1}{\gamma R}\frac{d}{dt}(P_b, V_b) \tag{9.11}$$

For small changes about a steady state initial condition, denoted by the subscript i, equations (9.10) and (9.11) become

$$T_a \Delta W_a - \frac{P_{ai}}{C_p}\frac{d}{dt}\Delta V_a = \frac{1}{\gamma R}\left[P_{ai}\frac{d}{dt}\Delta V_a + V_{ai}\frac{d}{dt}\Delta P_a\right] \tag{9.12}$$

$$T_b \Delta W_b - \frac{P_{ai}}{C_p}\frac{d}{dt}\Delta V_b = \frac{1}{\gamma R}\left[P_{bi}\frac{d}{dt}\Delta V_b + V_{bi}\frac{d}{dt}\Delta P_b\right] \tag{9.13}$$

If it is assumed that P_a, P_b, V_a and V_b do not vary greatly from their initial values and that

$$T_a = T_b = T_i = T_s;\ V_{ai} = V_{bi} = V_i;\ P_{ai} = P_{bi} = P_i,$$

then from equations (9.12) and (9.13)

$$T_s(\Delta W_a - \Delta W_b) = \frac{P_i}{C_p}\frac{d}{dt}(\Delta V_a - \Delta V_b) + \frac{P_i}{\gamma R}\frac{d}{dt}(\Delta V_a - \Delta V_b) + \frac{V_i}{\gamma R}\frac{d}{dt}(\Delta P_a - \Delta P_b)$$

$$= \left[P_i\left(\frac{1}{C_p} + \frac{1}{\gamma R}\right)\frac{d}{dt}(\Delta V_a - \Delta V_b) + \frac{V_i}{\gamma R}\frac{d}{dt}(\Delta P_a - \Delta P_b)\right] \tag{9.14}$$

But

$$C_p = \frac{\gamma R}{\gamma - 1} \text{ or } \frac{1}{C_p} + \frac{1}{\gamma R} = \frac{1}{R}$$

Therefore (9.14) reduces to,

$$\Delta W_a - \Delta W_b = \frac{1}{RT_s}\left[P_i \frac{\mathrm{d}}{\mathrm{d}t}(\Delta V_a - \Delta V_b) + \frac{V_i}{\gamma}\frac{\mathrm{d}}{\mathrm{d}t}(\Delta P_a - \Delta P_b)\right] \qquad (9.15)$$

At any particular instant, the volumes each side of the ram are,

$$V_a = V_{ai} + Ay$$
$$V_b = V_{bi} - Ay$$

Giving

$$V_a - V_b = 2Ay \qquad (9.16)$$

Substitution in (9.15) gives

$$\Delta W_a - \Delta W_b = \frac{1}{RT_s}\left[2P_iA\frac{\mathrm{d}}{\mathrm{d}t}\Delta y + \frac{V_i}{\gamma}\frac{\mathrm{d}}{\mathrm{d}t}(\Delta P_a - \Delta P_b)\right] \qquad (9.17)$$

Combining the valve (9.5) and ram (9.17) equations

$$2k_{1a}(\Delta x_v) + k_{2a}(\Delta P_a - \Delta P_b) = \frac{1}{RT_s}\left[2P_iA\frac{\mathrm{d}}{\mathrm{d}t}\Delta y + \frac{V_i}{\gamma}\frac{\mathrm{d}}{\mathrm{d}t}(\Delta P_a - \Delta P_b)\right]$$

or introducing the Laplace operator and rearranging

$$\left(\frac{V_i}{\gamma RT_s}s - k_{2a}\right)(\Delta P_a - \Delta P_b) = 2k_{1a}(\Delta x_v) - \frac{2P_iA}{RT_s}s(\Delta y)$$

and on further simplification

$$A_s(\Delta y) = k_1(\Delta x_v) - (k_2 + k_3 s)(\Delta P_a - \Delta P_t) \qquad (9.18)$$

where

$$k_1 = \frac{k_{1a}RT_s}{P_i}$$

$$k_2 = \frac{-k_{2a}RT_s}{2P_i}$$

$$k_3 = \frac{V_i}{2\gamma P_i}$$

It can be seen from Fig. 9.1 that the loading on the ram consists of a mass, viscous damping and external loading

$$A(\Delta P_a - \Delta P_b) = M\frac{\mathrm{d}^2}{\mathrm{d}_t^2}(\Delta y) + f\frac{\mathrm{d}}{\mathrm{d}t}(\Delta y) + \Delta L \qquad (9.19)$$

Combining (9.18) and (9.19), gives the open loop relationship for the actuator,

$$\left[\left(\frac{k_3M}{k_2f + A^2}\right)s^2 + \left(\frac{k_2M + k_3f}{k_2f + A^2}\right)s + 1\right]s(\Delta y)$$

$$= \frac{k_1A(\Delta x_v) - (k_2 + k_3s)\Delta L}{k_2f + A^2} \tag{9.20}$$

where

$$\omega_n = \sqrt{\left(\frac{k_2f + A^2}{k_3M}\right)} \text{ rad s}^{-1} \tag{9.21}$$

and damping ratio

$$c = \frac{k_3M + k_3f}{2\sqrt{\{(k_2f + A^2)k_3M\}}} \tag{9.22}$$

In a practical system f is kept small as it is wasteful of power, hence from (9.21)

$$(\omega_n)_{\text{pneu}} = \sqrt{\left(\frac{2\gamma P_iA^2}{V_iM}\right)} = \sqrt{\left(\frac{4\gamma P_iA^2}{VM}\right)}$$

where V is the swept volume.
For the equivalent hydraulic system

$$(\omega_n)_{\text{hyd}} = \sqrt{\left(\frac{4NA^2}{MV}\right)}$$

$$\frac{(\omega_n)_{\text{hyd}}}{(\omega_n)_{\text{pneu}}} = \sqrt{\left(\frac{N}{\gamma P_i}\right)}$$

If the fluid supply pressure in each system is assumed to be 1000 lbf in^{-2} then,

$$\frac{(\omega_n)_{\text{hyd}}}{(\omega_n)_{\text{pneu}}} = \sqrt{\left(\frac{1 \times 10^5}{1{\cdot}4 \times 10^3}\right)} = 8{\cdot}4$$

It is seen that, at normal temperatures, the natural frequency of the hydraulic system is much higher, giving a better dynamic response. The bulk modulus of a gas, operating under adiabatic conditions, is equal to the product of the pressure and the gas constant γ. Under isothermal conditions it equals the pressure.

The damping ratio can be calculated from equation (9.22), but experience indicates that this inherent contribution is very small for pneumatic servos, making it necessary to introduce additional stabilising methods.

9.4. Load sensitivity

Since the fluid is so compressible, the sensitivity to external loading is important. From equation (9.20) it is seen that the steady state relationship between ram velocity and valve stroke and external load will be,

$$[s(\Delta y)]_{ss} = \frac{k_1 A(\Delta x_v) - k_2(\Delta L)}{k_2 f + A^2} \tag{9.23}$$

Only when k_2 is zero is zero steady-state load sensitivity possible, otherwise to hold a steady velocity, an adjustment in (Δx_v) must be effected.

The sensitivity to load disturbances can also be found under dynamic conditions by considering equation (9.18)

$$As(\Delta y) = k_1(\Delta x_v) - (k_2 + k_3 s)(\Delta P_a - \Delta P_b) \tag{9.24}$$

If a position loop is arranged round the actuator, then,

$$\Delta x_v = \Delta x_i - \Delta y \tag{9.25}$$

The external loading can be expressed by

$$\Delta L = -(\Delta P_a - \Delta P_b)A \tag{9.26}$$

Combining these three equations and imposing the condition that the input is held fixed at $\Delta x_i = 0$

$$As(\Delta y) = -k_1(\Delta y) + (k_2 + k_3 s)\frac{\Delta L}{A} \tag{9.27}$$

$$\frac{\Delta L}{\Delta y} = \frac{A(k_1 + As)}{k_2 + k_3 s}$$

Converting into the frequency domain and reducing the expression to a single complex number,

$$\frac{\Delta L}{\Delta y} = \frac{A[k_1 k_2 + Ak_3\omega^2 + (Ak_2 - k_1 k_3)\omega i]}{k_2^2 + k_3^2\omega^2} \tag{9.28}$$

For small values of ω, equation (9.28) gives

$$\frac{\Delta L}{\Delta y} = A\frac{k_1}{k_2}$$

which is dependent on the valve characteristics. For large values of ω, equation (9.28) gives

$$\frac{\Delta L}{\Delta y} = \frac{A^2}{k_3} = \frac{2\gamma P_i A^2}{V_i}$$

which is the stiffness of the fluid in the ram.

9.5. Method of stabilisation

The open loop equation (9.20) for the system can be written $(\Delta L = 0)$

$$\frac{\Delta y}{\Delta x_v} = \frac{\{(k_1 A/(k_2 f + A^2)\}}{s\left(\frac{1}{\omega_n^2}s^2 + \frac{2c}{\omega_n}s + 1\right)} \tag{9.29}$$

Practical experience shows that $k_2 = f = c \approx 0$ is not unrealistic and therefore (9.29) reduces to

$$\frac{\Delta y}{\Delta x_v} = \frac{k_1 \omega_n^2/A}{s(s^2 + \omega_n^2)} = KG(s) \tag{9.30}$$

In order to improve the dynamic response, greater than 90° phase lead is needed in the feedback loop. This can be achieved by the addition of acceleration and velocity terms to the position feedback loop. From Fig. 9.3

$$\frac{\Delta y'}{\Delta y} = (K_a s^2 + K_v s + K_y) \tag{9.31}$$

Hence the loop transfer function is,

$$\frac{\Delta y}{\Delta x_i} = \frac{\dfrac{k_1 \omega_n^2}{A} \Big/ \{s(s^2 + \omega_n^2)\}}{1 + (K_a s^2 + K_v s + K_y) \dfrac{k_1 \omega_n^2}{A} \Big/ \{s(s^2 + \omega_n^2)\}} \tag{9.32}$$

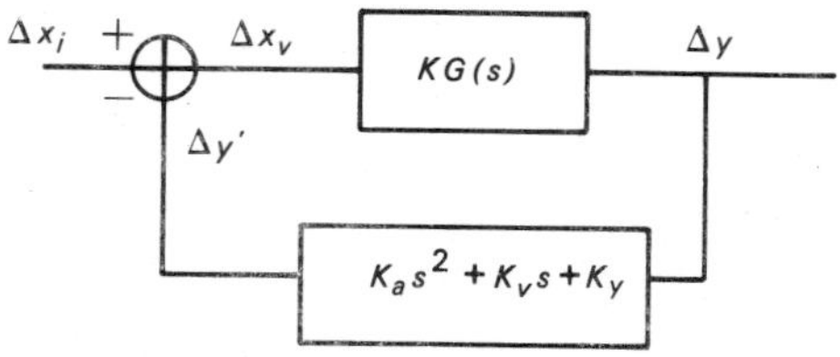

Fig. 9.3. Block diagram of pneumatic servo with position, velocity and acceleration feedback

A root locus plot can then be obtained from the characteristic equation,

$$\frac{k_1}{A} \omega_n^2 K_a \frac{\{s^2 + (K_v/K_a)s + K_y/K_a\}}{s(s^2 + \omega_n^2)} = -1 \tag{9.33}$$

The root locus plot for various pole-zero positions is shown in Fig. 9.4. It can be seen that 9.4(a) is completely unstable and 9.4(c) is conditionally unstable. This clearly confirms that the feedback loops need to be carefully tuned, otherwise any advantage is lost. The cases considered are obviously the worst condition since some inherent damping will be present in reality.

Pure acceleration feedback is not practical because accelerometers are not suitable and double differentiation of the position signal produces a signal with a high noise content. Since the pressure in the ram determines the acceleration of the output load, it can be used as an acceptable alternative. However, there are certain problems associated with this approach, particularly if the output load contains significant damping and spring forces. Also if an external load is applied to the ram, the initial pressure feedback would have the opposite effect of acceleration feedback.

If the load is pure mass,

$$A(\Delta P) = M \frac{d^2}{dt^2}(\Delta y) \tag{9.34}$$

and if the pressure feedback is $K_p \Delta p$, equation (9.33) can now be expressed

$$\frac{k_1}{A} \omega_n^2 \frac{M}{A} K_p \frac{\{s^2 + (K_v A/K_p M)s + (K_y A/K_p M)\}}{s(s^2 + \omega_n^2)} = -1 \tag{9.35}$$

which is similar to equation (9.33).

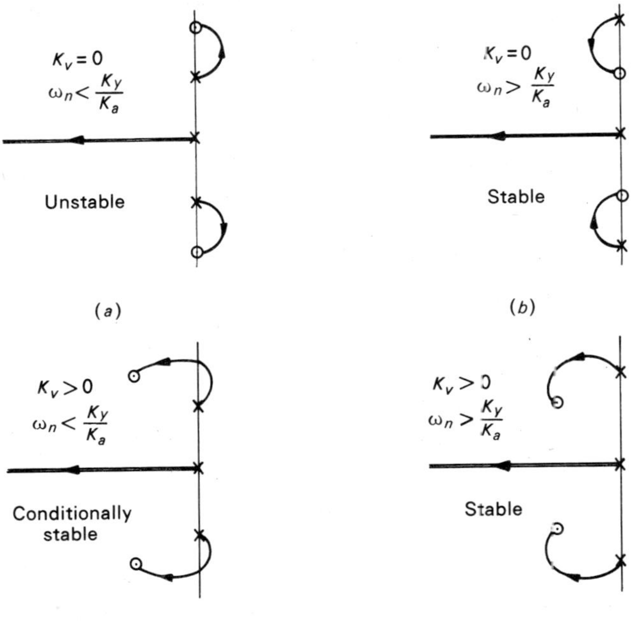

Fig. 9.4. Effect of feedback combinations on root locus (after Halstenberg [76])

One disadvantage of pressure feedback, in a real system is that for a small positional error, the system must build up a pressure differential across the ram to move the piston and thence eliminate the error. Pressure feedback will bend to prevent this process. Accelerator feedback on the other hand will not affect the system until the piston is actually moving.

Transient pressure feedback is an alternative method, in which the pressure signal is fed through a high pass filter so as to be only effective at the higher frequencies.

Incorporating the high pass filter transfer function in equation (9.35) gives,

$$\frac{k_1}{A} \omega_n^2 \frac{[K_p(M/A)s^2(\tau s/\tau s + 1) + K_v s + K_y]}{s(s^2 + \omega_n^2)} = -1$$

or

$$\frac{k_1}{\Delta}\omega_n^2\frac{[K_p(M/A)s^3 + K_v s^2 + (K_v/\tau + K_y)s + (K_y/\tau)]}{s(s^2 + \omega_n^2)(s + 1/\tau)} = -1 \tag{9.36}$$

In fact, this type of feedback is stiffer in relation to external load disturbances at the lower frequencies. Once more it is seen that careful selection of time constants is required to avoid instability. If τ is very big, then equations (9.35) and (9.36) become identical.

9.6. Stabilisation using auxiliary tanks

In his paper, Shearer shows that the same effect can be achieved using transient flow stabilisation, Fig. 9.5. Here flow from auxiliary tanks is diverted from or added to the valve flow during transient conditions.

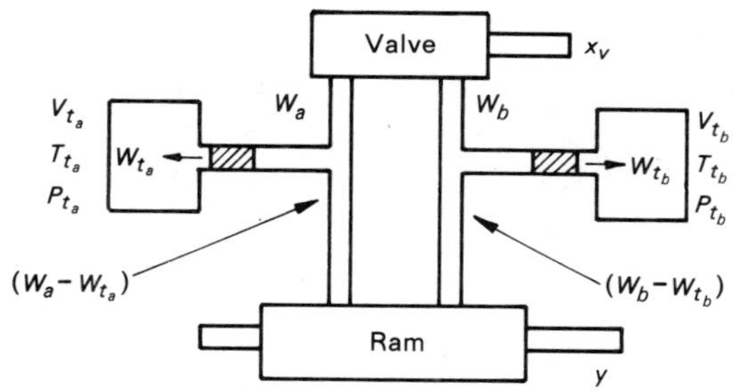

Fig. 9.5. Use of auxiliary tanks for stabilisation (after Shearer [70])

The transient flow rate W_t satisfies the equation (developed later) Shearer's analysis follows.

$$\frac{RT_s}{P_i}W_t = -k_2'\left(\frac{\tau s}{\tau s + 1}\right)(\Delta P_a - \Delta P_b) \tag{9.37}$$

This feedback is non-zero when $(\Delta P_a - \Delta P_b)$ is time variant, but for steady values of pressure drop across the piston, W_t is zero.

Capillary flow restrictors are used because their characteristics are more linear than orifices. The equations for the flow in these passages are,

$$W_{t_a} = \frac{ab^3}{24\mu RT_s l}(P_a^2 - P_{t_a}^2) \tag{9.38}$$

$$W_{t_b} = \frac{ab^3}{24\mu RT_s l}(P_b^2 - P_{t_b}^2) \tag{9.39}$$

where a = capillary passage width
b = capillary passage thickness
l = capillary passage length

Let

$$C_c = \frac{ab^3}{12\mu l}$$

and assuming small changes about the initial steady conditions such that $P_a = P_b = P_{a_t} = P_{b_t} = P_i$ then,

$$\Delta W_{t_a} = \frac{C_c P_i}{RT_s}(\Delta P_a - \Delta P_{t_a}) \quad (9.40)$$

$$\Delta W_{t_b} = \frac{C_c P_i}{RT_s}(\Delta P_b - \Delta P_{t_b}) \quad (9.41)$$

The energy equation applied to the identical tanks result in the equations

$$\Delta W_{t_a} = \frac{V_t}{\gamma R T_s}\frac{\mathrm{d}}{\mathrm{d}t}\Delta P_a \quad (9.42)$$

$$\Delta W_{t_b} = \frac{V_t}{\gamma R T_s}\frac{\mathrm{d}}{\mathrm{d}t}\Delta P_b \quad (9.43)$$

Combining equation (9.40) with (9.42) and (9.41) with (9.43) gives,

$$\Delta W_{t_a} = \frac{V_t}{\gamma R T_s}\frac{\mathrm{d}}{\mathrm{d}t}\left\{\left(\frac{-RT_s}{C_c P_i}\right)\Delta W_{t_a} + \Delta P_a\right\} \quad (9.44)$$

$$\Delta W_{t_b} = \frac{V_t}{\gamma R T_s}\frac{\mathrm{d}}{\mathrm{d}t}\left\{\left(\frac{-RT_s}{C_c P_i}\right)\Delta W_{t_b} + \Delta P_b\right\} \quad (9.45)$$

Rearranging terms,

$$\frac{V_t}{\gamma C_c P_i}\frac{\mathrm{d}}{\mathrm{d}t}(\Delta W_{t_a}) + \Delta W_{t_a} = \frac{V_t}{\gamma R T_s}\frac{\mathrm{d}}{\mathrm{d}t}(\Delta P_a) \quad (9.46)$$

$$\frac{V_t}{\gamma C_c P_i}\frac{\mathrm{d}}{\mathrm{d}t}(\Delta W_{t_b}) + \Delta W_{t_b} = \frac{V_t}{\gamma R T_s}\frac{\mathrm{d}}{\mathrm{d}t}(\Delta P_b) \quad (9.47)$$

The net flows to chambers *a* and *b* are,

$$(\Delta W_a - \Delta W_{t_a})$$

and

$$(\Delta W_b - \Delta W_{t_b})$$

The energy equations can now be applied to the chambers either side of the ram,

$$\Delta W_a - \Delta W_{t_a} - \frac{P_i A}{RT_s}\frac{\mathrm{d}}{\mathrm{d}t}(\Delta y) = \frac{V_i}{\gamma R T_s}\frac{\mathrm{d}}{\mathrm{d}t}(\Delta P_a) \quad (9.48)$$

$$\Delta W_b - \Delta W_{t_b} + \frac{P_i A}{RT_s}\frac{\mathrm{d}}{\mathrm{d}t}(\Delta y) = \frac{V_i}{\gamma R T_s}\frac{\mathrm{d}}{\mathrm{d}t}(\Delta P_b) \quad (9.49)$$

Equations (9.46) and (9.47) give,

$$\Delta W_{t_a} = \frac{(V_t/\gamma RT_s)s(\Delta P_a)}{(V_t/\gamma C_c P_i)s + 1} \tag{9.50}$$

$$\Delta W_{t_b} = \frac{(V_t/\gamma RT_s)s(\Delta P_b)}{(V_t/\gamma C_c P_i)s + 1} \tag{9.51}$$

Combining (9.48) with (9.50) and (9.49) with (9.51) gives

$$\Delta W_a - \frac{(V_t/\gamma RT_s)s(\Delta P_a)}{(V_t/\gamma C_c P_i)s + 1} - \frac{P_i A}{RT_s} s(\Delta y) = \frac{V_i}{\gamma RT_s} s(\Delta P_a) \tag{9.52}$$

$$\Delta W_b - \frac{(V_t/\gamma RT_s)s(\Delta P_b)}{(V_t/\gamma C_c P_i)s + 1} + \frac{P_i A}{RT_s} s(\Delta y) = \frac{V_i}{\gamma RT_s} s(\Delta P_b) \tag{9.53}$$

Subtract (9.53) from (9.52)

$$(\Delta W_a - \Delta W_b) - \frac{(V_t/\gamma RT_s)s(\Delta P_a - \Delta P_b)}{(V_t/\gamma C_c P_i)s + 1} - \frac{2P_i A}{RT_s} s(\Delta y)$$

$$= \frac{V_i}{\gamma RT_s} s(\Delta P_a - \Delta P_b) \tag{9.54}$$

But

$$\Delta W_a - \Delta W_b = 2k_{1a}(\Delta x_v) + k_{2a}(\Delta P_a - P_b)$$

giving

$$k_1(\Delta x_v) - k_2(\Delta P_a - \Delta P_b) - k_2' \frac{\tau s(\Delta P_a - \Delta P_b)}{\tau s + 1} - As(\Delta y) = k_3 s(\Delta P_a - \Delta P_b) \tag{9.55}$$

to be compared with (9.18)
where

$$k_2' = \frac{C_c}{2}$$

$$\tau = \frac{V_t}{\gamma C_c P_i}$$

Combining with the load equation 9.19

$$k_1(\Delta x_v) - \left(k_2 + k_2' \frac{\tau s}{\tau s + 1} + k_3 s\right)\left[\left(\frac{M}{A}s + \frac{f}{A}\right)s(\Delta y) + \frac{\Delta L}{A}\right]$$

$$= As(\Delta y) \tag{9.56}$$

Further information about pneumatic servos can be obtain in [71, 72, 73].

9.7. Hot gas servomotors

Hot gas servomotors, which operate around 1000°F and 1000 lbf in^{-2}, are particularly useful for space applications where propellant exhaust gases can be usefully recirculated. One particular difference between this type of system and the more conventional pneumatic actuator [74, 75, 76], is the dirtiness of the fluid medium. This aspect and the higher temperature create special problems in actuator design.

One possible configuration [75] is shown in Fig. 9.6. Several useful features of this assembly are:

1. The application of metal bellows in place of the more conventional ram eliminates sealing friction and lubrication problems. Their use will also reduce inertia. In addition, the thin walled bellows with their large corrugated heat transfer area in contact with ambient air, will lower the temperature of the actuator chamber, thus reducing material creep and fatigue problems.
2. The use of a thin walled metal bellows has advantages over the pipe of the jet pipe valve, since the reduced inertia will give a faster response.
3. The use of pivoting push rods rather than the normal connecting rods, will help to eliminate binding and frictional effects.
4. Dual bellows actuator will eliminate backlash and frictional hysteresis. Metal bellows with a spring stiffness of 160 lbf in^{-1} and operating pressures of 3000 lbf in^{-2}, capable of withstanding 4000 reversals at 1000°F are available. It is also possible to spray the hot contact side with a ceramic coating, thus reducing the metal temperature.

Another possible arrangement is shown in Fig. 9.7 which is suitable for long stroke actuators or large flows. The valve has a jet pipe first stage followed by a poppet arrangement in the second stage. The poppet valve, which uses seating rather than shear action, has considerable advantage over the spool valve in high temperature, dirty fluid applications.

Since hot gas is very compressible, the natural frequency is low and control of the dynamic response is more difficult. Compressibility allows considerable excess accelerating forces at the instant of release of breakaway friction. The dynamic response can be sensitive to the amplitude of the command signal as well as the magnitude of the friction load. Dither is a possible alternative to reduce friction effects.

Gas systems are very expensive and their use can only be justified in aero space applications, where minimum weight, reliability and high environmental temperatures are important. Cost of propellants are expensive, although in time the system may become feasible as an industrial application.

There are a number of suitable propellants, for example solid propellants for short duration applications and liquid propellants for systems requiring large quantities. Bipropellants, having liquid fuel and oxidisers are normally not used because of the storage tank complication. Monopropellants such as hydrazine and hydrogen peroxide, are unstable compounds which decompose

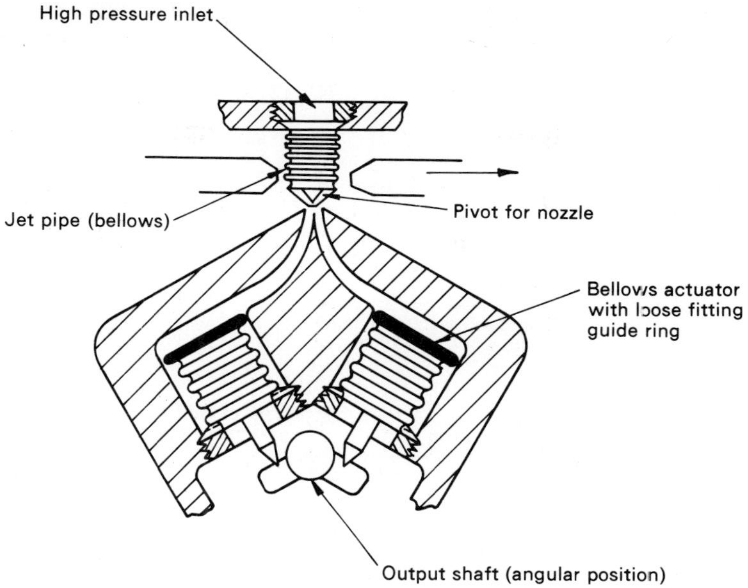

Fig. 9.6. Single stage hot gas jet pipe control valve with tandem bellows actuator

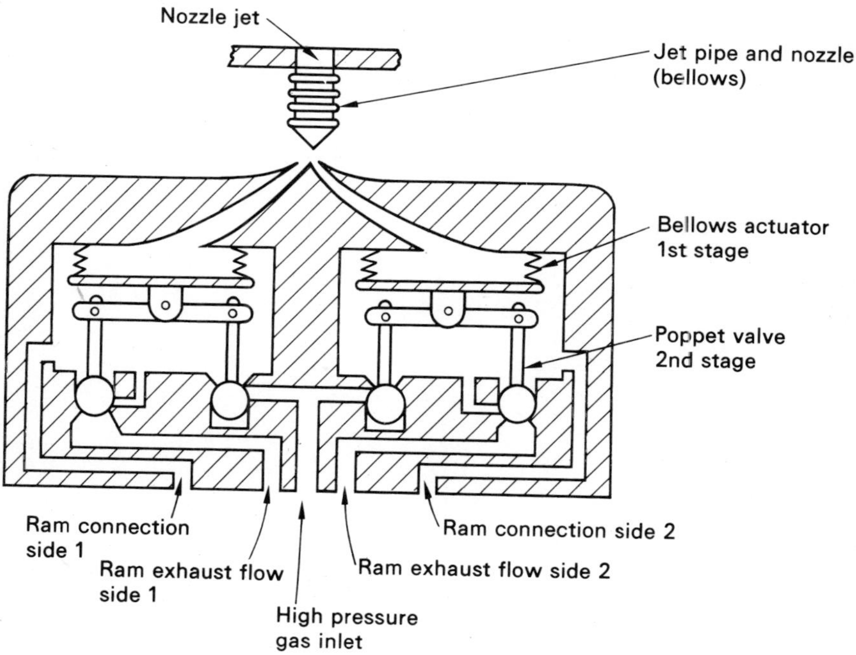

Fig. 9.7. Two-stage hot gas control valve, first stage jet pipe, second stage poppet

spontaneously at high temperature and pressure to produce hot gas around 1100–2000°F. A typical early type gas generator in which a liquid propellant is converted to a gas is shown in Fig. 9.8.

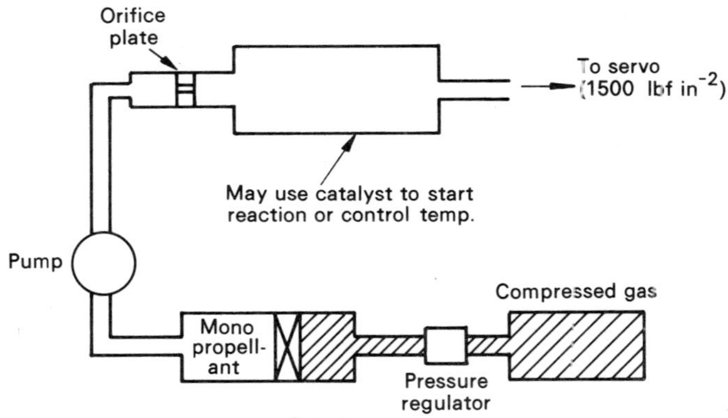

Fig. 9.8. One of the earlier methods of generating gas

Chapter 10

Some practical aspects of servo testing and design

10.1. Introduction

The information discussed so far, has in the main been of a theoretical nature. However, while this information is invaluable for design, a stage is reached when the actual components must be fitted together and made to satisfy a specification.

The formulation of a specification should give information under the following headings:

1. What the system is required to do, and how accurately it should do it.
2. Information about existing components adjacent (in the signal transmission sense) to the system and the output loading conditions. Also information about any external loading disturbances which may be injected at some other point in the loop.
3. The cycle of operation the system is likely to experience during its working life.
4. Reliability, including operation under extreme conditions of temperature and humidity.
5. Built-in safety precautions.

Obviously all of these points will not apply to every system.

While the specification should contain sufficient information to leave the engineer in no doubts as to the requirements, it should not be inflexible. Many hours of development have been wasted in the past, trying to meet unrealistically stringent specifications.

No matter how detailed and precise an investigation has been carried out using a theoretical approach, it is very rarely indeed that a system will meet its performance requirements without some development. A systematic analysis will get the project in the right direction but practical testing is needed for final adjustments. On the other hand, it should be understood that the system cannot be successfully developed by practical testing alone, and a trial and error approach along these lines should not be attempted. More than often it is uneconomical and generally unsuccessful. The two approaches are complementary to each other and, with sensible application, lead to an efficient solution of the problem.

The approach to testing a new system could be summarised as follows:

1. Static tests on each component intended for use in the system loop.
2. Dynamic tests on each component, where necessary or practical. This avoids penalising the whole system by the performance of one component in the loop.
3. Static tests on the complete system in the open loop condition.
4. Dynamic tests on the complete system, to confirm stability and fix the loop gain. Open loop responses are often not practical, such as when the controller is an integrator. In these cases the loop must be closed and reliance placed on the theoretical value of loop gain. The approach to this gain value must be made in small steps, to avoid unnecessary damage to the system.
5. Inclusion and testing of additional loops and networks where necessary to achieve the required performance.
6. Simulated component wear and fault tests.
7. Duty cycling, environmental and reliability testing, where necessary.

It is worth noting that in most cases, development testing will impose, by far, the most severe operating conditions which a system is likely to experience during its working life.

10.2. Position transducers

One of the simplest and most useful devices for converting displacement into an electrical signal is the potentiometer. It has the advantage of being cheap and reliable, while providing good accuracy within the limits of its resolution. There are a wide range of potentiometers available, employing wire wound cores, carbon tracks and film deposits [77]. Their mode of operation can be linear or non-linear. Multi-turn potentiometers are formed using a helix configuration and give much better accuracy and resolution than the single turn units.

Wire-wound potentiometers have the disadvantage that when used in a high gain loop, the output signal can show a stepped effect. This results in a detectible 'roughness' of motion in the control system.

One of the main problems the designer must look out for in a system is mismatch of components resulting in loading errors. The potentiometer is a particularly convenient component for demonstrating this problem.

Consider the simple circuit shown in Fig. 10.1, where a potential divider circuit using a potentiometer is loaded by a circuit resistance R_L, which could be the input to a pen recorder for example.

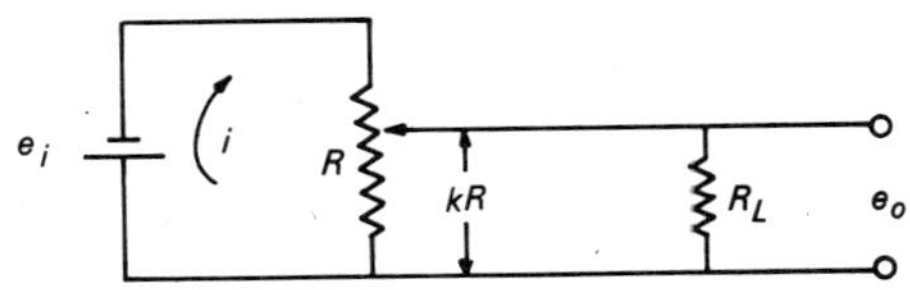

Fig. 10.1. A potentiometer loaded with resistance R_1

For the input circuit,

$$e_i = Ri$$

When the output takes no current, $R_L = \infty$ and

$$e_o = kRi$$

giving

$$\frac{e^o}{e^i} = k$$

Where R_L is finite, the output circuit is composed of two resistances in parallel. The system is further complicated if the loading is inductive or capacitive. For the simpler resistive case,

$$\frac{1}{R_T} = \frac{1}{kR} + \frac{1}{R_L}$$

or

$$R_T = \frac{kRR_L}{R_L + kR}$$

The total resistance as seen by e_i is now,

$$R_T' = R(1 - k) + \frac{kRR_L}{R_L + kR}$$

and the current in the circuit is

$$i = \frac{e_i}{R_T'} = \frac{R_L + kR}{kR^2(1 - k) + RR_2} e_i$$

but,

$$e_o = R_T i = \frac{kRR_L}{R_L + kR} i$$

and hence,

$$\frac{e_o}{e_i} = \frac{k}{k(1 - k)(R/R_L) + 1}$$

It can easily be shown that R_L must be 15 times as large as the potentiometer resistance R to limit the error in linearity to 1 per cent. These are cases where it is not possible to satisfy this criterion. Fortunately the cathode follower [78] can overcome this problem, as it has the characteristic of transforming a high impedance source at the input, into a low impedance source at the output.

To reduce the problems associated with potentiometer resolution and wear, the a.c. pick-off is often introduced. This does, however, increase the cost and more equipment is required. In one arrangement an armature, made from a ferromagnetic material, rotates in a stator consisting of a four arm inductance bridge. The movement increases the inductance of two arms and reduces the inductance of the other two. The a.c. pick-off requires excitation from an oscillating voltage generally either 400 Hz, 1200 Hz or 2400 Hz. The desired signal is super imposed on the exciting frequency, referred to as the carrier frequency. A high excitation frequency is preferred when the system is required to work over a fairly wide bandwidth.

Since the required output signal exists as the envelope of the carrier signal, it is often necessary to demodulate the output to recover the d.c. voltage signal. This task can be accomplished by a phase sensitive rectifier, which rectifies the signal and maintains the correct polarity of the resultant d.c. signal.

A simple form of phase sensitive rectifier is shown in Fig. 10.2, which shows also the carrier wave and the demodulated d.c. signal. The phase of the input signal is sensed by comparison with the reference supply voltage.

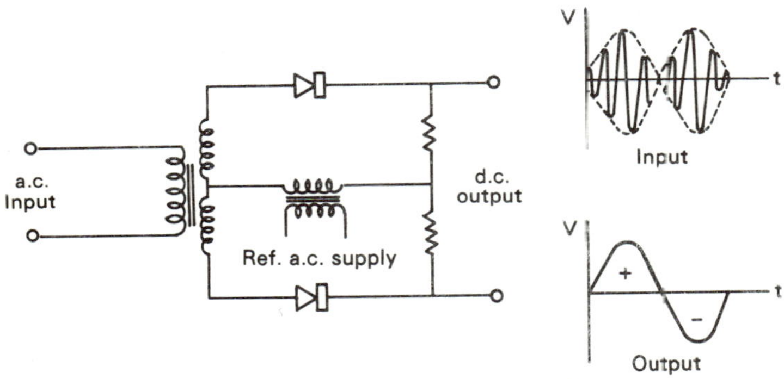

Fig. 10.2. A phase sensitivity rectifier and its modulated and demodulated signals

10.3. Recording instruments

There are many methods of recording information in a system, either as a permanent or a semi-permanent record. The method selected depends on a number of factors, such as whether or not the information is steady or rapidly changing.

The most reasonably priced chart recorders use ink on paper and a mechanical pen, while in the higher price range a high voltage is used to burn a trace on specially prepared paper. Recorders of this type have response times in the region of 0·01 s.

For rapidly fluctuating data, the ultra-violet recorder can be used with great effect. In this case the bandwidth extends beyond 1·5 kHz (response time 0·001 s). The trace in this case is obtained by directing the light beam onto a photosensitive paper.

Various systems of chart drives are available covering speed ranges from 0·5 in hr^{-1} up to 160 in s^{-1}. The chart drive is usually selected to suit the dynamic response of the pen.

10.4. System testing

Initially it is usually necessary to carry out those tests which prove the basic engineering standard of the mechanical and electrical components and circuits, e.g., hydraulic pressure tests and electrical insulation tests. The British Standard Specifications give excellent guidance in this direction.

With regard to the performance of the system, the procedure is usually as follows:

1. Alignment of feedback potentiometers, valves and torque motors. Some of this work can be started in the sub-assembly stages.
2. Checking forward and feedback path gains, and circuits providing derivative terms. This includes making sure that feedback signals are of the correct sign.
3. Checking the loop characteristics.

With regard to loop performance, the following areas need to be investigated.

1. Long term inaccuracies of the controlled process, involving the examination of hysteresis effects, thresholds, lack of linearity and drift.
2. Short term inaccuracies such as velocity lags, phase relationships, limiting velocity and acceleration, overshoots, etc.
3. Stability.

These features of the system performance are investigated by subjecting the system to known input functions and recording the output behaviour.

The most commonly used input functions are [79, 80]:

Step function.
Ramp function.
Simple harmonic function.
Impulse function.

The impulse function [81] has until recently been only of academic interest, but it now is assuming greater practical significance.

10.4.1 The step function

This is useful for all three general stages of testing, providing that for the routine test it is not complicated by spurious effects that need special interpretation. The effects of non-linearities are not usually masked by this approach.

In addition to maximum rates and acceleration figures (indicating saturation of the system), the step response gives an indication of stability. This makes the approach very effective in the initial stages of development for quick comparative tests.

For routine testing, the step response record can be checked for such points of interest as first cross-over point, percentage overshoot, and response time.

10.4.2 The ramp function

This type of test input is particularly important when the expected command signals in the field are of a similar nature, e.g. a radar aerial tracking. The required information is often best obtained by recording the error signal at some convenient point, usually in electrical terms and converting it by means of the electrical gains of the system into the final output movement. The error signal gives a measure of the friction in the system.

10.4.3 Simple harmonic functions

This is the traditional method of analysis developed originally from work on electronic devices and is useful provided that the output waveform is sinusoidal. Most control systems analysis equipments reduce distorted output waveforms to the fundamental harmonic and if the system is non-linear, it is difficult to assess the real value of this approach.

The conventional presentation of the results is to plot amplitude ratio and phase angle against frequency. This can be either in the form of a polar or a Bode diagram. Charts are available to assist in the conversion of the instrument readings to amplitude and angle (in fact some instruments present the results digitally).

It should also be noted that results of a closed loop system presented in this form can be readily converted to open loop using the Nichols chart.

10.4.4 Impulse function [81]

Modern computing techniques now enable the impulse response of a system to be calculated if a 'white noise' random signal is used. True 'white noise' has a power spectral density which is constant for all frequencies, but the signal used for testing is a generated approximation to this, known as a 'pseudo-random binary sequence'.

The computer calculates the cross correlation function, that is, the relationship between the random input and the output. For a 'white noise' signal it can be shown that the cross correlation function is the impulse response of the system.

This method can be organised into a go/no-go system suitable for routine testing, and a test then only occupies a matter of seconds. Alternatively, for analysis purposes, the actual impulse response can be plotted.

For routine test purposes this method relies on the development test to establish the tolerable extremes of performance. This would be fully documented with regard to frequency response, stability margin, step response and, if necessary, ramp function response. The impulse responses of these two extremes are incorporated in programme units which are plugged into the impulse tester. The response of the system under test is compared with these units on a go/no-go basis by the test console.

10.5. Frequency response equipment

Commercial equipment [82, 83] available for frequency response measurements use different operating principles, but the information obtained is basically the same. The general arrangement is shown in Fig. 10.3. An oscillator drives the comparator both directly and via the system under test. Information from the comparator is then displayed on a suitable indicator.

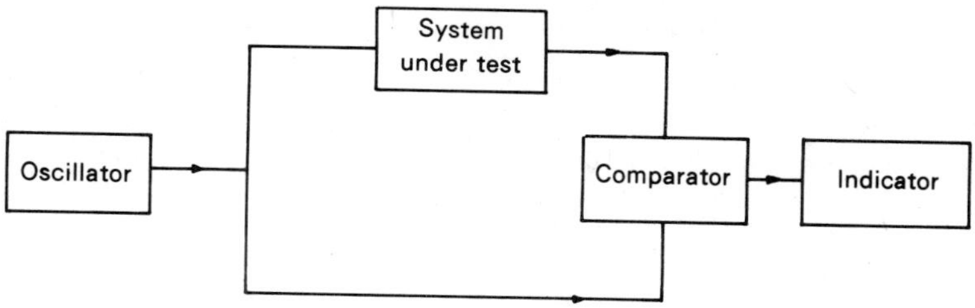

Fig. 10.3. Block diagram of a frequency response test rig

The response of any linear system can be determined from the system transfer. This is the ratio of two polynomials,

$$\frac{\theta_o}{\theta_i}(s) = \frac{b_o + b_1 s + b_2 s^2 + \cdots}{a_o + a_1 s + a_2 s^2 + \cdots}$$

If the input is a pure sine wave of frequency ω then in the steady state, the relationship between the input and output signals may be obtained from the transfer function by putting $s = \omega j$. The resultant expression is in the form,

$$\frac{\theta_o}{\theta_i}(\omega j) = R(\omega) + Q(\omega)j$$

The terms $R(\omega)$ and $Q(\omega)$ of the complex number are termed the reference and quadrature components of the transfer function. Hence, if the information is presented at the indicator in this form, then

$$\text{amplitude} = \sqrt{\{(R(\omega))^2 + (Q(\omega))^2\}} = G$$

and

$$\text{phase angle} = \tan^{-1}\frac{Q(\omega)}{R(\omega)} = \phi$$

Referring now to Fig. 10.3, if the output from the oscillator is $\sin \omega t$, and the system under test modifies this signal, introducing an amplitude factor G and phase angle ϕ, then

$$G \sin(\omega t + \phi) = \text{system output}$$

These signals are multiplied in the comparator, using either synchro resolvers, or watt meters, or quarter square multipliers, giving,

$$G \sin \omega t \sin(\omega t + \phi) = \frac{G}{2}[\cos\phi - \cos(2\omega t + \phi)]$$

At the same time, the system output is multiplied by another signal in quadrature, that is cos ωt,

$$G \cos \omega t \sin (\omega t + \phi) = \frac{G}{2} [\sin \phi - \sin (2\omega t + \phi)]$$

When the excitation frequency is above 0·5 Hz, the harmonic terms of the above equations may be removed by either electrical filtering or by the mechanical properties of the indicator. Hence, knowing $(G/2) \cos \phi$ and $(G/2) \sin \phi$, G and ϕ can be determined.

In the low frequency region, the harmonic terms begin to take effect and result in difficulty of obtaining a steady reading from the indicator. In this area, better results can be achieved if the average values are computed over a finite time, usually one cycle. The average of the harmonic terms are then zero.

10.6. Amplitude response

Although a servo system may be statically linear, that is, it may follow the linear law of its feedback potentiometer with great accuracy, in a dynamic sense it will probably be very non-linear for movements up to 50 times threshold [66]. This is illustrated by the frequency response shown in Fig. 8.1. The frequency response results show a considerable worsening of phase angle with reduction of amplitude. Part of this is due to the effective reduction of loop gain, as a result of the valve area characteristic, and part due to the factors which contribute to the threshold.

At low frequencies, the small amplitude waveform may show distinct flat tops and in the worst cases there may be large steps in the movement. Even a small degree of flat topping of the waveform will be discernible both by sight and touch. This is due to the system having to operate through the dead zone at each reversal of movement, whilst the steps are caused by the dead zone allowing overshooting of the signal.

An interesting and confusing side effect can occur when frequency response testing a system with a considerable dead zone which is due mainly to seal friction. At low frequencies the seal friction causes flat topping and consequently reduces the recorded amplitude. With increase of frequency this effect disappears, and the amplitude attains its correct value. If the convenient, but incorrect procedure is followed of using the low frequency measurement as the reference amplitude, then the test results will show a much higher resonance peak than is the true condition.

10.7. The servo amplifier

In control systems where electrical signals are present, some form of amplification is generally necessary, especially for the error signal. A d.c. amplifier is normally used, which is capable of amplifying signals at low frequencies, including zero frequency or direct current signals. The d.c. amplifier is much more prone to drift than its a.c. counterpart. Another problem arises from the fact that to handle d.c. signals, interstage coupling must be resistive.

There are three main causes of drift [84]:

1. Variation in power supplies.
2. Change of component characteristic, due to ageing and temperature changes.
3. The effect of heater voltage variation.

Obvious methods of reducing drift due to conditions 1 and 2 are to use stabilised power supplies and a careful choice of components. Various circuit arrangements are also readily available for reducing drift and temperature effects.

The dynamic response of this type of amplifier is normally well in excess of the requirement of the hydraulic servo motor. However, matching of the torque motor, used to drive the spool valve, to the amplifier can be critical and this is discussed in detail in the next section.

10.8. Matching the torque motor to the servo amplifier

The torque motor is an important component in the electro-hydraulic servo motor, as it converts the error signal into mechanical movement at the spool valve. This is achieved by the use of a simple permanent magnet motor with limited rotational movements. In this type of torque motor the differential current across their coils produces a resultant torque and movement. The characteristic of this device is therefore dependent on two variables and the output torque T_o is given by

$$T_o = k_g i - k_s \theta_r$$

where

k_g = torque motor gain (ft lbf A^{-1})
k_s = torque motor stiffness, lbf ft rad^{-1}
i = input differential current (mA)
θ_r = torque motor shaft movement (rad)

A plot of some typical characteristics is shown in Fig. 10.4. It is seen that for any particular input differential current, the torque motor shaft position is dependent on the external load at the valve assembly. The external loading T_L consists of flow forces, friction and centring spring forces.

An approximation to the dynamic performance of the torque motor can be found by summing the torques acting on such a system. Apart from having to accelerate the rotor of inertia J, the torque developed by the action of a differential current must overcome the viscous damping term with coefficient f, the relay stiffness and the external load torque.

Hence,

$$J\frac{d^2\theta_r}{dt^2} + f\frac{d\theta_r}{dt} + k_s\theta_r + T_L = k_g i$$

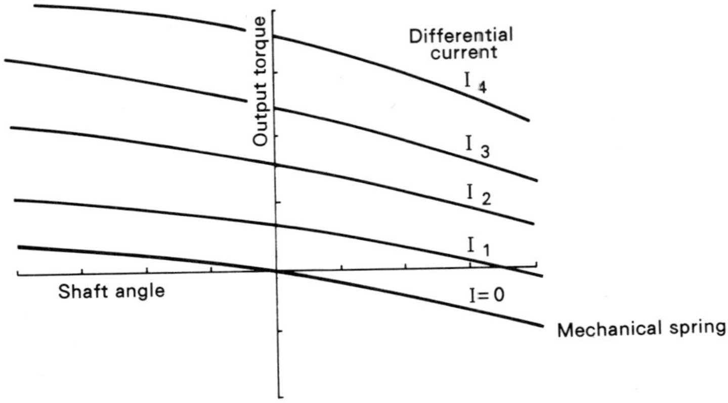

Fig. 10.4. Static characteristics of a torque motor

giving a transfer operator, if $T_L = 0$, of

$$\theta_r = \left[\frac{k_g/J}{s^2 + (f/J)s + k_s/J}\right] i$$

A typical torque motor would have the following parameters,

Rotor inertia (J) $= 1{\cdot}2 \times 10^{-5}$ lb ft^2
Damping Coefficient (f) $= 1 \times 10^{-4}$ lbf ft s rad^{-1}
Gain (k_g) $= 10{\cdot}2$ lbf ft A^{-1}
Stiffness (k_s) $= 0{\cdot}196$ lbf ft rad^{-1}

Hence, the natural frequency of the torque motor would be,

$$f_n = \frac{1}{2\pi}\sqrt{\frac{k_s}{J}} = \frac{1}{2\pi}\sqrt{\left(\frac{0{\cdot}196 \times 32{\cdot}2}{1{\cdot}21 \times 10^{-5}}\right)} = 115 \text{ Hz}$$

and the damping ratio

$$c = \frac{f}{J} \times \frac{1}{2\omega_n} = \frac{f}{2\sqrt{(Jk_s)}} = 0{\cdot}184$$

The bandwidth of this component would not normally limit the response of most hydraulic systems.

Although the torque motor and amplifier have acceptable bandwidths as separate components, it is often the case that in combination, severe limitations on the overall frequency response can occur if matching of the components is ignored.

The amplifier is effectively a voltage generator driving current through its own impedance and the combined impedance, resistive and inductive, of the torque motor coils. Thus the amplifier must drive the current against the back e.m.f. developed in the torque motor. Due to the magnitude and phase relationship of the back e.m.f., this voltage increases with frequency and amplitude, tending to cancel the operating voltage of the servo amplifier. This results in attenuation of the output. High impedance output valves and current feedback in the servo amplifier are effective methods of overcoming these problems.

10.9 Natural frequency calculations in the servo actuator

Much development time and expense can be saved by an appreciation at the design stage of the effect of a low natural frequency on servo performance. The natural frequency of the system is closely related to the bandwith, which in turn determines the speed of response of the system. Hence the higher the natural frequency, the faster the system response. Since the natural frequency is dependent on both inertia and stiffness, both of these should be kept to an optimum value. However, even an intensive design effort will be limited by the stiffness-weight characteristics of the materials available and by the hydraulic fluid bulk modulus.

In a response test on a servo system, frequencies of the same order as the resonant frequencies contribute large angles of lag and any attempt to correct this by increasing the servo gain only accentuates the resonant peak. However, if the resonant frequency can be kept outside the working bandwidth the situation is much improved. This is summarised in Fig. 10.5.

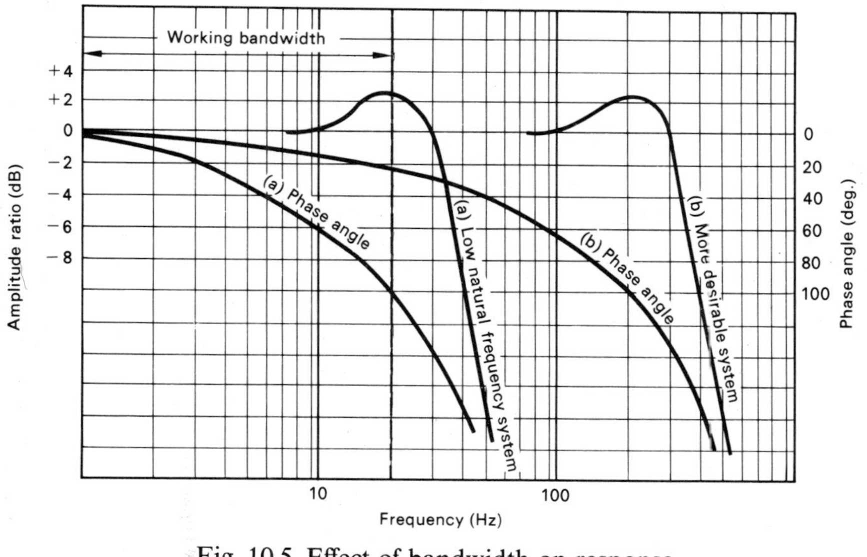

Fig. 10.5. Effect of bandwidth on response

The problem of designing for a high natural frequency must be approached simultaneously from several different aspects. The structure itself must have minimum weight, a feature which also helps to minimise the kinetic energies to be dissipated in the event of a runaway failure. In addition the structure should be designed for maximum stiffness, in this respect box type structures sometimes offer greatest possibilities.

In order to increase the stiffness of the oil column, the actuator should be designed for maximum piston area and minimum stroke.

An example of the assessment of stiffness for the arrangement shown in Fig. 10.6 can now be developed. It is assumed that the push rods and rocker arm of

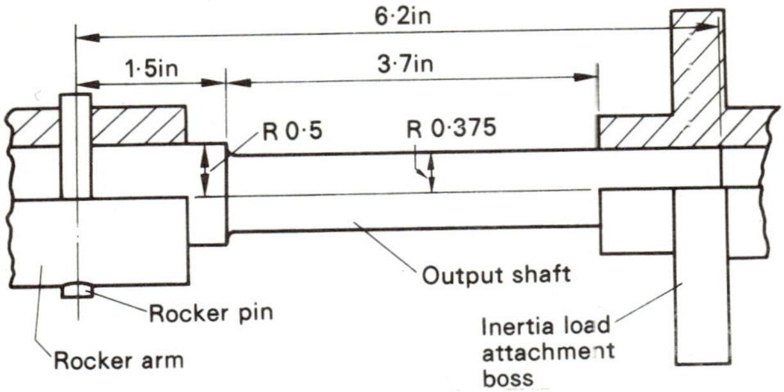

Fig. 10.6. Mechanical connection between actuator and load

the servo motor are stiff and the main contribution to flexibility is from the output shaft.

The stiffness of a shaft in torsion is given by

$$\frac{T}{\theta} = \frac{GJ}{l}$$

where T = applied torque
θ = angular defection
G = modulus of rigidity (12×10^6 lbf in^{-2})
J = polar moment of inertia
l = lengths of the sections

For the first section of Fig. 10.6

$$J_1 = \frac{\pi}{2}\left(\frac{d}{2}\right)^4 = \frac{3{\cdot}14}{2} \times (0{\cdot}375)^4 = 0{\cdot}0312 \text{ in}^4$$

where d = diameters of the sections.
Hence,

$$\frac{T}{\theta} = \frac{12 \times 10^6 \times 0{\cdot}0312}{3{\cdot}7} = 0{\cdot}102 \times 10^6 \text{ lbf in rad}^{-1}$$

also for the second section,

$$J_2 = \frac{3{\cdot}14}{2} \times (0{\cdot}5)^4 = 0{\cdot}97 \text{ in}^4$$

and hence

$$\frac{T}{\theta} = \frac{12 \times 10^6 \times 0{\cdot}097}{1{\cdot}5} = 0{\cdot}78 \times 10^6 \text{ lbf in rad}^{-1}$$

Since these two stiffnesses are in series, the overall mechanical stiffness λ_s is,

$$\lambda_s = \frac{1}{9{\cdot}8 + 1{\cdot}28} \times 10^6 = 0{\cdot}091 \times 10^6 \text{ lbf in rad}^{-1}$$

The oil stiffness can be assessed using the dimensions shown for the piston arrangement of Fig. 10.7.

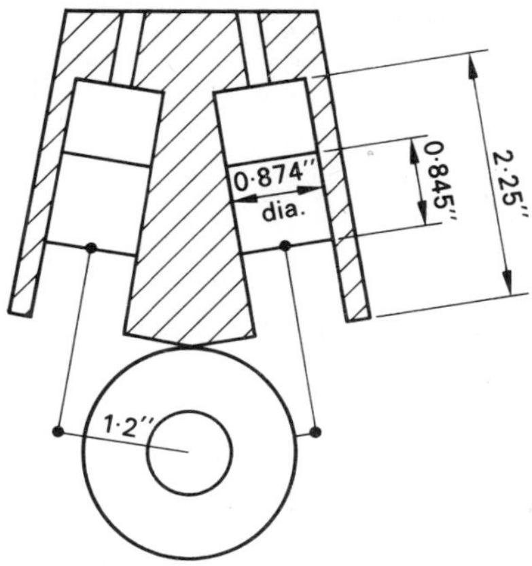

Fig. 10.7. The rotary actuator

The volume of the piston is,

$$= 0{\cdot}845 \times (0{\cdot}874)^2 \times \frac{3{\cdot}14}{4} = 0{\cdot}51 \text{ in}^3$$

and the volume of the cylinder is,

$$= (0{\cdot}874)^2 \times 2{\cdot}25 \times \frac{3{\cdot}14}{4}$$

$$= 1{\cdot}36 \text{ in}^3$$

The resultant volume of a chamber (in mid position) is

$$0{\cdot}5\,(1{\cdot}36 - 0{\cdot}51) = 0{\cdot}425 \text{ in}^3$$

The oil stiffness λ_o is given by

$$\lambda_o = \frac{NA^2}{V}$$

Taking a value of $N = 2 \times 10^5$ lbf in^{-2} for a common hydraulic oil,

$$\lambda_o = \frac{2 \times 10^5 \times /0{\cdot}6)^2}{0{\cdot}425} = 1{\cdot}7 \times 10^5 \text{ lbf in}^{-1}$$

since the radius is 1·2 in., and there are two chambers, the total hydraulic stiffness is

$$\lambda_o = 2 \times 1{\cdot}7 \times 10^5 \text{ X } 1{\cdot}2^2 = 4{\cdot}9 \times 10^5 \text{ lbf in rad}^{-1}$$

Since the oil and structural stiffness are in series, the total overall stiffness is given by,

$$\lambda_T = \frac{1}{1{\cdot}1 + 0{\cdot}204} \times 10^5 = 0{\cdot}768 \times 10^5 \text{ lbf in rad}^{-1}$$

Oil volumes in the passages between the control valve and the pistons have been ignored, since the volume of these passages is very small compared with the swept volume.
In this particular system, the rotational inertia is 23·2 lb ft^2 from the specification and hence the natural frequency is

$$f_n = \frac{1}{2\pi}\sqrt{\frac{\lambda_T}{J}} = \frac{1}{2\pi}\sqrt{\left(\frac{0{\cdot}768 \times 10^5 \times 32{\cdot}2}{12 \times 23{\cdot}2}\right)} = 15 \text{ Hz}$$

If the natural frequency had been based solely on the oil stiffness, f_n would be 38 Hz. Hence the mechanical structure can considerably affect the final value of the natural frequency.

10.10. Performance assessment

For this particular system the specification is given as follows:

Supply pressure = 1500 lbf in^{-2}
Output load inertia = 23·2 lb ft^2
Required steady state rate = 600°s^{-1}
Required response $\pm \frac{3}{4}°$ output at 15 Hz
External damping load = 5·04 lbf in s rad^{-1}
Load torque = 180 lbf in

From this information the engineer can assess the required natural frequency, piston sizes and information regarding the choice of the control valve. The process of assessment must follow a number of paths, the results are then compared in order to choose the correct system parameters.

The following method can be used to determine the piston area in relation to the frequency response requirements. Assuming sinusoidal excitation of the output, then,

$$\theta_o = \hat{\theta}_o \sin \omega t$$

where θ_o is the maximum output amplitude required at a frequency of ω rad s^{-1}. Double differentiation of this equation will produce the expression for output acceleration, hence,

$$\frac{d^2\theta_o}{dt^2} = -\hat{\theta}_o\omega^2 \sin \omega t$$

or in the case of maximum acceleration

$$= -\hat{\theta}_o\omega^2$$

From the specification the values for $\hat{\theta}_o$ and ω can be substituted and hence the maximum acceleration required in order to meet the specification can be determined.

$$\left(\frac{d^2\theta_o}{dt^2}\right)_{max} = \tfrac{3}{4} \times (2 \times 3{\cdot}14 \times 15)^2 = 6630^\circ\ s^{-2}$$

Since the inertia of the output load is known, the necessary torque required to be produced by the actuator can be calculated.

$$\text{Torque} = J\frac{d^2\theta_o}{dt^2} = \frac{23{\cdot}2 \times 12}{32{\cdot}2} \times \frac{6630}{57{\cdot}3} = 1000\ \text{lbf in}$$

To this figure an additional 10 per cent is added as an estimate of internal friction, seals, etc., hence

$$\text{Torque} = 1100\ \text{lbf in}$$

The supply pressure is given as 1500 lbf in^{-2}, and therefore the piston area can be calculated,

$$\text{Piston area required} = \frac{1100/1{\cdot}2}{1500} = 0{\cdot}61\ \text{in}^2$$

This calculation assumes that all the supply pressure is available for acceleration.

Having considered the conditions necessary to produce adequate acceleration to satisfy the frequency response, it is now necessary to look at the conditions required to satisfy the velocity requirements.

At steady velocity the force to be overcome is,

$$\text{Force} = \frac{180 + 5{\cdot}04 \times 600/57{\cdot}3}{1{\cdot}2} = 194\ \text{lbf}$$

For most efficient work conditions it can be shown that two-thirds of the supply pressure should be used on this case, hence,

$$\text{Piston area required} = \frac{194}{(1500 \times 2/3)} = 0{\cdot}194\ \text{in}^2$$

Clearly the acceleration requirement provides sufficient piston area. The next step is to determine the maximum flow requirement from the steady rate condition so that the valve can be designed. For the steady rate condition,

$$\frac{d\theta_o}{dt} = 600^\circ\ s^{-1} = 10{\cdot}42\ \text{rad}\ s^{-1}$$

The linear piston rate will be,

$$\frac{dx}{dt} = r\frac{d\theta_o}{dt} = 1{\cdot}2 \times 10{\cdot}42 = 12{\cdot}6\ \text{in}\ s^{-1}$$

and hence the maximum flow rate must be,

$$Q_{max} = A\frac{dx}{dt} = 12{\cdot}6 \times 0{\cdot}61 = 7{\cdot}7\ \text{in}^3\ s^{-1}$$

The maximum oscillatory rate required is

$$\left(\frac{d\theta_o}{dt}\right)_{max} = \tfrac{3}{4} \times 2 \times 3{\cdot}14 \times 15 = 70{\cdot}5^\circ\ s^{-1}$$

$$\left(\frac{dx}{dt}\right)_{max} = 1{\cdot}2 \times \frac{70{\cdot}5}{57{\cdot}3} = 1{\cdot}47\ in\ s^{-1}$$

$$Q_{max} = 0{\cdot}61 \times 1{\cdot}47 = 0{\cdot}89\ in^3\ s^{-1}$$

This figure of flow rate is much less than that required by the steady rate condition.

10.11 Choice of control valve

The simplest single stage valve for controlling both the direction of the flow and the flow rate is the three-way valve, but this will control the flow to only one side of the load, and a differential area ram must be used, Fig. 10.8. A more common form of the single stage valve, is the four-way valve shown in Fig. 10.9.

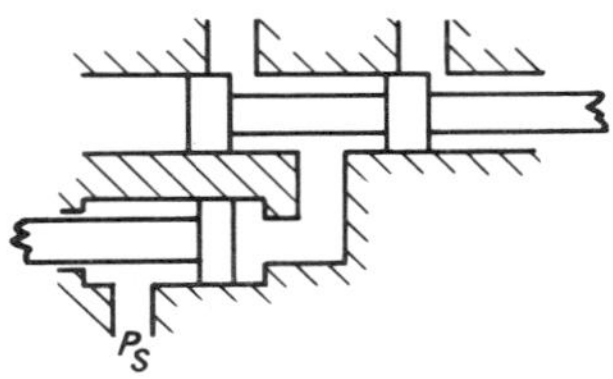

Fig. 10.8. Three-way valve and single acting cylinder

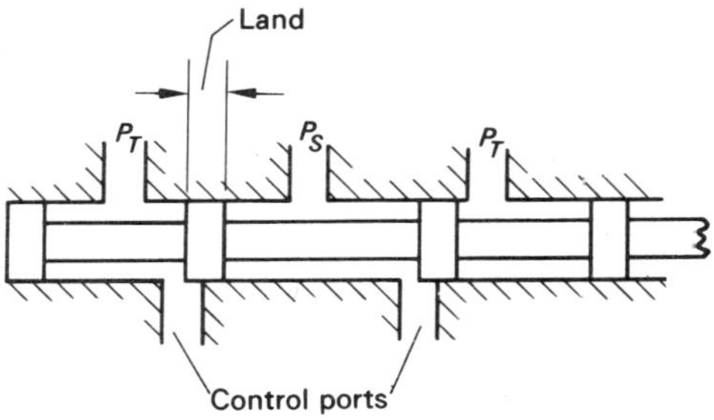

Fig. 10.9. The four-way servovalve

The valve is made up of a spool, the sliding portion and a liner or sleeve. Two sets of orifices are bored symmetrically around the sleeve, together with three other sets of holes to provide passages for the inlet pressure and the outlet exhaust to tank. The orifices must be symmetrically arranged around the sleeve and should not be less than two, although three orifices are preferable. This is in order that the pressure be distributed evenly around the spool, reducing stiction effects.

Fluid is metered through the control valve ports when the spool uncovers a quadrant or sector of the orifice. The uncovered area exposed for each value of the spool displacement gives a non-linear area characteristic. An example of

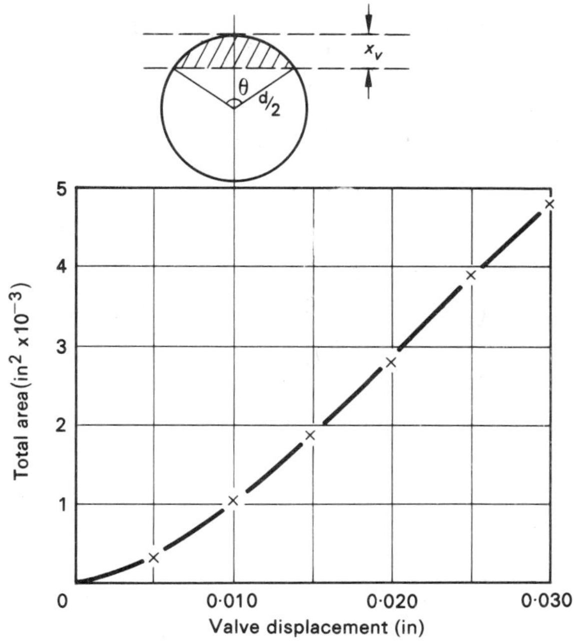

Fig. 10.10. The area–displacement characteristic of a segmental orifice

this is shown in Fig. 10.10 in which the uncovered area has to be calculated from the relationship

$$a = \frac{d^2}{8}(\theta - \sin\theta)$$

$$\cos\frac{\theta}{2} = \frac{d - 2x_v}{d}$$

where

a = exposed orifice area
d = orifice diameter
x_v = spool displacement

In this case there are three orifices, each of diameter 0·076 in.

It can be seen from Fig. 10.10 that operating the valve at larger displacements helps to linearise the characteristic.

The section of the spool which controls the uncovering of the orifice area is termed the land. A valve has underlap or overlap if the land is smaller or larger respectively than the orifice diameter, when the spool is central.

An overlapped valve saves fluid loss when the spool is central but tends to reduce the accuracy of the servo. Its use also tends to reduce servo stability.

An underlapped valve gives much more sensitive control at the expense of fluid loss. Its use tends to improve the servo stability. Although in most cases the ideal valve would have zero lap, in practice this is difficult to achieve without the use of very expensive production techniques. Typical flow characteristics are shown in Fig. 10.11.

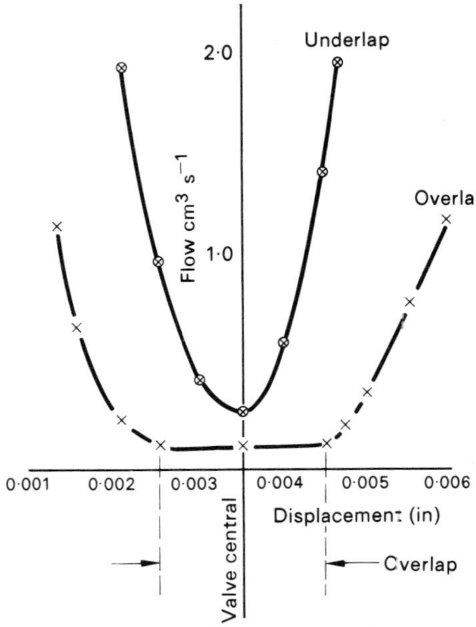

Fig. 10.11. Measured variation of flow rate with spool displacement for underlapped and overlapped valves

Experience shows that for small spool valves of the type used in this example, and for a typical hydraulic oil, the flow through the control orifices can be calculated from

$$Q = 103a\sqrt{(P_s - P_1)}$$

In the example, the flow required is 7·7 $\text{in}^3\ \text{s}^{-1}$ and for optimum conditions the pressure drop across one set of orifices will be (see chapter 3)

$$\begin{aligned} P_s - P_1 &= \tfrac{1}{3} \times \tfrac{1}{2} \times 1500 \\ &= 250 \text{ lbf in}^{-2} \end{aligned}$$

This takes into account the maximum efficiency criterion, i.e. $\frac{1}{6}\, P_s$ drop across each of the valves two orifice sets, leaving $\frac{2}{3}\, P_s$ for driving the load.

Hence the total orifice area required must be,

$$\begin{aligned} a_T &= \frac{7{\cdot}7}{103 \times \sqrt{250}} = \frac{7{\cdot}7}{1630} \\ &= 47{\cdot}2 \times 10^{-4} \text{ in}^2 \end{aligned}$$

This should be developed at a valve displacement of a little less than half the orifice diameter. Assuming three orifices are to be used in each control port set, then for each orifice,

$$a = \tfrac{1}{3} \times 47{\cdot}2 \times 10^{-4} = 15{\cdot}8 \times 10^{-4}\ \text{in}^2$$

and since,

$$a = \frac{\pi d^2}{4} \text{ or } d^2 = \frac{15{\cdot}8 \times 10^{-4} \times 4 \times 2}{3{\cdot}14}$$

assuming maximum uncovered area developed at 50 per cent displacement.

$$d = 6{\cdot}34 \times 10^{-2}\ \text{in}$$

The nearest practical orifice size is 0·076 in diameter, and the experimental results from a valve with three orifices of this diameter in each control port set are in Fig. 10.12. Also shown in this figure are the corresponding flow values calculated from the flow equation.

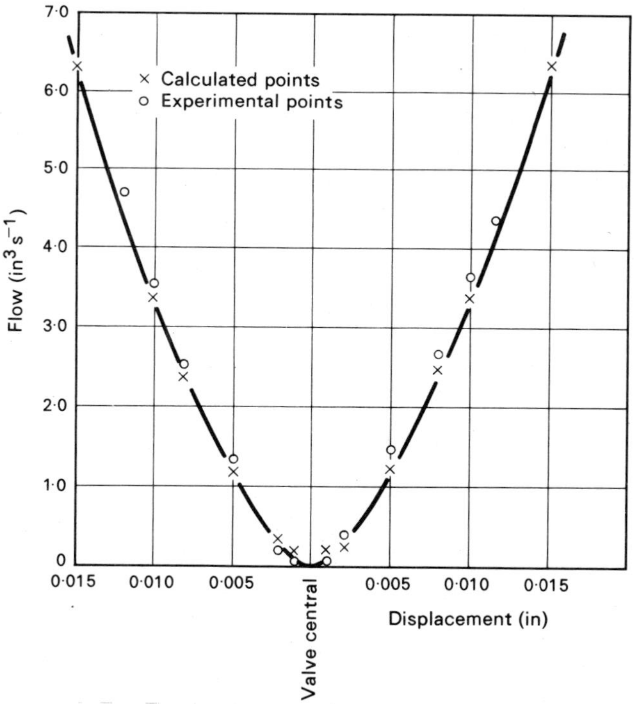

Fig. 10.12. Measured and predicted flow rates through a spool valve

Valves are often assessed in terms of horsepower, which in some cases is defined by,

$$\text{Horsepower} = \frac{P_s \times Q}{12 \times 550}$$

where

$$P_s = \text{supply pressure (lbf in}^{-2})$$
$$Q = \text{maximum flow (in}^3\ \text{s}^{-1})$$

Another way to define the horsepower is in terms of the load flow and pressure characteristics.

These few simple calculations allow the engineer to get a feel of the system and should enable components to be selected. If the components are being developed from scratch, then a much more detailed study is necessary to ensure proper performance and reliability. Fairly large inertia loading conditions can cause trouble and need careful study. This can result from both flow force effects, in the case of electro-hydraulic servo, and also from cavitation.

Chapter 11

Control of pressure

11.1 Introduction

The control of pressure is equally important in pneumatics as in hydraulics. Pressure control valves act to regulate pressure in a circuit. In hydraulic circuits these may function as either relief valves or as reducing valves.

In the relief valve mode of operation the valve is normally closed and opens at a predetermined pressure, by-passing the excess flow. This is shown in Fig. 11.1. This type of valve is a main source of heat generation in the circuit, while flow is being by-passed to tank.

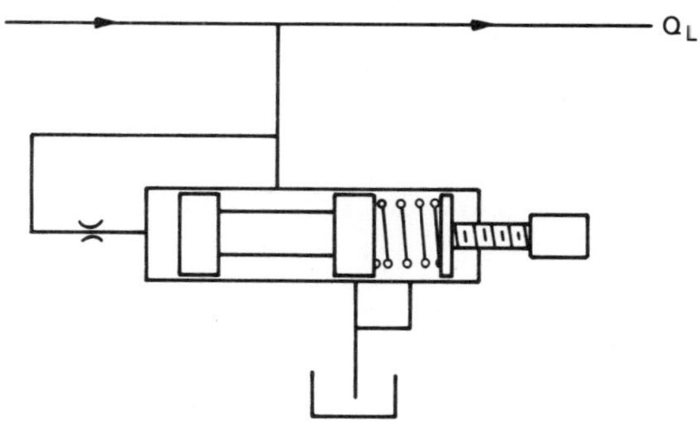

Fig. 11.1. A relief valve

In the reducing valve mode of operation, shown in Fig. 11.2, the valve is normally open. It closes to maintain a minimum pressure by restricting flow in the line.

The relief valve is used to control the pressure in the system by rapidly expelling small amounts of fluid. These valves are used for example to protect the hydraulic pump, and in turn the electric motor, from overload. They are also used to maintain the pressure in a circuit at a predetermined level.

Pressure reducing valves reduce the pressure in a section of the circuit. For example, in a hydraulic press circuit, the full pressure that the pump set develops

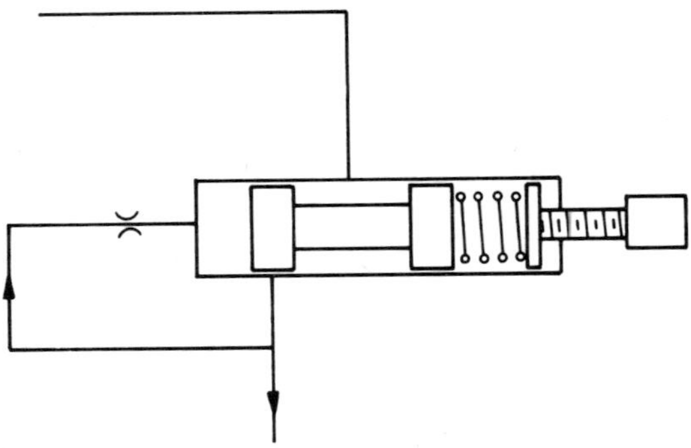

Fig. 11.2. A pressure reducing valve

may be needed at the press cylinder, but a much lower pressure may be needed to operate cylinders used for ejecting the work piece.

From these two basic modes of operation stem a wide range of specialised pressure control valves, shown in Table 11.1.

Table 11.1

Valve	Application
Safety relief valves	Preset valve used to provide a pressure limit in event of system failure.
Counterbalance valve	Used to maintain a constant back pressure on a cylinder, to hold a load for example
Unloading valve	A relief valve used to control pump pressure and opens under the application of a pilot pressure to unload the pump
Sequence valve	A relief valve which allows flow to pass to another circuit once a set pressure is achieved
Load dividing valve	Used to proportion pressure between two pumps in series
Decompression valve	Controls the rate at which the pressure of a compressed fluid is released

Pneumatic pressure control valves are more commonly termed pressure regulators and are used to safeguard the pneumatic system. Normal inlet pressures are around 150 lbf in^{-2} and these can be reduced to zero by the regulator. The unit also compensates for pressure fluctuations between the source and the regulator inlet.

A simple type of regulator is shown in Fig. 11.3. Commercial units are normally combined with a lubricator and filter. The regulator consists of a spring which can be adjusted externally. The spring force acts on a diaphragm and as long

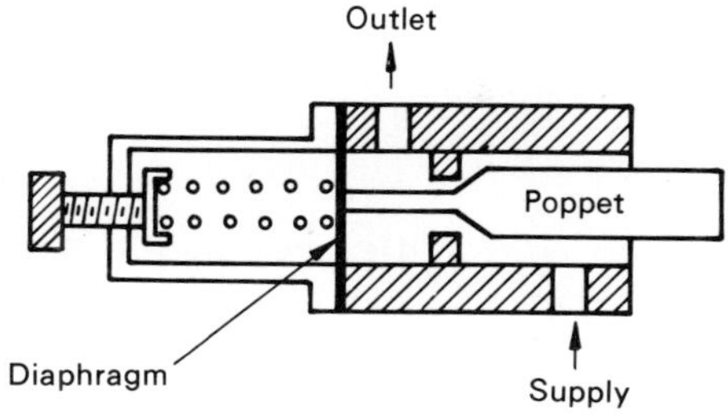

Fig. 11.3. A simple non relieving pneumatic pressure regulator

as the downstream pressure is less than the spring pressure, the valve will remain open. This allows the downstream pressure to rise until it equals the spring pressure. At this point the pressure differential across the diaphragm is zero and the valve remains closed. If the controlled pressure rises above the desired value, the diaphragm rises reducing the flow to the system and hence the pressure.

In some lubricators, the action of the passing air stream causes a reduction in the pressure in the top of the dome and the oil from the reservoir is forced up the feed pipe where it can drip through a constriction into the main air stream atomising into a micro fog of lubricated oily air. A needle valve restriction is normally incorporated in the feed tube to regulate the drip feed rate.

The filter unit relies on a vortex motion to separate the water content in the air. The moisture laden air enters at the inlet, passes over vanes at the deflector section and receives a spinning motion. Moisture separates and falls to the

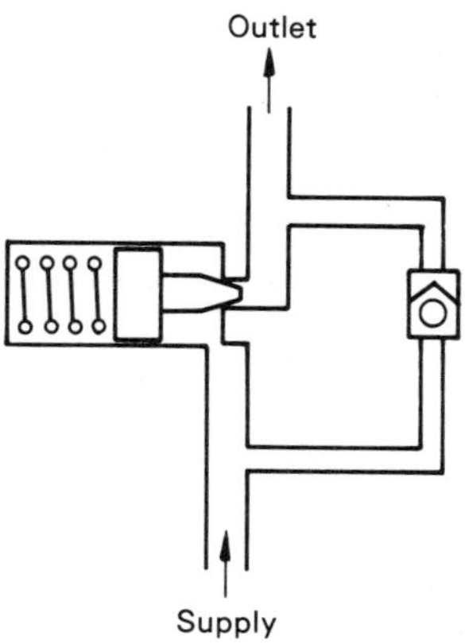

Fig. 11.4. A sequence valve, with non-return valve in parallel

bottom of the bowl. The exit path for the air is through a sintered bronze filter element to remove any small dirt particles. Some units have a silica-gel purifier element to dry the air.

It is essential that the air be lubricated if moving part pneumatic units are used. It is important to note that for non-moving part fluidic elements, the air must be dry and oil-free and in these applications it is necessary to fit a filter to remove any oil.

Other types of pneumatic pressure control valves are,

Sequence valves
Safety valves

The sequence valve directs air from the one circuit to another when there is a pressure build up in the former. It is seen from Fig. 11.4 that there is no flow from supply to outlet until the supply pressure is sufficient to overcome the spring force. Flow from outlet to inlet is unrestricted.

Safety valves are used where excessive pressure build up is likely to occur.

11.2. Applications—hydraulic

1. *Relief valve* Hydraulic relief valves are normally placed in the circuit to prevent the system exceeding a predetermined pressure. Fig. 11.5 shows the relief valve *A* arranged to blow-off excess pressure to tank.

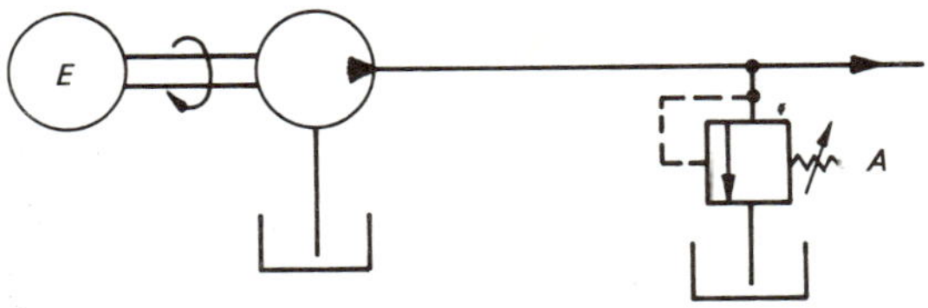

Fig. 11.5. A relief valve in the delivery line

2. *Reducing valve* In Fig. 11.6, the relief valve *A* controls the pump pressure, while reducing valve *B* allows a low pressure supply to be available. Notice that the reducing valve is in series.

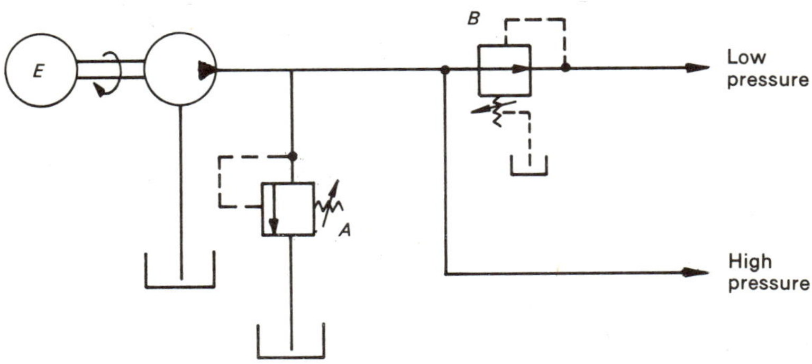

Fig. 11.6. A relief valve *A* and a reducing valve *B* in a circuit

3. *Sequence valve* The two position four-way directional valve *A*, Fig. 11.7, is moved so that oil pressure is directed to the end of cylinder *B*. The piston of cylinder *B* advances to the end of its stroke and then pressure builds up causing the sequence valve *C* to open. Oil pressure is now directed to cylinder *D* and its piston moves to the end of its stroke.

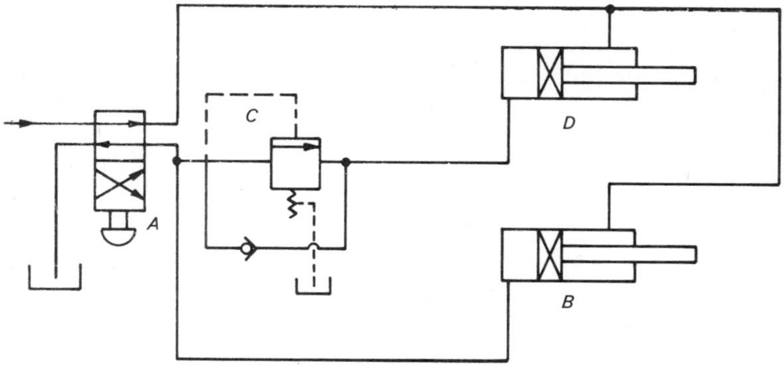

Fig. 11.7. Sequencing of cylinders *B* and *D* by means of a sequence valve *C*

If the directional valve *A* is moved back to its original position, the oil pressure is directed simutaneously to both cylinders *B* and *D*, causing movement.

4. *Decompression circuit* In some hydraulic circuits it is necessary to release the oil at a controlled rate until the pressure has dropped to a reasonable level. Oil flows into the cylinder, Fig. 11.8, through the main ports of the

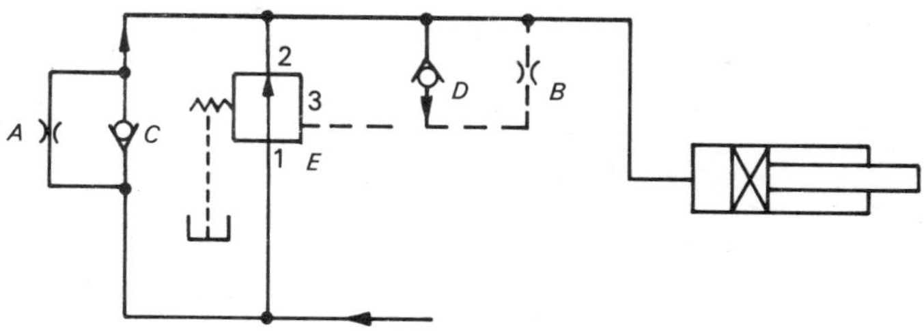

Fig. 11.8. A decompression circuit

valve *E*, until the pressure rises to a point where the predetermined pressure is reached at port 3. Flow from 1 to 2 is then stopped. Further flow continues into the cylinder via the non-return valve *C* until maximum pressure is reached. When the applied pressure is removed, the only flow path from the cylinder to tank is through *A*. This allows the cylinder to decompress gradually, preventing excessively high flows in the pipes. Orifice *B* is used to damp the valve. At a point determined by the force of the spring, the valve opens, allowing unrestricted flow back to the tank.

11.3. Applications—pneumatic

1. *Regulators* Cut in and cut out of a compressor tends to result in an unsteady supply pressure to a system. A pressure regulator is needed if a steady supply is essential. Such a valve is also useful if a pressure smaller than the main supply is needed. The regulator referred to in Fig. 11.3 is non-relieving. If this type of regulator were controlling the pressure in a fixed volume, there would be no means of relieving the controlled pressure if it became too high. The symbols for the 'non-relieving', Fig. 11.9(*a*), and self relieving pressure regulator, Fig. 11.9(*b*), are shown.

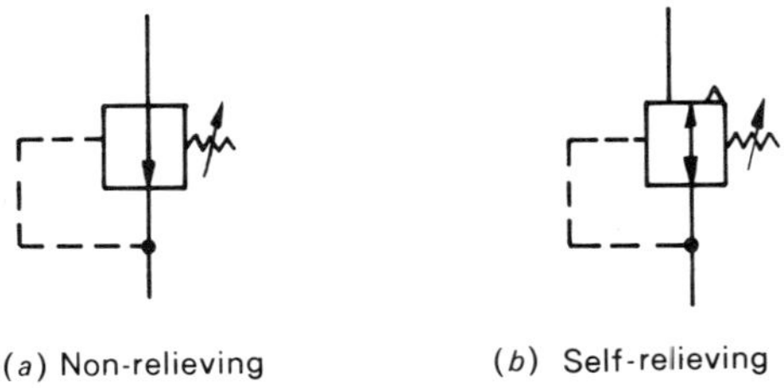

Fig. 11.9. C.E.T.O.P. symbols for pressure regulators (*a*) non-relieving (*b*) relieving

In Fig. 11.10, regulator *A* sets a different circuit pressure to that of regulator *B*. These pressures are then used to control the two cylinders via the directional valves.

2. *Sequence valves* These valves direct air from the main circuit to a second circuit when there is a pressure build up in the former. In Fig. 11.11(*a*), there is no flow from supply to outlet until the supply pressure is sufficient to overcome the spring force. Flow from outlet to inlet is unrestricted. In Fig. 11.11(*b*), the four-way valve *A* is actuated and cylinder rod *B* extends. When it is fully extended pressure builds up in the left side of the cylinder allowing flow

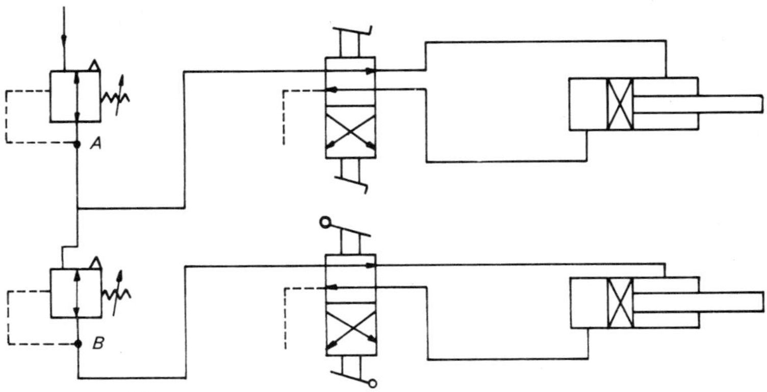

Fig. 11.10. Use of pressure regulators in practice

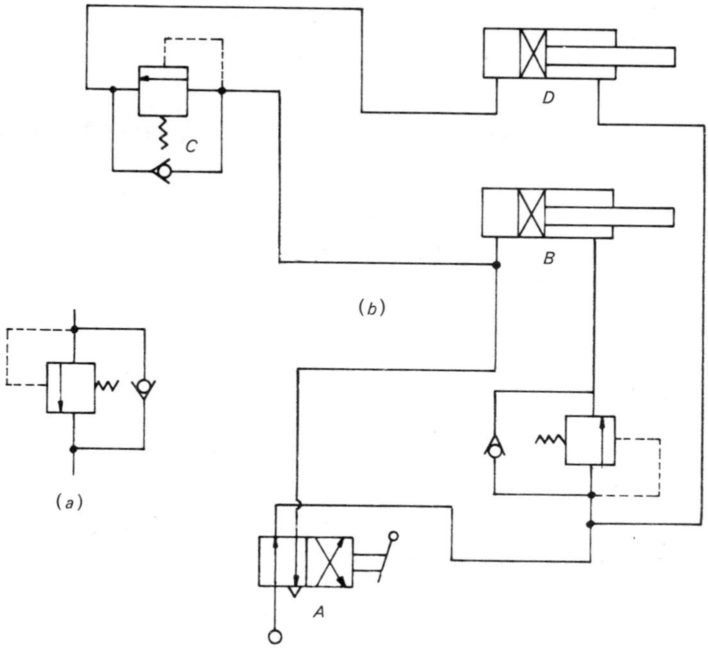

Fig. 11.11. (*a*) sequence valve and non-return valve in parallel (*b*) pneumatic circuit using sequence valves

through sequence valve *C*. Cylinder rod *D* now extends. When valve *A* is returned to its original position, *D* retracts and then *B* retracts.

11.4. Response of spool valves

Pressure control valves use feedback and hence can be treated as simple control loops. Because of this need for feedback, it is possible for the valves to become unstable. Another form of instability can result from lack of damping in even the simplest of arrangements such as a ball and spring pressure relief valve. It is also important that valves work quickly to correct for errors in the circuit. This requires good dynamic response, but in most cases a balance must be sought between stability and rapid response.

Spool valves can be actuated in a number of ways, mechanically, electrically or for large power units, by a pilot system. These valves are popular in hydraulics because they allow large amplifications but the time taken to open or close them may contribute significantly to the overall time constant of the equipment.

The response of electrically operated small valves can be improved by replacing the conventional solenoid with a coil moving in the field of a permanent magnet. The response times can be reduced below 10 ms by this method, an improvement by a factor of four or more over conventional a.c. or d.c. solenoids. The moving coil arrangement has proved its worth in vibration generators,

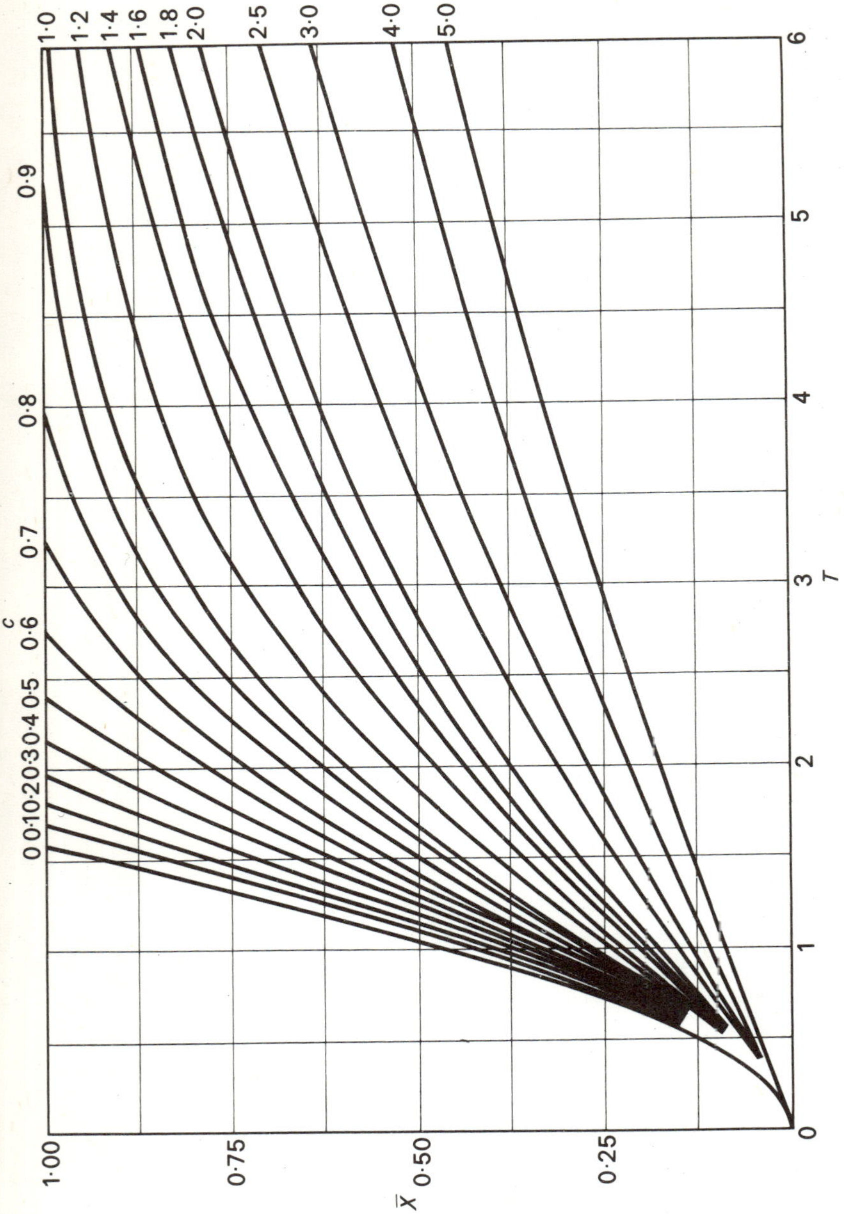

Fig. 11.12 Solutions of the equation

$$M\ddot{x} + f\dot{x} + k(x + x_1 + x_2) = F$$

$$\bar{x} = \frac{k(x + x_2)}{F - kx_1}, \quad T = t\sqrt{\frac{k}{m}}, \quad c = \sqrt{\frac{f^2}{4kM}}$$

(after Lorenz and Stringer [85])

their low inertia and low electrical inductance being a particular advantage.

The response time of a spool valve will depend upon the instantaneous values of the forces acting in the axial direction of the spool. The force applied to the spool must overcome the inertia force due to the moving parts, the friction force due to shearing of the oil layer in the radial gap between spool and bore, the return spring force and any force due to the fluid flowing through the valve.

A balance of force on the spool gives the equation,

$$M\ddot{x} + f\dot{x} + k(x + x_1 + x_2) + F_\beta \pm \rho L Q = F \tag{11.1}$$

where x is the spool displacement from the zero lap position
x_1 is initial spring compression
x_2 is the valve overlap
M is mass of moving parts
$f = \mu\pi d L_1/\varepsilon$ sliding friction coefficient
μ is fluid viscosity
d is bore diameter
L_1 is length of spool in sliding contact with the bore
ε is radial clearance between spool and bore
k is return spring stiffness

Lorenz and Stringer [85] have derived a solution for this equation, shown in Fig. 11.12, which predicts the time taken for the spool to slide a given distance,

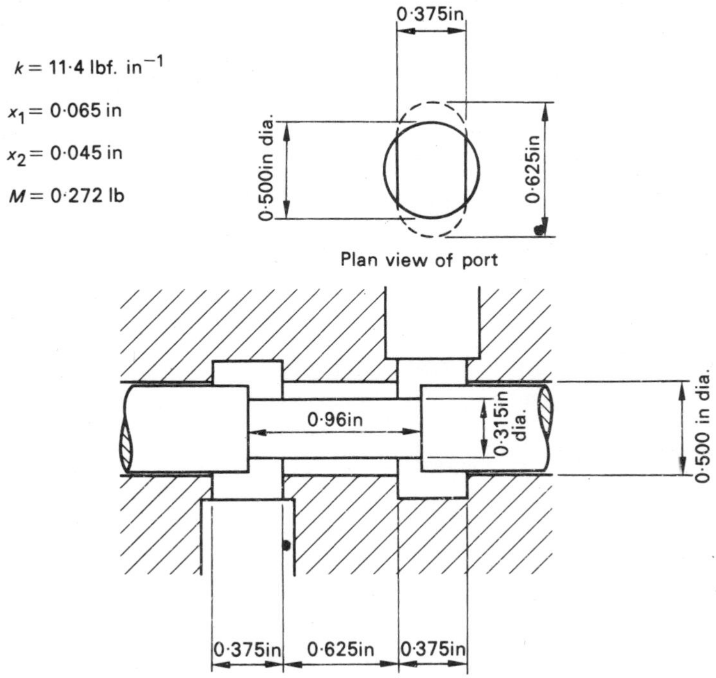

Fig. 11.13. Chamber and port dimension for valve (after Lorenz and Stringer [85])

after a particular value of force F has been applied in the form of a step. For this result the flow force effects have been ignored.

Example

The example given in [85] is particularly revealing. For the valve arrangement shown in Fig. 11.13, with no flow and an applied force F of 8.14 lbf, equation (11.1) is

$$0{\cdot}000705\ddot{x} + 0{\cdot}15\dot{x} + 11{\cdot}4(x + 0{\cdot}11) = 8{\cdot}14 \tag{11.2}$$

represented by the central full line of Fig. 11.14.

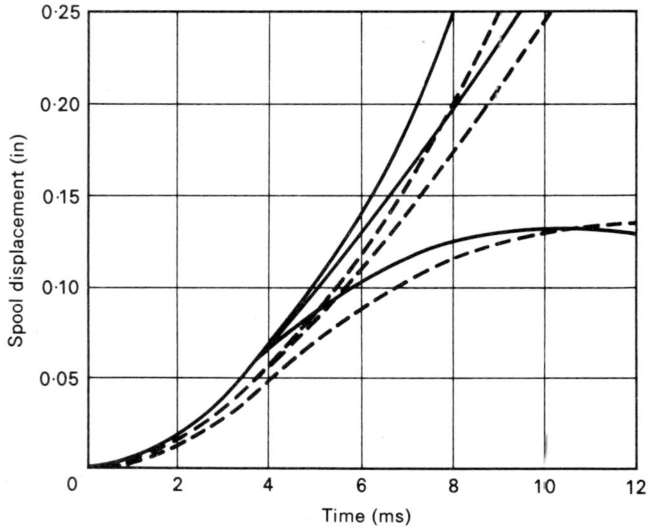

Fig. 11.14. Spool motion for valve solenoid current 3 amp., and pressure drop 250 lbf in^{-2} (After Lorenz and Stringer [85]). Acknowledgement to the Institution of Mechanical Engineers

When flow is present it is possible to express (11.1) in terms of the pressure drop across the valve and the valve displacement. In this example the pressure drop is 250 lbf in^{-2}.

For the same case, (11.1) becomes,

$$0{\cdot}000705\ddot{x} + 0{\cdot}15\dot{x} + 11{\cdot}4(x + 0{\cdot}11)$$
$$+ 208\frac{(1 - 6{\cdot}2x \pm 19{\cdot}4x)}{(1 + 10{\cdot}8x)^2}x \pm \frac{0{\cdot}176\dot{x}}{(1 + 10{\cdot}8)^2} = 8{\cdot}14$$

These results are given by the upper and lower full lines in Fig. 11.14, which are for different directions of flow through the valve. The dotted lines are experimental results.

11.5. Use of flow forces

Although flow forces have a considerable adverse effect on spool response time, in certain applications they can be arranged to improve the speed of response. Work carried out by Carrington [40] shows that rapid response of a three-way valve can be achieved by using both shaping and angled port force compensation, and is best used when the flow in the centre port is bi-directional, Fig. 11.15. It

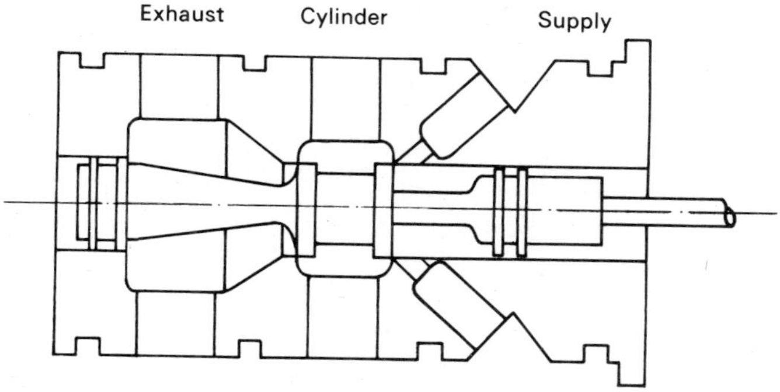

Fig. 11.15. Fast-acting three-way valve (after Carrington [40])

can be seen from Fig. 11.16, that the angled ports have reduced the closing flow force when flow is from supply to cylinder. When the flow is from cylinder to exhaust the use of the shaped spool results in the flow tending to open the valve over most of its travel.

With the spool actuated using a moving coil device and this type of compensation, response times around 10 ms have been achieved at 5000 lbf in^{-2} and 136 gal min^{-1}.

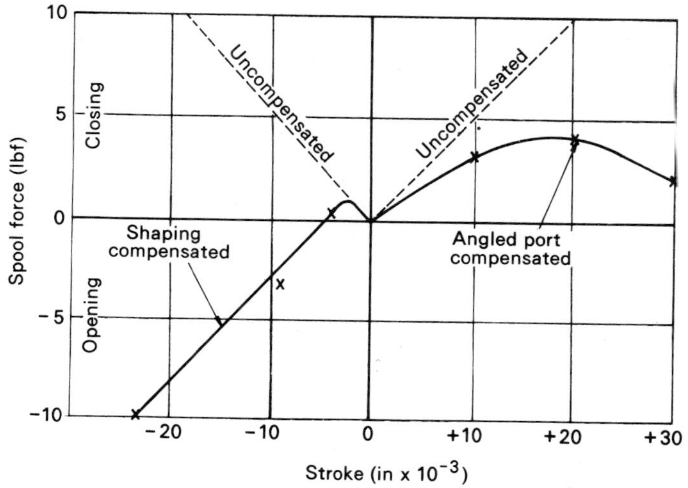

Fig. 11.16. Flow forces in fast-acting three-way valve (after Carrington [40])

11.6. Dynamics of pressure control valves

The analysis of the relief valve is similar to that developed by Ma [86] in which the dynamic behaviour of the pressure reducing valve is studied. A typical arrangement is shown in Fig. 11.17 where V_s is the volume of the pressure

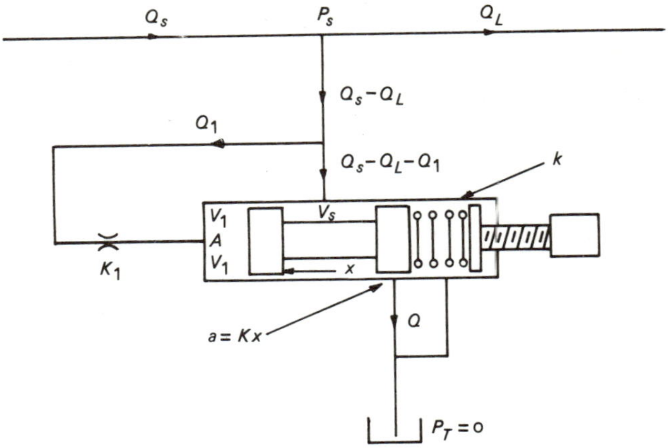

Fig. 11.17. Schematic of a relief valve

sensing chamber and V_1 the volume of the pressure control chamber. The spring preload is set to determine the desired relief pressure and this results in a force kx_o acting on the spool. The difference between the preload force and the pressure in the pressure sensing chamber will cause the spring to deflect an amount x, the movement of the spool. Under dynamic conditions force balance is

$$kx_o - AP_1 = M\frac{d^2x}{dt^2} + f\frac{dx}{dt} + kx + F_\beta \tag{11.3}$$

where M is the mass of the moving parts and x is the valve opening.

Flow force effects will contain steady state and transient force components,

$$F_\beta = 2C_qKx(P_s - P_T)\cos\phi - \rho L\frac{dQ}{dt} \tag{11.4}$$

Hence, combining (11.3) and (11.4)

$$kx_o + \rho L\frac{dQ}{dt} = M\frac{d^2x}{dt^2} + f\frac{dx}{dt} + kx + 2C_vKxP_s\cos\phi + P_1A \tag{11.5}$$

The flow through the main orifice, Q, can be approximated by considering small flow increments as shown in chapter 3.

$$Q = \frac{\partial Q}{\partial x}x + \frac{\partial Q}{\partial P}P_s$$

$$= C_xx + C_pP_s \tag{11.6}$$

where

$$C_p = \frac{C_q K x \sqrt{(2/\rho)}}{\sqrt{(P_s - P_T)}}$$

$$C_x = C_q K \sqrt{\left\{\frac{2}{\rho}(P_s - P_T)\right\}}$$

Differentiation of (11.6) gives

$$\frac{\partial Q}{\partial t} = C_x \frac{\partial x}{\mathrm{d}t} + C_p \frac{\partial P_s}{\partial t} \tag{11.7}$$

These results can now be substituted into (11.5) giving,

$$kx_o + \rho L c_x \frac{\mathrm{d}x}{\mathrm{d}t} = P_1 A - \rho L C_p \frac{\mathrm{d}P_s}{\mathrm{d}t} + M \frac{\mathrm{d}^2 x}{\mathrm{d}t^2} + f \frac{\mathrm{d}x}{\mathrm{d}t} + kx + \mathrm{K}' P_s x$$

where $K' = 2C_q K \cos \phi$
Now let,

$$(K' P_s + k) = k_e \text{ the equivalent spring stiffness}$$

and

$$(f - \rho L c_x) = f_e \text{ the equivalent friction coefficient.}$$

Then

$$kx_o - P_1 A + \rho L C_p \frac{\mathrm{d}P_s}{\mathrm{d}t} = M \frac{\mathrm{d}^2 x}{\mathrm{d}t^2} + f_e \frac{\mathrm{d}x}{\mathrm{d}t} + k_e x \tag{11.8}$$

The term $\rho L C_p(\mathrm{d}P_s/\mathrm{d}t)$ can be considered small. Then,

by letting $c = f_e/\sqrt{(2Mk_e)}$, (11.8) becomes, when transformed

$$kx_o - P_1 A = k_e \left(\frac{s^2}{\omega_n^2} + \frac{2c}{\omega_n} s + 1\right) \tag{11.9}$$

where $\omega_n = \sqrt{(k_e/M)}$ is the natural frequency of the spool and equivalent spring.
Flow into and out of the pressure sensing chamber can be defined by

$$Q_1 = K_1(P_s - P_1) = \frac{V_1}{N} \frac{\mathrm{d}P_1}{\mathrm{d}t} - A \frac{\mathrm{d}x}{\mathrm{d}t} \tag{11.10}$$

This assumes the orifice to be small enough for laminar flow to exist and takes into account compressibility effects in this chamber. Equation (11.10) can be expressed in terms of P_1, the pressure in the sensing chamber,

$$P_1 = \frac{K_1 P_s + Asx}{K_1(s/\omega_s + 1)} \tag{11.11}$$

where $\omega_s = NK_1/V_1$.

Flow through the pressure control chamber can be defined by,

$$(Q_s - Q_L) - Q_1 - K_e P_s = \frac{V_s}{N}\frac{dP_s}{dt} - Q \qquad (11.12)$$

where K_e is the leakage coefficient. Again compressibility effects in this chamber are included.

Substitution from (11.6) and (11.10) into (11.12) gives

$$(Q_s - Q_L) - K_1 P_s + K_1\left[\frac{K_1 P_s + Asx}{K_1(s/\omega_s + 1)}\right] - K_e P_s - C_p P_s + C_x x = \frac{V_s}{N}\frac{dP_s}{dt}$$

or

$$(Q_s - Q_L)\left(\frac{s}{\omega_s} + 1\right) + C_x\left[1 + \left(\frac{1}{\omega_s} + \frac{A}{C_x}\right)s\right]x$$

$$= \left[\left(\frac{s}{\omega_s} + 1\right)\left(1 + \frac{K_1}{K_s} + \frac{s}{\omega_p}\right)K_s - K_1\right]P_s \qquad (11.14)$$

where $\omega_p = NK_s/V_s$ and

$$K_s = C_p + K_e.$$

Expanding the right-hand side of (11.14) gives

$$(Q_s - Q_L)\left(\frac{s}{\omega_s} + 1\right) + C_x\left[1 + \left(\frac{1}{\omega_s} + \frac{A}{C_x}\right)s\right]x$$

$$= K_s\left[\frac{s^2}{\omega_s\omega_p} + \frac{1}{\omega_p}\left(1 + \frac{\omega_p}{\omega_s} + \frac{V_1}{V_s}\right)s + 1\right]P_s \qquad (11.15)$$

This equation is essentially the same as that derived by Merritt [87]. The right-hand side of (11.15) can be further simplified if it is assumed that $V_1/V_s \ll 1$, giving

$$(Q_s - Q_L)\left(\frac{s}{\omega_s} + 1\right) + C_x\left[1 + \left(\frac{1}{\omega_s} + \frac{A}{C_x}\right)s\right]x$$

$$= K_s\left(1 + \frac{s}{\omega_s}\right)\left(1 + \frac{s}{\omega_p}\right)P_s \qquad (11.16)$$

Equations (11.9) and (11.11) can be combined to give,

$$kx_o - \frac{AP_s}{\left(\frac{s}{\omega_s} + 1\right)} - \frac{A^2 sx}{K_1\left(\frac{s}{\omega_s} + 1\right)} = K_e\left(\frac{s^2}{\omega_n^2} + \frac{2c}{\omega_n}s + 1\right)x \qquad (11.17)$$

and this allows the block diagram in Fig. 11.18 to be drawn.

If the orifice controlling flow into the pressure sensing chamber were removed, then $K_1 = \infty$ and $s/\omega_s = 0$. Equations (11.16) and (11.17) then reduce to,

$$(Q_s - Q_L) + (C_x + As)x = K_s\left(1 + \frac{s}{\omega_p}\right)P_s \qquad (11.18)$$

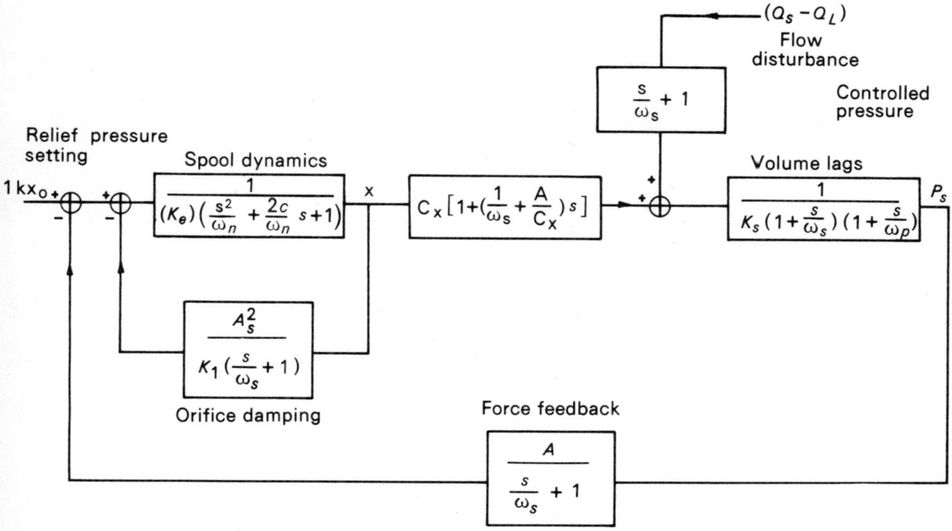

Fig. 11.18. Block diagram of relief valve

and

$$kx_o - AP_s = K_e\left(\frac{s^2}{\omega_n{}^2} + \frac{2c}{\omega_n}s + 1\right)x \tag{11.19}$$

which gives the new block diagram of Fig. 11.19.

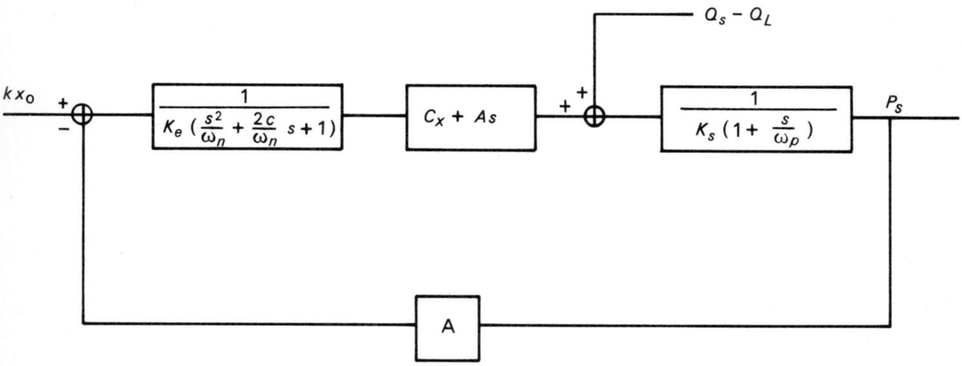

Fig. 11.19. Block diagram of relief valve without orifice damping

Examination shows that stability will be affected since a secondary damping loop around the spool has been removed. This system now relies on the inherent damping of the valve which is usually small. This type of instability results in valve chatter.

Consider the loop shown in Fig. 11.19, when a load disturbance has settled out so that $(Q_s - Q_L) = 0$. Then

$$\frac{P_s}{kx_o} = \frac{(C_x + As)/\left\{K_sK_e\left(\dfrac{s^2}{\omega_n{}^2} + \dfrac{2c}{\omega_n}s + 1\right)\left(1 + \dfrac{s}{\omega_p}\right)\right\}}{1 + (C_x + As)A/\left\{K_sK_e\left(\dfrac{s^2}{\omega_n{}^2} + \dfrac{2c}{\omega_n}s + 1\right)\left(1 + \dfrac{s}{\omega_p}\right)\right\}} \tag{11.20}$$

From the characteristic equation of (11.20)

$$\frac{s^3}{\omega_n{}^2\omega_p} + \left(\frac{1}{\omega_n{}^2} + \frac{2c}{\omega_n\omega_p}\right)s^2 + \left(\frac{1}{\omega_p} + \frac{2c}{\omega_n} + \frac{A^2}{K_sK_e}\right)s + \left(1 + \frac{AC_x}{K_sK_e}\right) = 0$$

The Routh–Hurwitz criterion requires that

$$\left(\frac{1}{\omega_p} + \frac{2c}{\omega_n} + \frac{A^2}{K_sK_e}\right) - \frac{(1/\omega_n{}^2\omega_p)/(1 + AC_x/K_sK_e)}{1/\omega_n{}^2 + 2c/\omega_n\omega_p} \geqslant 0 \text{ for stability}$$

or

$$\frac{2c\omega_p}{\omega_n} + \frac{2c\omega_n}{\omega_p} + \frac{A^2}{K_sK_e}(\omega_p + 2c\omega_n) \geqslant \frac{AC_x}{K_sK_e} \tag{11.21}$$

If the damping term $c = 0$, then for stability

$$\omega_p \geqslant C_x/A$$

A pressure relief valve will respond rapidly to load changes if no restrictor is present, but will have poor stability and pressure control accuracy. The poor stability characteristics makes the valve noisy due to valve chatter. Use of a restrictor will alleviate the situation, but will give a slower response. Two-stage pressure control valves have much better stability characteristics.

More information about relief valve dynamics can be obtained in [88] and [89].

Chapter 12

Speed control

12.1. Pneumatic actuators

The most commonly used method of controlling the speed of pneumatic motors and cylinders is flow restriction. The needle valve, Fig. 12.1, is one of the simplest

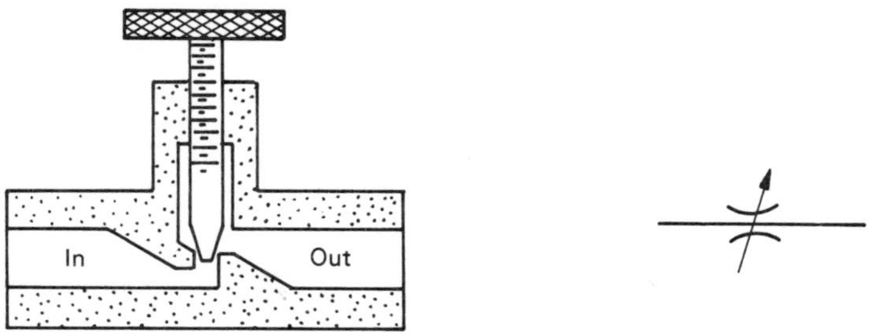

Fig. 12.1. A needle flow restrictor and its C.E.T.O.P. symbol

variable restrictors available. The orifice area is easily varied between zero and fully open by turning the adjusting screw. If this restrictor, which is basically a two-way valve, is placed in the supply line to a pneumatic motor, the motor will achieve maximum speed when the needle valve is wide open and will stop after the valve is closed. Intermediate speeds can be obtained by setting the restrictor at intermediate openings. With the restrictor in the supply line the motor is said to be under meter-in speed control. Figure 12.2(*a*) shows a meter-in control of cylinder speed. The rod speed is controlled when it is extending, but the non-return valve in parallel with the restrictor allows the rod to retract quickly. Figure 12.2(*b*) shows a meter-out system, in which the restrictor is placed in the return or exhaust line from the cylinder. Here again the non-return valve allows rapid cylinder rod retraction. In many cases the variable restrictor and non-return valve are available as an integral unit. Figure 12.3 shows a typical construction. Flow from left to right must pass through the restriction whereas flow in the opposite direction lifts the poppet off its seat and can bypass the restriction.

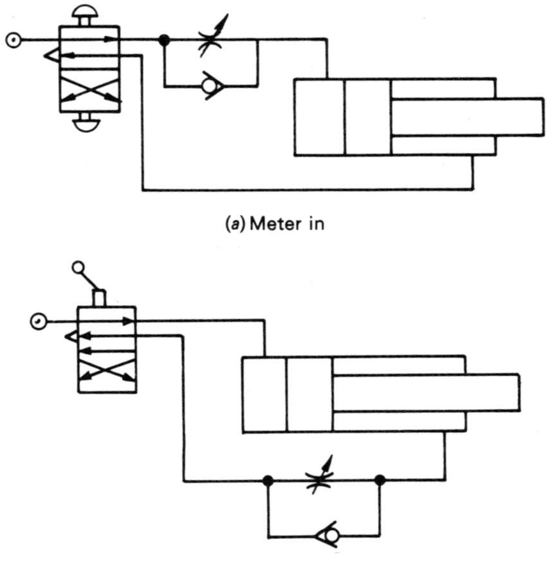

Fig. 12.2. Speed control using (*a*) meter-in (*b*) meter-out

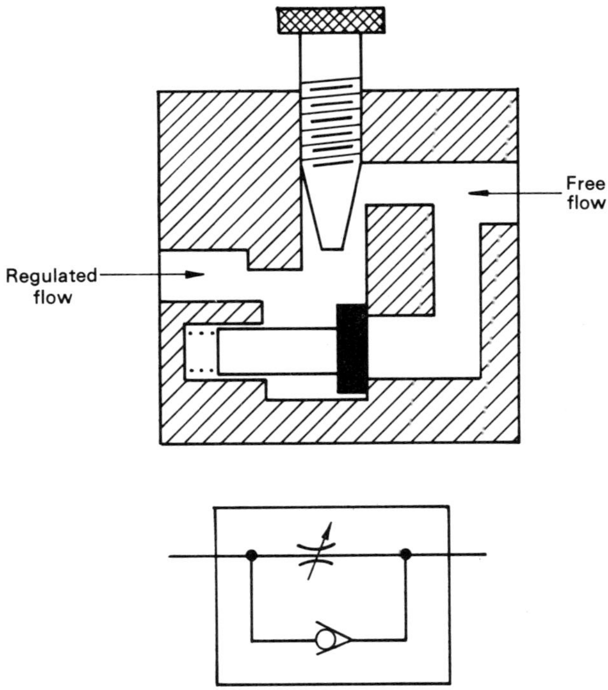

Fig. 12.3. Variable restrictor and non-return valve as an integral unit

In order to better understand the relative advantages and disadvantages of meter-in and meter-out controls, it is necessary to look at system dynamics. Figure 12.4, for example, shows a typical response of a pneumatic cylinder

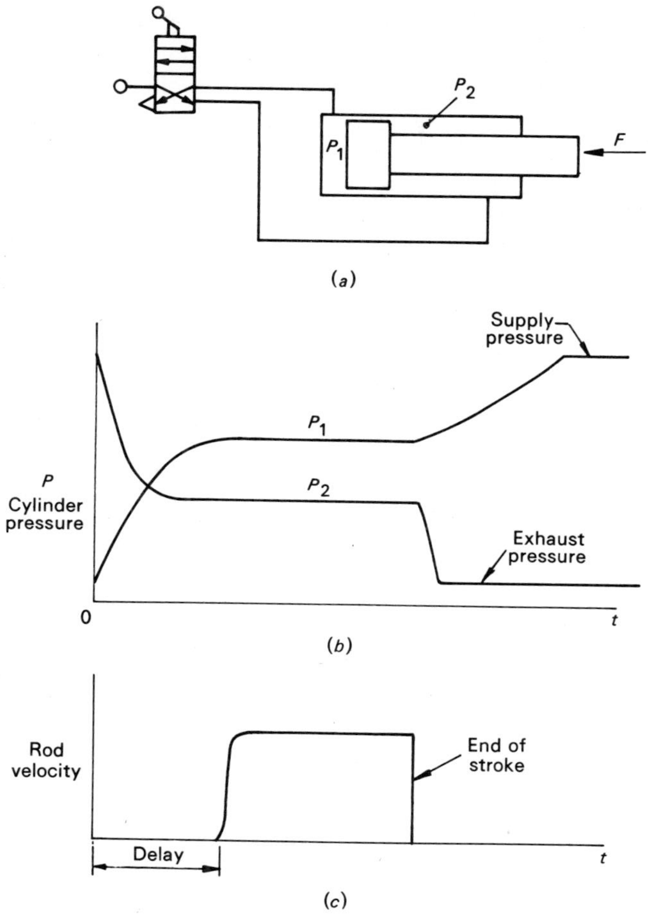

Fig. 12.4. Typical response of a pneumatic cylinder loaded with a constant force

loaded with a constant force F. At zero time the valve is actuated to make the rod extend. The blank end of the cylinder is connected to supply and the pressure P_1, which was initially at exhaust pressure, rises. At the same time the rod end is connected to exhaust and pressure P_2 which was initially at supply pressure, falls. The speed of the pressure changes depends on the compressibility of the air, the size of the valve ports, supply pressure and cylinder size. The cylinder rod will not move until the differential in the pressures P_1 and P_2 is sufficient to provide a force F to equal the load on the rod. The rod then extends at constant velocity, the pressures P_1 and P_2 remaining constant since the incoming and outgoing volumetric flow rates are equal. When the rod stops at the end of its stroke the blank end pressure rises to supply pressure while the rod end pressure falls to exhaust pressure.

How do restrictions in the supply and exhaust lines affect this picture? In order to best answer this question, it is convenient to use a quantitative comparison. Consider for example a single rod double acting cylinder with piston rod diameter $D_2 = 0{\cdot}5$ in and piston diameter $D_1 = 2{\cdot}0$ in. If there is a constant force F opposing piston rod extension then the force balance equation is

$$P_1A_1 - P_2A_2 = F \tag{12.1}$$

Consider a meter-in speed control of this cylinder. If the demanded speed is small then it is reasonable to assume that the rod end pressure P_2 will be very

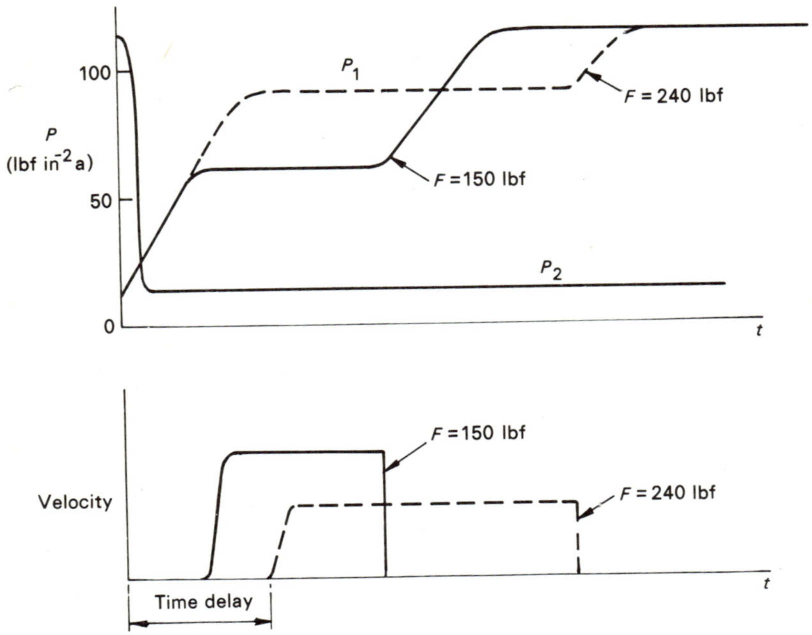

Fig. 12.5. Effect of load change on speed of pneumatic cylinder with meter-in speed control

nearly equal to exhaust pressure in the steady state, Fig. 12.5. Equation (12.1) can then be written

$$P_1 = \frac{F}{A_1} + \frac{A_2}{A_1} P_e \tag{12.2}$$

If the exhaust is to atmosphere, P_e can be taken as atmospheric pressure and for this case

$$P_1 = [F/\pi + 13{\cdot}8] \text{ lbf in}^{-2} \text{ a} \tag{12.3}$$

If the supply pressure is 100 lbf in^{-2} g, equation (12.3) shows

1. The maximum load that the cylinder can drive, the stall load $\hat{F}$, is

$$\hat{F} = (114{\cdot}7 \times \pi) - (14{\cdot}7 \times 15\pi/16) = 317 \text{ lbf}$$

2. The minimum load that the cylinder can control is

$$\check{F} = -\pi \times 13{\cdot}8 = -43 \text{ lbf}$$

At this force, the pressure P_1 is absolute zero. A larger negative force or pull would cause the cylinder to 'run away'. This inability to cope with overhauling loads is a serious disadvantage of the meter-in system and does not allow its use in drilling or reaming operations.

The dynamic response of a meter-in control is sketched in Fig. 12.5. Two loads have been considered, F = 150 lbf and 240 lbf. For the smaller load, the steady state P_1 is given by equation (12.3) as 62 lbf. No rod movement occurs until P_1 has risen from atmospheric pressure to this value. If a low rod speed is demanded the metering orifice will be adjusted to present a large resistance to inflow and under these conditions the delay required for the build up of P_1 can be considerable. This is another disadvantage of meter-in control in pneumatic circuits. If the load is increased to 240 lbf, the steady state value of P_1 is increased to 90 lbf in^{-2} a and the pressure build-up time is further increased. An important point to note is that the steady state velocities for the two loads are different. The speed is load dependent and this is a serious disadvantage of a speed control system. The ratio of the two speeds can be calculated. The incoming mass flow rate W is

$$W = W_r f(P_1/P_s)$$

when W_r is constant and is given by equation (3.5). If the steady state mass flow rates at F = 150 lbf, and 240 lbf, are respectively W_a and W_b then

$$\frac{W_b}{W_a} = \frac{f(P_1/P_s)_b}{f(P_1/P_s)_a} = \frac{f(0{\cdot}79)}{f(0{\cdot}54)}$$

Figure 3.7 can be used to give $W_b/W_a = 0{\cdot}837$. The mass flow rate into the cylinder has been reduced at the higher loading because the pressure P_1 has risen. The volumetric flow rate into the cylinder dictates the speed of the rod in the steady state. Mass flow rate W and volumetric flow rate Q are related by

$$W = \rho Q$$

$$\therefore \frac{Q_b}{Q_a} = \frac{W_b}{W_a}\frac{\rho_a}{\rho_b} \tag{12.4}$$

If the process is isothermal then

$$\frac{\rho_a}{\rho_b} = \frac{P_{1a}}{P_{1b}} \tag{12.5}$$

$$\therefore \frac{Q_b}{Q_a} = \frac{W_b}{W_a}\frac{P_{1a}}{P_{1b}} \tag{12.6}$$

$$= 0{\cdot}837 \times 62/90 = 0{\cdot}577$$

The speed is nearly halved at the higher loading.

If meter-out control is used and the steady state blank end pressure is assumed to equal the supply pressure, equation (12.1) gives the rod end pressure P_2 as

$$P_2 = (122 - 0{\cdot}34F) \text{ lbf in}^{-2} \text{ a} \tag{12.7}$$

The stall load is found by putting $P_2 = 0$ in this equation. It is 370 lbf. A very important feature of meter-out control is that negative loads can be dealt with. Equation (12.7) shows that the more negative F, the larger is pressure P_2.

The dynamic response of a meter-out control is sketched in Fig. 12.6. On operation of the directional control valve, P_1 rises to near supply pressure.

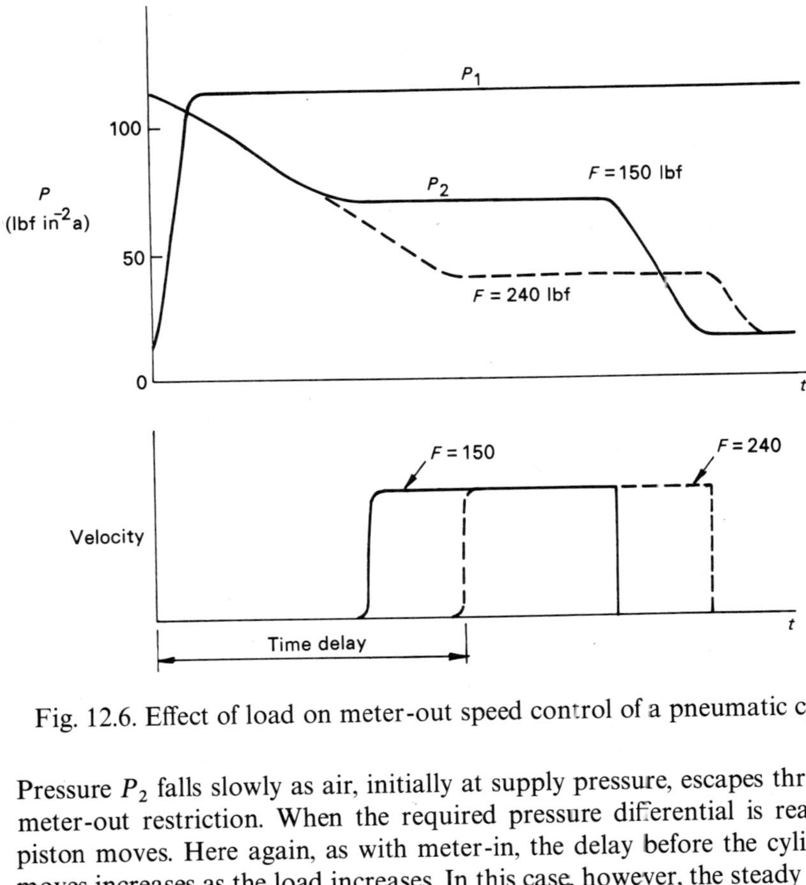

Fig. 12.6. Effect of load on meter-out speed control of a pneumatic cylinder

Pressure P_2 falls slowly as air, initially at supply pressure, escapes through the meter-out restriction. When the required pressure differential is reached the piston moves. Here again, as with meter-in, the delay before the cylinder rod moves increases as the load increases. In this case, however, the steady state rod velocity is the same under both loading conditions. This is due to the fact that the flow through the restriction is

$$W = W_r(P_2/P_r)f(P_e/P_2) \tag{12.8}$$

where W_r and P_r are constants (see equation (3.5)). Choked flow occurs when $P_e/P_2 < 0{\cdot}532$ for air and hence in this case the outflow is choked when $P_2 >$

28 lbf in^{-2} a. The load equation, equation (12.7), then indicates that choking occurs if the load is less than +277 lbf (75 per cent of stall load). If again mass flow rates at loads of 150 lbf and 240 lbf are W_a and W_b then since the flow is choked

$$\frac{W_a}{W_b} = \frac{P_{2a}}{P_{2b}} \tag{12.9}$$

Equation (12.4) then gives

$$\frac{Q_b}{Q_a} = \frac{P_{2b}\,\rho_a}{P_{2a}\,\rho_b} = 1 \tag{12.10}$$

Hence provided that the outflow is choked, the volumetric flow rates and hence the steady state velocities are independent of load. The meter-out control of pneumatic cylinders has thus two decided advantages over meter-in.

1. Speed is less affected by load changes.
2. It will not run away under the influence of overhauling loads. These factors make meter-out control by far the more popular for pneumatic circuits.

Note that in Fig. 12.6, it was assumed that the rod end pressure was initially supply pressure. If a two-position directional control valve is used with a transient closed centre position, this is nearly the truth. If the valve is three position with an open centre, a different response can occur. The initial value of both P_1 and P_2 would be atmospheric pressure if the valve were initially moved from its open mid-position to its actuating position. Under these conditions an initial fast lunge of the piston can occur. Consider for example the case when the load is zero and the meter-out restriction is high. When the valve moves from the mid-position, P_1 will rise and the rod will move quickly. Equilibrium conditions, for this zero load, are reached when pressure P_2 is 122 lbf in^{-2} a. If the small flow rate through the restriction is neglected it is seen that, for an isothermal process, P_2 will rise to 122 lbf in^{-2} a from 14·7 lbf in^{-2} a when the rod end volume has been reduced by a factor 14·7/122 = 0·12. In other words the rod will move 88 per cent of its stroke before equilibrium is reached. Open centre valves should therefore, not be used with meter-out speed control circuits.

Both meter-in and meter-out controls are poor if stiction forces are present and low speeds are required. Figure 12.7(*a*) shows a stiction characteristic which could be partly caused by cylinder seal friction as well as by the system load. 20 lbf is needed to start the load moving, but only 5 lbf is needed to keep it in motion. When the pressure P_1 in a meter-in circuit, Fig. 12.7(*b*), has built up sufficiently to overcome the 20 lbf stiction load, the piston lunges since as soon as it moves there is more force available than that required for steady motion (5 lbf). The lunge forward reduces the pressure in the blank end and when this has fallen below that required to give a 5 lbf thrust, the rod stops. Pressure P_1 then builds up again and the cycle repeats. The frequency of the stick slip oscillation increases as the rod extends as a result of the increasing blank end volume, Fig. 12.7(*c*).

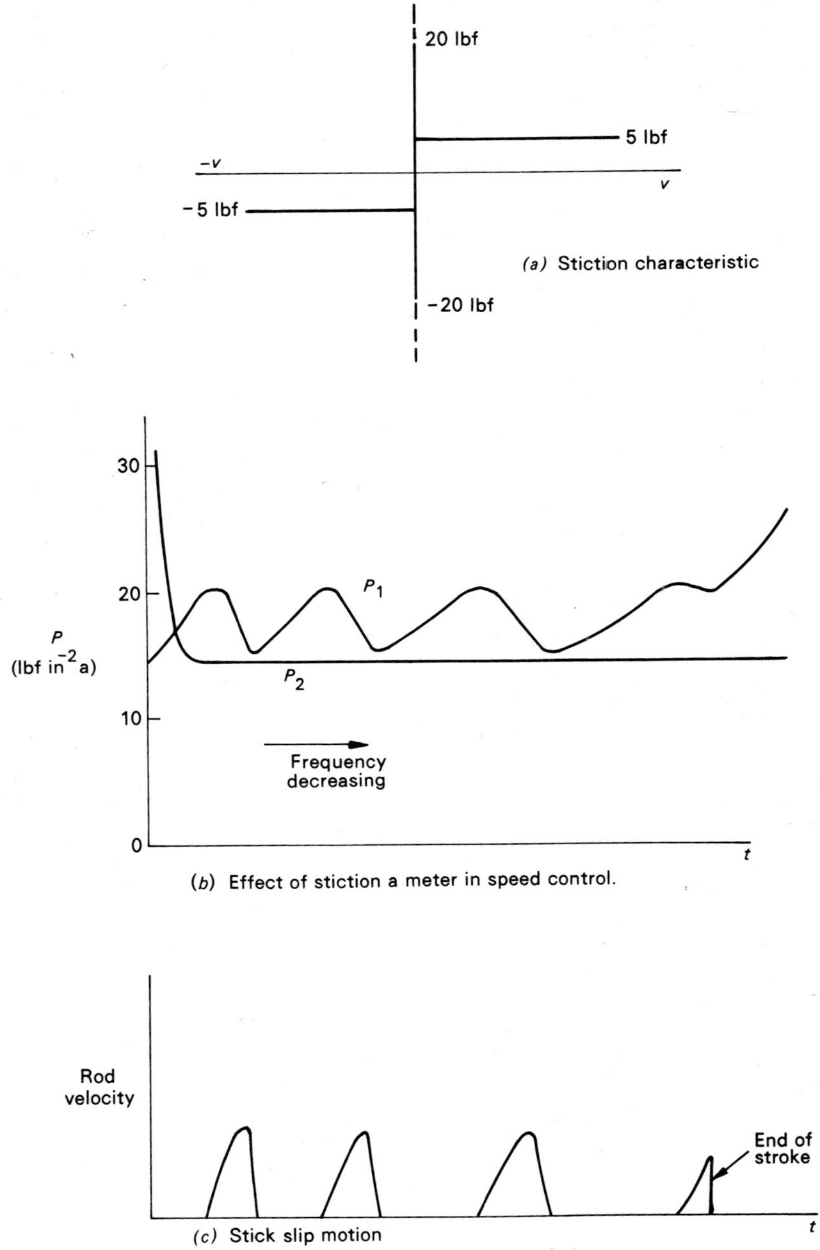

Fig. 12.7. Effect of stiction on a meter-in speed control

12.2. Dynamic analysis of meter-out speed control

The above discussion of speed control assumed the simplest of loads. In practice, however, inertia loads, friction and spring loads are encountered and it is sometimes necessary to predict their effects on system performance. The following analysis shows how inertia load effects can be determined. For the system shown

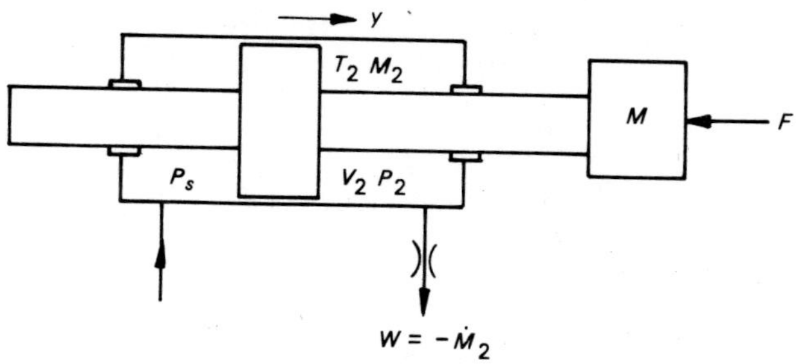

Fig. 12.8. Schematic of pneumatic cylinder with meter-out control.

in Fig. 12.8, the mass M_2 of air in the exhaust side of the cylinder is given by the equation of state as

$$M_2 = P_2V_2/RT_2 \tag{12.11}$$

The rate of change of mass is, assuming for simplicity an isothermal process

$$\dot{M}_2 = \left(\frac{P_2}{RT_2}\right)\dot{V}_2 + \left(\frac{V_2}{RT_2}\right)\dot{P}_2 \tag{12.12}$$

or

$$\frac{\dot{M}_2}{M_2} = \frac{\dot{V}_2}{V_2} + \frac{\dot{P}_2}{P_2} \tag{12.13}$$

Equation (9.12) can be applied if adiabatic conditions exist. If the flow is choked, as it usually is in meter-out circuits,

$$W = -\dot{M}_2 = kP_2 \tag{12.14}$$

where $k = W_r/P_r$

The load equation is

$$P_s - P_2 = (M/A)y + F/A \tag{12.15}$$

Note that the pressure in the left-hand side of the cylinder is assumed to remain constant at P_s (infinite capacity) F is a step load applied at time $t = 0$

$$\text{Since } V_2 = V_2(0) - Ay \tag{12.16}$$

equation (12.15) becomes

$$P_s - P_2 = F/A - (M/A^2)\dot{V}_2 \tag{12.17}$$

Equations (12.13), (12.14) and (12.17) combine to give

$$\ddot{V}_2 + \frac{kP_2}{M_2}\ddot{V}_2 + \frac{P_2A^2}{MV_2}\dot{V}_2 = \frac{kP_2A^2}{MM_2}\left[\frac{F}{A} - P_s\right] \tag{12.18}$$

This is a now linear differential equation; the fact that the coefficients are functions of the dependent variables makes it difficult to solve. Note, however, that in the new steady state following the application of the step load F, the velocity is V_{2s} given by

$$\dot{V}_{2s} = \frac{MV_{2s}}{P_{2s}A^2}\frac{kP_{2s}A}{MM_{2s}}[P_s - P_{2s} - P_s]$$

$$= \frac{-kV_{2s}P_{2s}}{M_{2s}} = -kRT_{2s} \tag{12.19}$$

Initially equation (12.14) gives

$$W(0) = -\dot{M}_2(0) = kP_2(0)$$

and equation (12.13) gives

$$\dot{V}_2(0) = [V_2(0)/M_2(0)]\dot{M}_2(0) = -[V_2(0)/M_2(0)]kP_2(0)$$
$$= -kRT(0) \tag{12.20}$$

If the process is isothermal $T_s = T(0)$ and the steady state velocities are the same before and after the load application. This agrees with conclusions drawn earlier in this chapter.

Some indication of system behaviour can be obtained quickly if the variables are assumed to have only small changes about their initial values. In this case equation (12.18) can be approximated by

$$\ddot{V}_2 + \frac{kP_2(0)}{M_2(0)}\ddot{V}_2 + \frac{P_2(0)A^2}{MV_2(0)}\dot{V}_2 = \frac{kP_2(0)A^2}{MM_2(0)}\left[\frac{F}{A} - P_s\right] \tag{12.21}$$

or

$$\ddot{V}_2 + 2c\omega_n\ddot{V}_2 + \omega_n^2\dot{V}_2 = \omega_n^2\dot{V}_2(0) \tag{12.22}$$

The natural frequency ω_n is given by

$$\omega_n^2 = P_2(0)A^2/(MV_2(0)) \tag{12.23}$$

Equation (12.23) shows also that the natural frequency tends to increase as the rod moves to the right reducing V_2.

The damping ratio c is

$$c = -\tfrac{1}{2}\frac{\dot{V}_2(0)}{A}\sqrt{\left\{\frac{2M}{P_sV_2(0)}\right\}} = \tfrac{1}{2}\dot{y}(0)\sqrt{\left\{\frac{2M}{P_sV_2(0)}\right\}} \tag{12.24}$$

The damping depends on the initial rod velocity or valve mass flow. The response of the system is now reasonably well defined. Equations (12.19) and (12.20) show that the steady state velocities are the same before and after load application. Equation (12.22) shows that there is a transient oscillation after the load application and equations (12.23) and (12.24) enable the damping and frequency of the oscillation to be calculated.

The same results can be found by using the linearised orifice flow equation developed earlier in chapter 3. As an exercise in using valve coefficients it is worthwhile repeating the analysis by this method. The equation of state, equation (12.11), can be written

$$\frac{\delta M_2}{M_2} = \frac{\delta V_2}{V_2} + \frac{\delta P_2}{P_2} - \frac{\delta T_2}{T_2} \tag{12.25}$$

where δ is a small increment. If a polytropic relationship exists between T_2 and P_2 then

$$\frac{\delta T_2}{T_2} = \frac{n-1}{n}\frac{\delta P_2}{P_2} \tag{12.26}$$

where $n = 1$ for an isothermal process and $n = 1{\cdot}4$ for air for an adiabatic process. Equation (12.25) and (12.26) give

$$\frac{\delta M_2}{M_2} = \frac{\delta V_2}{V_2} + \frac{1}{n}\frac{\delta P_2}{P_2} \tag{12.27}$$

Assume small perturbations about the initial conditions and put

$$\left.\begin{aligned} M_2 &= M_2(0) + \delta M_2 \\ M_2 &= M_2(0) + \delta M_2 \\ P_2 &= P_2(0) + \delta P_2 \\ V_2 &= V_2(0) + \delta V_2 \end{aligned}\right\} \tag{12.28}$$

Equation (12.27) is then

$$\frac{\delta M_2}{M_2(0)} = \frac{\delta V_2}{V_2(0)} + \frac{1}{n}\frac{\delta P_2}{P_2(0)} \tag{12.29}$$

The linearised orifice flow equation is

$$\delta W = W(0)(1+\zeta)\frac{\delta P_2}{P_2(0)} - \tfrac{1}{2}W(0)\frac{\delta T_2}{T_2(0)} \tag{12.30}$$

and using equation (12.26)

$$\delta W = W(0)\frac{\delta P_2}{P_2(0)}\left\{(1+\zeta) - \tfrac{1}{2}\left(\frac{n-1}{n}\right)\right\} \tag{12.31}$$

But

$$\delta W = -\delta M_2 \tag{12.32}$$

Equation (12.29) also gives

$$\frac{\dot{M}_2}{M_2(0)} = \frac{\dot{V}_2}{V_2(0)} + \frac{1}{n}\frac{\dot{P}_2}{P_2(0)} \tag{12.33}$$

or

$$\frac{\delta \dot{M}_2}{M_2(0)} = \frac{\delta V_2}{V_2(0)} + \frac{1}{n} \frac{\delta \dot{P}_2}{P_2(0)} \tag{12.34}$$

The load equation is

$$\delta P_2 = -(M/A)\,\delta \ddot{y} - F/A \tag{12.35}$$

and

$$\delta V_2 = V_2(0) - A\delta y$$

Equations (12.31), (12.32), (12.34), (12.35) combine to give for $n = 1$

$$\dddot{y} + \frac{W(0)\,(1 + \zeta)}{M_2(0)}\ddot{y} + \frac{P_2(0)A^2}{MV_2(0)}\dot{y} = \frac{W(0)\,(1 + \zeta)F}{MM_2(0)} \tag{12.36}$$

This equation is second order in y and has the same natural frequency and damping ratio as equation (12.22). Experimental results of pneumatic speed control are presented in [90].

12.3. Speed control of hydraulic motors

A variable displacement pump can be used for varying the speed of a hydraulic cylinder or motor. The hydrostatic transmission, Fig. 12.9, takes advantage of this and provides an infinitely variable transmission ratio. The high pressure relief valve and is associated non-return valves in Fig. 12.9 protect the system from overloads. Motor and pump leakage is made up by the small capacity pump. Such a system has a high overall efficiency since the pump can be adjusted to give only the required flow rate. There is no wastage of power like that arising in systems using fixed displacement pumps where that portion of pump output

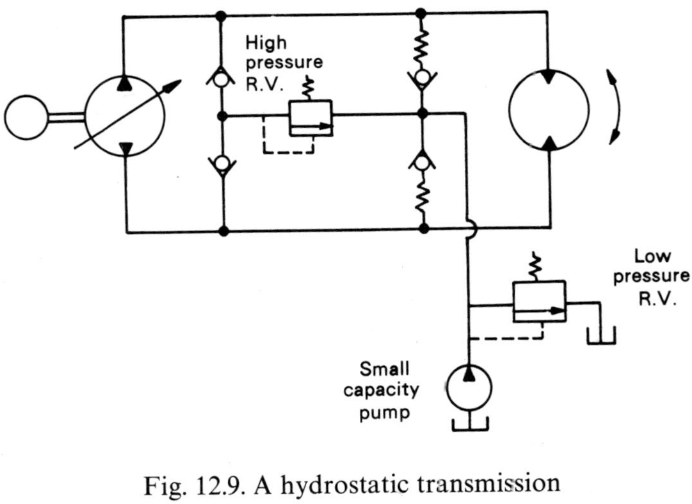

Fig. 12.9. A hydrostatic transmission

not needed by the motor must be dumped via a relief valve. Hydrostatic transmissions have other important advantages such as,

1. Speed is nearly independent of load.
2. Torque can be maintained constant as speed varies.
3. High torque to inertia and torque to weight ratios.

In general, however, such systems are expensive and their use is generally restricted to high power applications. For the smaller power range (0–20 hp) fixed displacement pumps are commonly used.

Since the volumetric flow rate through a hydraulic motor determines its speed, then some means for diverting part of the pump's output are necessary if the speed of a motor driven by a fixed displacement pump is to be varied. The circuit in Fig. 12.10(*a*) uses meter in and is so designed that excess flow is dumped to tank via a relief valve. This of course is inefficient. Figure 12.10(*b*)

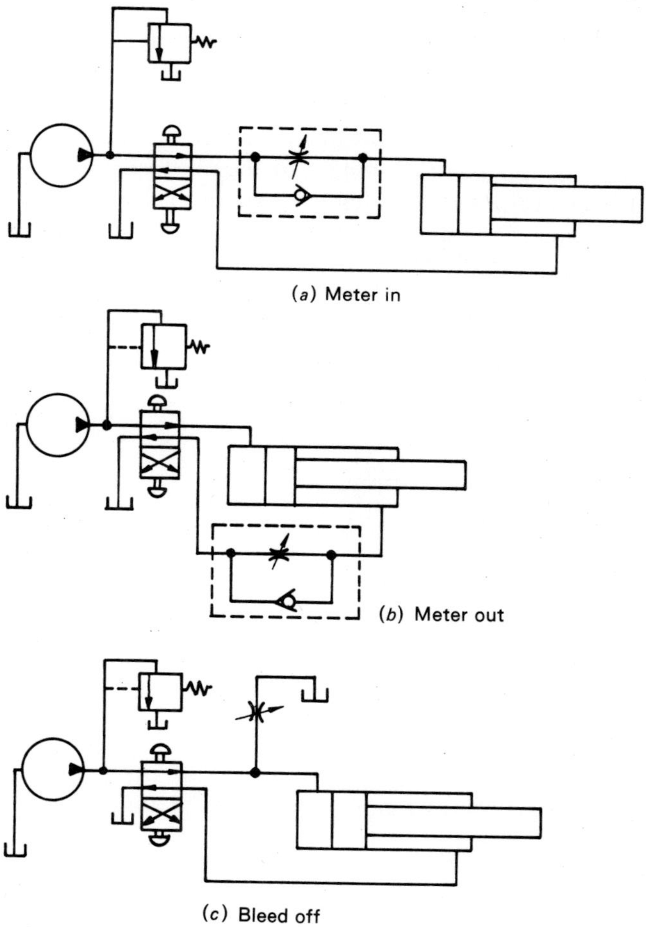

Fig. 12.10. Means of controlling speed in a hydraulic system (*a*) meter-in (*b*) meter-out (*c*) bleed off

shows a hydraulic meter-out circuit and here again the relief valve passes excess flow back to tank. In hydraulic circuits, in particular, meter-out control can lead to high pressure intensification at the rod end of a single rod double acting cylinder. Cylinders can have blank end to rod end area ratios as high as 2 to 1 and even under a zero load condition this would cause rod end pressures to rise to twice supply pressure. This could result in damaged seals or piping and is a strong argument in favour of meter-in control in some applications. Other than this, meter-out control is as popular in hydraulic circuits as in pneumatic circuits. Its ability to control overhauling loads makes it popular for reaming, drilling and milling applications when the tool tends to pull through the work. The delays that arise in pneumatic metering circuits due to compressibility are practically non-existent in hydraulic circuits in which the high oil bulk modulus allows cylinder pressures to adjust quickly to the value required by the load.

Where the inefficiency of meter-in or meter-out circuits cannot be tolerated, the bleed off method may be used, Fig. 12.10(*c*). In this case the flow rate to the cylinder is the difference between the pump flow rate and the bleed off or bypass flow rate. The big advantage is that the excess flow is forced to tank at load pressure and not at the high pressure set by the relief valve. Variation of pump leakage with pressure, results in cylinder speed varying slightly with load.

Earlier discussion of pneumatic systems showed how choked flow in a restrictor could be used to give a load independent speed control. Unfortunately there is no such phenomenon in the case of liquid flow. Liquid flow rate through an orifice is determined by the orifice area and the orifice drop. The system load dictates the pressure drop and hence the flow rate and speed are load dependent in the circuits shown in Fig. 12.10. The pressure compensated flow control valve has been designed to overcome this difficulty. In Fig. 12.11 the setting of the

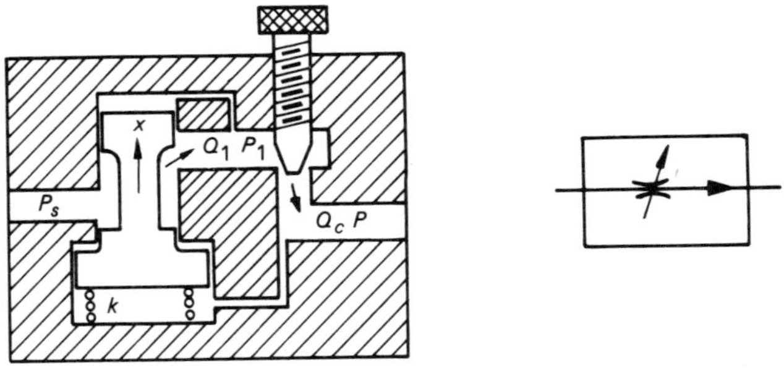

Fig. 12.11. Schematic of a pressure compensated hydraulic flow control valve

needle valve fixes the flow rate. If the outlet pressure varies the spool moves in a direction to compensate for the change in outlet pressure. For example, if pressure P falls, the spool moves downwards restricting the flow Q_1 and reducing the pressure P_1. By this means the pressure drop across the needle valve is maintained constant irrespective of load changes.

A static analysis [89] of this valve brings to light the important parameters.

$$\text{Flow rate } Q_1 = C_{q1} f(x) \sqrt{\left\{\frac{2}{\rho}(P_s - P_1)\right\}} \tag{12.37}$$

where the orifice area $a_1 = f(x)$ and x is measured from the closed position. The flow rate Q_c through the control orifice is

$$Q_c = Q_1 = C_{qc}\, a_c \sqrt{\left\{\frac{2}{\rho}(P_1 - P)\right\}} \tag{12.38}$$

where a_c is the area of the control orifice. Consideration of the forces acting on the spool gives .

$$\text{Pressure force} + \text{Bernoulli force} = \text{Spring force}$$

$$A(P_1 - P) + \frac{\rho Q_1^{\,2}}{f(x)} \cos 69° = k(l_o - x) \tag{12.39}$$

where l_o is the initial spring compression when x is zero.
Equation (12.38) gives

$$P_1 - P = \rho Q^2/(2C_{qc}^{\;2} a_c^{\;2}) \tag{12.40}$$

Substitution of equation (12.40) in equation (12.39) gives

$$Q^2 = \frac{2C_q^{\;2} a_c^{\;2} k(l_o - x)}{[\rho A + 2C_{qc}^{\;2} a_c^{\;2} \rho \cos 69°/f(x)]} \tag{12.41}$$

The dependency of Q on x can be eliminated if l_o is made much greater than the spool movement x and if $(C_{qc}a_c)^2$ is much less than $f(x)$, the spool orifice area. Equation (12.41) then reduces to

$$Q^2 = 2(C_{qc}a_c)^2 k l_o/\rho A \tag{12.42}$$

and the flow rate is independent of load. Chapter 2 showed that the flow coefficient for an orifice varies with flow number. If a flow control valve is operating at low flow numbers, where C_q varies rapidly with flow number ($\lambda < 300$), i.e. laminar flow, then some means must be provided for offsetting temperature effects C_q increases with temperature if the flow is laminar. Temperature compensation can be achieved by making the stem of the adjusting screw of a material with a high coefficient of thermal expansion. As temperature increases the increasing C_q will be offset by the reduction in orifice area caused by the spool expansion.

The dynamic performance of pressure compensated flow control valves is described in [89]. A typical commercial valve is there shown to have an oscillatory flow response to a step input in load pressure P. For example, a pressure step of 22 kg cm^{-2} gave rise to a damped oscillating flow rate Q, of initial amplitude 30 cm^3 s^{-1} and natural frequency 25 Hz. The flow settled out after 0·12s.

12.4. Dynamic analysis of hydraulic meter-out control

For simplicity's sake it is assumed that the flow control valve's response is much faster than that of the rest of the system so that its dynamics can be neglected. The continuity equation then gives, Fig. 12.12

$$Q = Av(0) = Av - \frac{V_2}{\mathrm{N}}\frac{\mathrm{d}P_2}{\mathrm{d}t} \tag{12.43}$$

where $v(0)$ is the demanded speed of the cylinder rod and Q is the constant flow rate through the flow control valve. The system is loaded with a constant force F and inertia M. Hence

$$A(P_s - P_2) = AP_l = M\frac{\mathrm{d}v}{\mathrm{d}t} + F \tag{12.44}$$

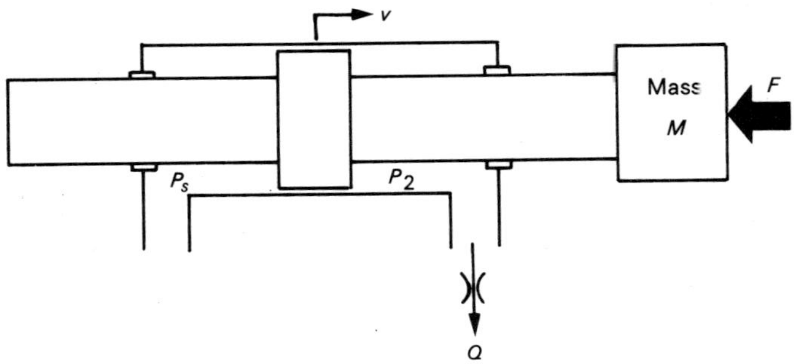

Fig. 12.12. Schematic of hydraulic meter-out control

The pressure in the left-hand side of the cylinder is assured to remain constant at P_s.
Equations (12.43) and (12.44) combine to give

$$\frac{\mathrm{d}^2P_l}{\mathrm{d}t^2} + \omega_n{}^2P_l = \omega_n{}^2\frac{F}{A} \tag{12.45}$$

$$P_l = F/A + [P_l(0) - F/A] \cos \omega_n t \tag{12.46}$$

This solution represents an undamped oscillation of frequency $\omega_n = \sqrt{(NA^2/MV_2)}$. System damping is zero because the flow control valve, by maintaining a constant flow rate, eliminates valve damping. In practice, seal friction and load friction will attenuate the oscillation. Chapter 6 discussed the stability of the hydraulic servo and this was seen to depend on valve damping—the decrease of flow rate with increase of load pressure. The meter-in control of a pneumatic cylinder discussed earlier in this chapter also depended on valve damping for its stability.

Equation (12.46) can also be written as

$$P_2{}^* = P_s{}^* - F/A - \{P_s{}^* - P_2{}^*(0) - F/A\} \cos \omega_n t \tag{12.47}$$

Where as before the asterisk denotes that the pressure is measured relative to tank pressure P_e. Equations (12.47) and (12.43) give

$$\Delta v = v - v(0) = \frac{V_2\omega_n}{NA}\{P_s{}^* - P_2{}^*(0) - F/A\}\sin\omega_n t \tag{12.48}$$

The response of a cylinder controlled by meter-out is shown in Fig. 12.13. Here the initial load is zero before a step load $F = 0{\cdot}4\, P_s^*$ is applied to the cylinder rod.

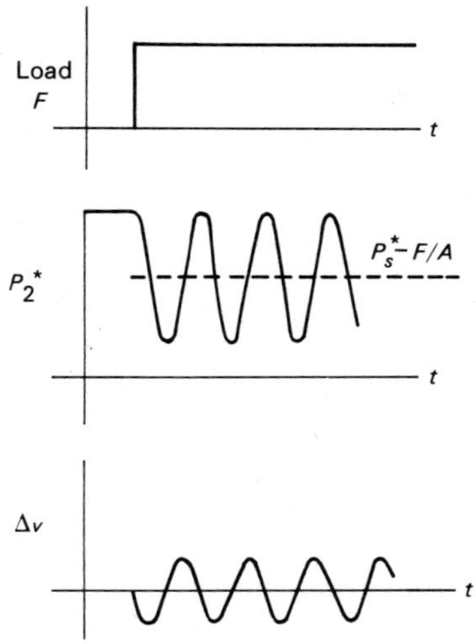

Fig. 12.13. Response of a hydraulic cylinder with meter-out speed control

On the whole, speed control by hydraulics is much superior to that by pneumatics. Time delays are minimal in hydraulic systems where compressibility is low and the pressure compensated flow control valve makes possible load independent speed control. At very low demanded speeds stick slip oscillation [91] can also cause problems in hydraulic systems.

12.5. Hydropneumatics

Although hydraulic systems are much better than pneumatic for fine feed control, there are two disadvantages that may preclude their use in some cases. These are (1) the expense of hydraulics and (2) the fact that higher speeds can be achieved with pneumatic cylinders. Because of this, hydropneumatic or 'air and oil' systems [92] have been developed to take advantage of some of the good points of both hydraulic and pneumatic systems. They basically use air for power and oil for control. Figure 12.14 shows oil and air cylinders in tandem. With the pneumatic four-way directional control valve as shown, the cylinder rod extends. Oil is forced through the meter-out flow control valve which controls the speed of the whole unit. The oil reservoir makes up leakage and is also necessary in this case to make up the difference in volumes of the blank end and rod end of the single rod double acting cylinder. This circuit is also controlled during retraction of the rod. The two position two-way valve may be operated manually or by cam to bypass the flow control valve and allow fast rod extension over parts of the stroke. Figure 12.15 shows another method of 'skip' feed in which the pneumatic cylinder rod can travel quickly over a distance determined by the setting of the two adjusting nuts on the rod of the hydraulic cylinder. This might be used on a drilling machine where the drill is brought quickly to the workpiece before being slowed down for the actual drilling operation. A pneumatic cylinder can provide this fast movement more quickly and efficiently than a hydraulic cylinder.

The tandem cylinder of Fig. 12.14 tends to be long and bulky. Use of an air hydraulic cylinder, which is basically a cylinder with air on one side of the piston and oil on the other, provides a more compact unit by means of which feed rates down to $\frac{1}{8}$ in min- can be achieved.

Another arrangement for the hydropneumatic control of a cylinder is shown in Fig. 12.16. Here only one cylinder—a hydraulic cylinder—is necessary. Each side of the cylinder has its own air pressurised oil reservoir. When the control valve is actuated air is supplied, to the left-hand reservoir, pressurising the oil and causing the rod to extend. The exhausting oil forces its way through the meter-out control into the right-hand reservoir which is at exhaust pressure. This system is not quite so bulky as the tandem cylinder and has the additional advantage that the designer has some freedom in the positioning of the reservoirs in relation to the cylinder.

In systems using air and oil, it is essential that the two fluids be kept apart at all times. Air oil reservoirs should use bags or pistons, as in hydraulic accumulators, to separate the air and oil. If this precaution is not taken the oil may become excessively aerated and may froth, resulting in a very 'spongy' system.

12.6. Synchronisation

Exact synchronisation of the movements of two or more cylinders is a difficult problem. It is required, in practice, when distributed loads must be applied at a uniform rate as, for example, in long power shears, presses and aircraft control surfaces. At first glance, one might think that two cylinders could be synchronised simply by supplying them from a common line. Unfortunately no two cylinders are alike and differences in seal friction, bore size and leakage all tend to result in lack of synchronisation. If two nominally identical cylinders are driven from the same source, the cylinder rod with the least friction will move first and, in some cases, may even complete its stroke before the other cylinder rod moves. If both rods are moving, the cylinder with the greater leakage will have the slower rod speed and so too will the one with the slightly larger bore.

Flow control valves provide the least expensive method of achieving approximate synchronisation. If the cylinders, for example, use meter-out controls, then it is possible to adjust each control until the best possible synchronisation is attained. Extremely slow feeds are difficult to achieve by this method since dirt and temperature changes can lead to variations of speed in this region.

A more direct method is to couple the cylinders mechanically. If a rack is

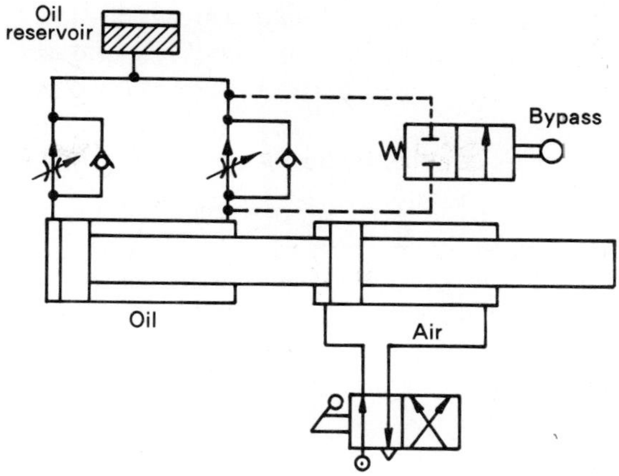

Fig. 12.14. A hydropneumatic system

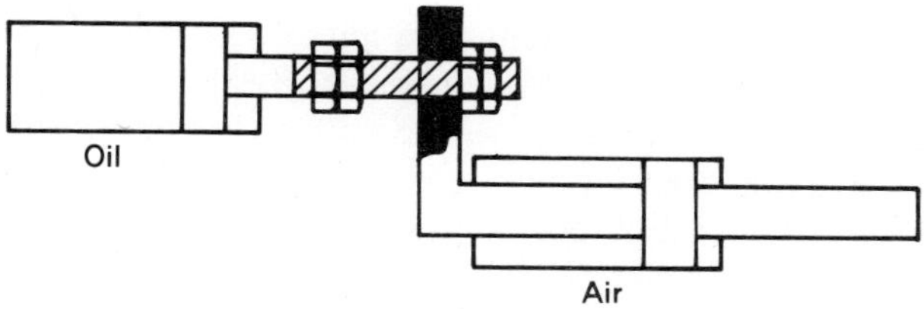

Fig. 12.15. Mechanical arrangement providing skip feed

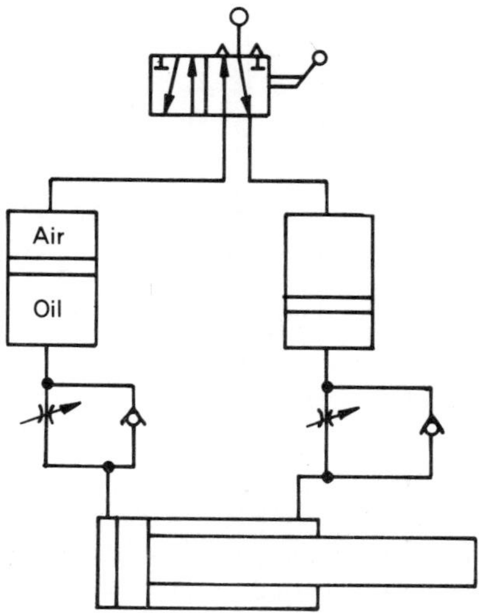

Fig. 12.16. Hydropneumatic system using pressurised reservoirs

attached to each piston rod and if the two gear wheels that roll on these racks are connected by a common shaft, very good synchronisation is possible in both directions. This system tends, however, to be bulky especially when strokes are long.

Another method uses coupled hydraulic motors as a flow divider, Fig. 12.17.

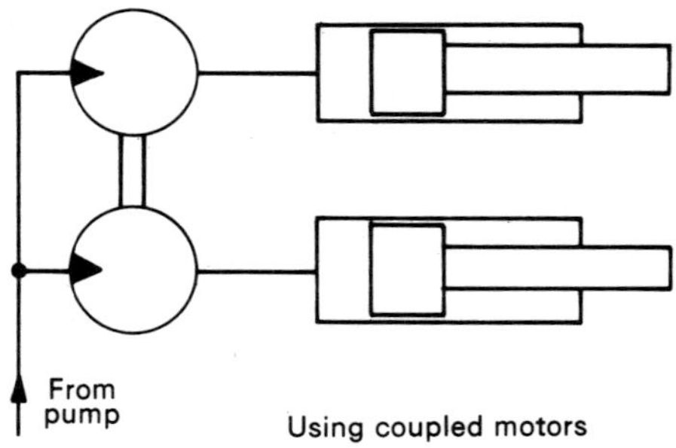

Fig. 12.17. Synchronisation of hydraulic cylinder speed using coupled motors

Two identical motors are driven by a pump and the flow divides equally between them before passing to the cylinders. Leakage in the cylinders should be kept to a minimum in this circuit since it affects synchronisation. A big advantage of this method over the previous one is that the cylinders need not be parallel; they can have any relative positions. Unfortunately hydraulic motors are expensive for the higher power ranges.

When two similar hydraulic cylinders are moving at the same speed they both displace the same flow rate (neglecting leakage). Hence, if the cylinders are connected in series, Fig. 12.18, their movements will be synchronised. Because

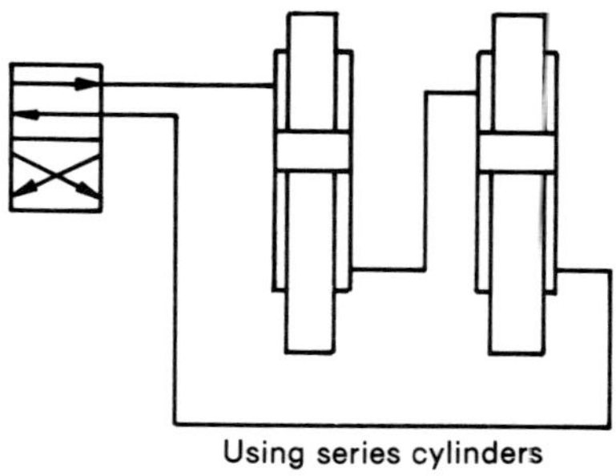

Fig. 12.18. Synchronisation of cylinder speeds using identical cylinders in series

of the need for similar displaced flow rates this method can only be used with through rod double acting cylinders. Leakage is again a critical factor but, in addition, it is essential that all air be eliminated from the cylinders and the connecting lines. In such a system, any small error in synchronisation could be corrected at the end of the stroke, Fig. 12.19. Assume that cylinder *B* has more leakage than *A* so that as the rods move downwards piston *A* bottoms first. Rod B would stop if the simple circuit of Fig. 12.18 were used. In Fig. 12.19, rod A at the bottom of its stroke operates limit switch LA. This energises the solenoid on the synchronising valve C which, when actuated, connects the supply directly to the top of cylinder B and makes it complete its stroke. When limit switch LB is operated the valve C is reset and the rods can be raised.

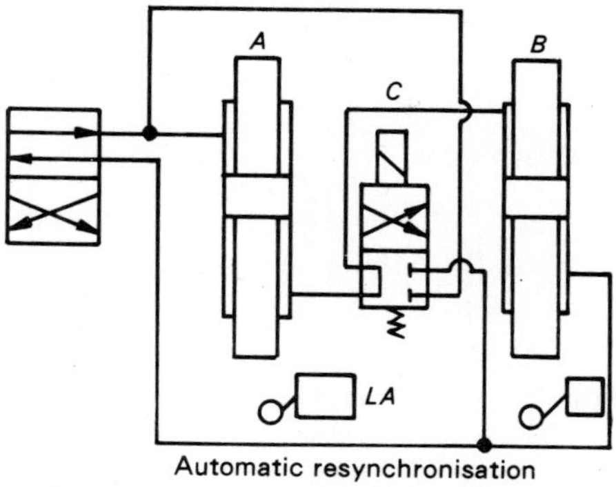

Fig. 12.19. Circuit for automatic resynchronisation of two cylinders

As a last example, consider the flow dividing valve shown in Fig. 12.20. This consists of a hollow spool which is free to slide in a sleeve. Orifices in the ends of the spool connect the supply to the cylinders A and B. The spool is in equilibrium only if the flow rates to A and B are equal. If, for example, more flow goes

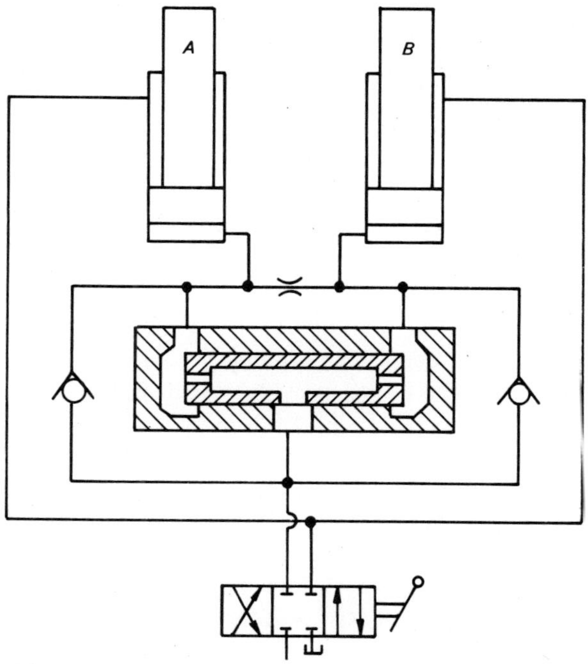

Fig. 12.20. Use of a flow dividing valve for synchronisation

to *A* than to *B*, the pressure on the left-hand end of the spool will be less than that on the right. The resulting force moves the spool to the left, restricting the flow to *A* and attempting to achieve equal flow rates. The restrictor in the line between the cylinders is necessary to ensure that if one cylinder rod bottoms early then the other will be able to complete its stroke also Of the examples discussed, this last one is the only closed loop control. The spool is actuated by errors in flow rate and is thus an error-actuated controller.

12.7. Speed control servos

Except for the last example, all the above techniques of speed control and synchronisation are essentially open loop systems. There is no measurement of the actual speed. Such control is not good enough for applications like gun turret and radar drives. High precision speed control requires measurement so that actual speed and demanded speed can be compared. Any errors are used to drive the system in the sense to reduce the errors. For example, accurate speed control can be achieved by using a servo valve, Fig. 12.21, to modulate

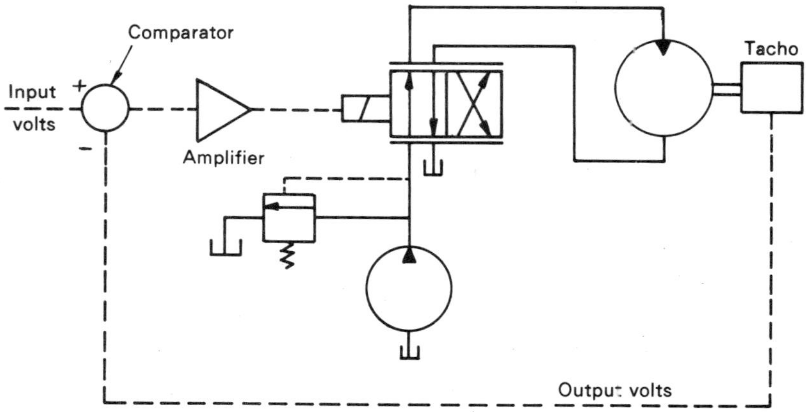

Fig. 12.21. Speed control servo

the flow rate from a fixed displacement pump to a motor or cylinder. A tacho-generator on the output shaft measures the speed. If, for example, this is less than the desired value, the positive signal from the comparator will cause more flow to be passed to the motor, thereby increasing its speed. Since this is basically a meter-in speed control, it suffers from the disadvantage of inefficiency caused by excess flow rate being dumped to tank via the relief valve. Its response is fast, since the servovalve and motor can be mounted close together providing a compact drive with a minimum of fluid inertia and compressibility lag. The student should verify that this servo is type O and will exhibit steady state errors to step inputs.

Motor speed can also be controlled by means of a variable displacement pump with servo driven swash plate, Fig. 12.22. In this case, errors in the speed are amplified to drive a servo valve which changes the angle of the swash plate on the pump. This is sometimes known as the servo pump system. It is an efficient system in that the pump only supplies the flow rate required at any instant and as such is popular in high power installations such as winding control.

Because of its bulk and the amount of fluid contained, its response is slower than that of the constant flow system of Fig. 12.21.

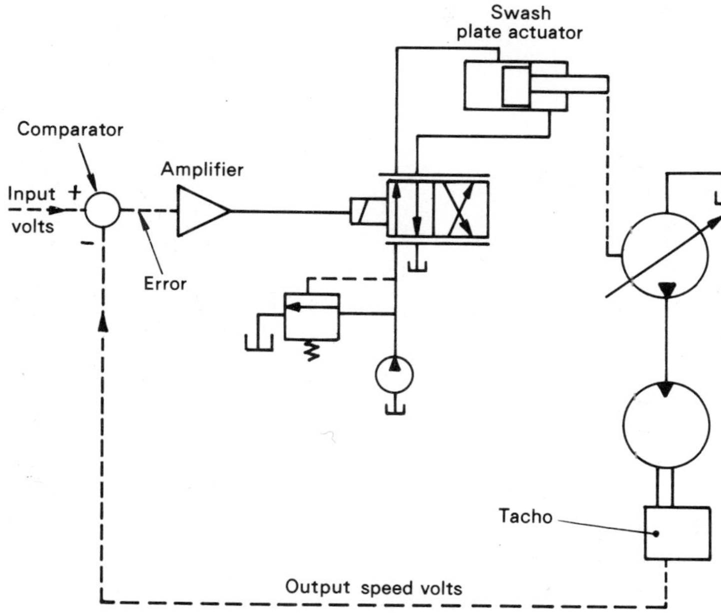

Fig. 12.22. Speed control servo using a variable displacement pump

Chapter 13

Fluidics

13.1. Introduction

The word fluidics has been used [93] to describe engineering systems in which fluid motion or displacement contains significant information as well as power. The conventional fluid power control valves fall into this category, but over the last ten to fifteen years tremendous progress has been made in the design and application of fluidic devices that contain no moving mechanical parts. Many attractive advantages stem from the fact that they have no moving parts—they are reliable, have long life and have high environmental tolerance.

Although interest in the subject is great at the moment, the basic principles of these devices were known many years ago. For example, Coanda's work on reattached flow was carried out around 1932 and was published in 1938. The most significant advances however did not come until 1959 when the results of research at the Diamond Ordnance Fuze Laboratories and at the Massachusetts' Institute of Technology came to light. Today's enthusiasm for the subject is reflected in the already enormous literature on fluidics. Several well attended international conferences have been organised by the British Hydromechanics Research Association and by the International Federation of Automatic Control.

Fluidics now offers a viable alternative to the other 'ics'—electronics, hydraulics and pneumatics. Although it has not quite lived up to its promise of fifteen years ago, it has nevertheless in many cases provided a solution where other methods failed.

13.2. Devices without moving parts

13.2.1 The proportional, analogue or beam-deflection amplifier

If two fluid streams impinge, an exchange of momentum occurs and a resultant flow is developed with a direction and velocity such as to maintain the momentum flux of the system. This is the principle of operation of the proportional amplifier, Fig. 13.1, in which control jets are used to deflect a power jet. The flow is usually contained between parallel top and bottom plates. The figure illustrates how the power jet is captured by receivers which lead directly to the

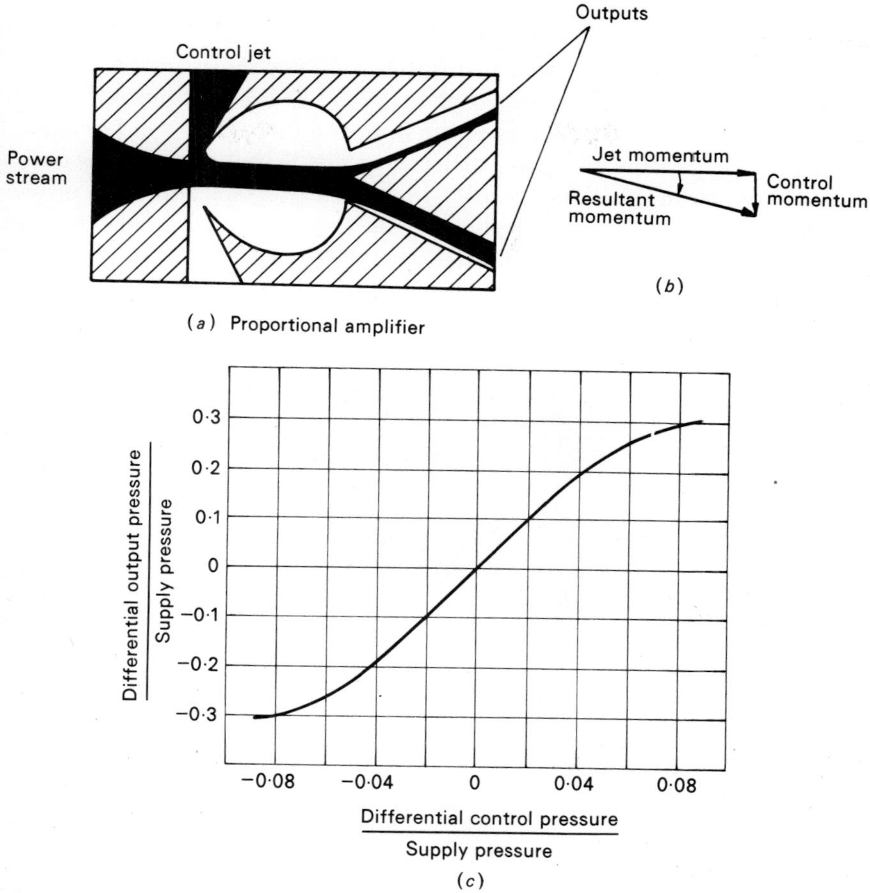

Fig. 13.1. (*a*) a proportional amplifier (*b*) addition of momenta gives power jet deflection (*c*) output pressure against control pressure of a typical element

load. If the jet is deflected from the central or neutral position, in which conditions in both receivers are the same, its proportioning between the receivers is changed and a differential in the receiver pressures and flows will be established. The device acts as an amplifier since a low energy control jet can be used to vary the amount of a high energy power stream entering a receiver. The performance can be described by the pressure gain G_p, which is the difference in the output pressures divided by the difference in the control pressures when the outputs are blocked. Typical pressure gains lie in the range 2–10. Higher gains can be achieved by staging or cascading several proportional amplifiers. A typical performance curve is shown in Fig. 13.1(*c*). Note that the maximum recovered differential pressure is only about one-third of the supply pressure, due to losses within the element. Flow gain G_q and power gain G_h are also used as performance figures. Flow gain is the difference in output flows divided by the difference in control flows when there is no load on the device. Its value is typically 5–15. Power gain is the difference in output powers (defined as the product of output flow and output pressure) divided by the difference in control powers.

Proportional amplifiers can be sensitive to load changes. For example, if the load resistance at outlet *A* increases, the flow rate through *A* will be reduced. In order to stop the flow then being deflected to side *B*, thus changing the output of the device and in some cases leading to instability, it is necessary to provide vents through which flow not required by the load can be diverted. The vents also supply the fluid entrained by the jet. Improved stability, at the cost of decreased gain, can be achieved by using a centre dump as shown in Fig. 13.2.

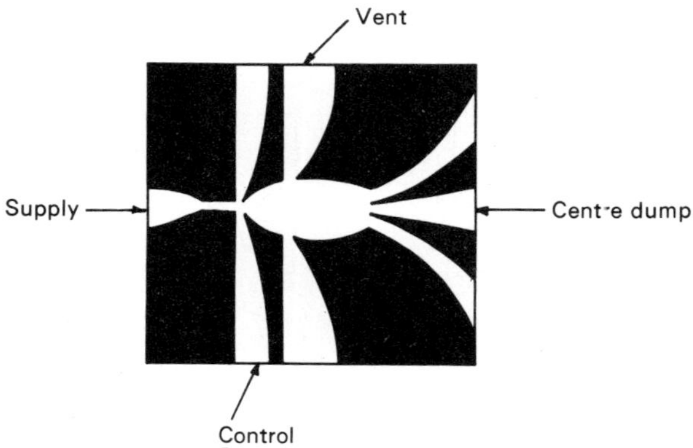

Fig. 13.2. A proportional amplifier with a centre dump

Because of their proportional characteristics, the beam deflection amplifier has found application in closed loop control systems, particularly in the aerospace industry. As with all analogue devices, noise can cause problems and great care must be taken to ensure that the signal to noise ratio is kept at an acceptable value, particularly in cascaded amplifiers. Research has shown that noise is minimised by reducing the aspect ratio, i.e. by reducing the ratio of the distance between top and bottom plates to the supply orifice width.

Aspect ratio is normally about 2 and nozzle widths lie in the range 0·1–1 mm. Operating pressure can be anything from 10 to 100 kN m^{-2}.

It is important to note that the principle of momentum interaction applies whether the flow is laminar or tubulent. Commercially available elements operate with turbulent flow but research has shown that devices using deflection of laminar jets can be more efficient [155].

13.2.2 The bistable amplifier—the Coanda effect

Whilst the proportional amplifier can operate with either laminar or turbulent flow, the bistable amplifier requires turbulent flow for fast switching times.

The Coanda effect can be explained with the aid of Fig. 13.3 which shows a

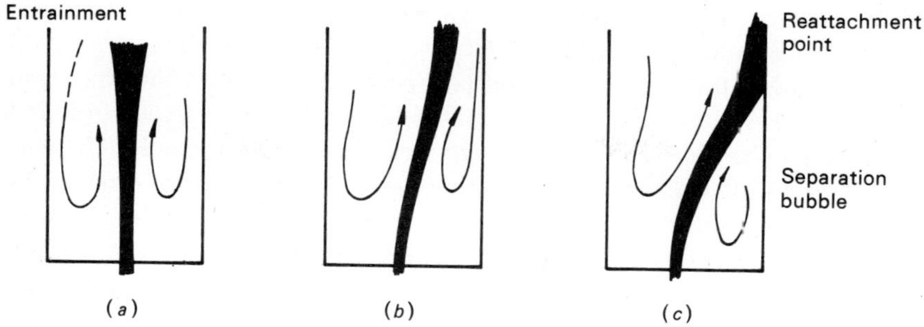

Fig. 13.3. The Coanda effect

turbulent jet of air discharging through a slot into a chamber with top and bottom plates. On entering the chamber the jet picks up or entrains air and this removal of air from the chamber causes the pressure on both sides of the jet to fall. This is an unstable situation. If the jet is disturbed and moves to the right, Fig. 13.3(*b*), the air entrained on the right of the jet must pass through a smaller area and hence the pressure falls on the right side of the jet. There is now a resultant force tending to push the jet further to the right and ultimately the jet will reattach (at the reattachment point) to the right-hand side wall, forming a separation bubble, Fig. 13.3(*c*). If there is a lack of symmetry in the device, e.g. if one wall of the chamber is nearer to the orifice than the other, then the jet will spontaneously reattach to the nearer wall. This phenomenon of reattachment is put to good use in the bistable amplifier, Fig. 13.4. This device is symmetrical

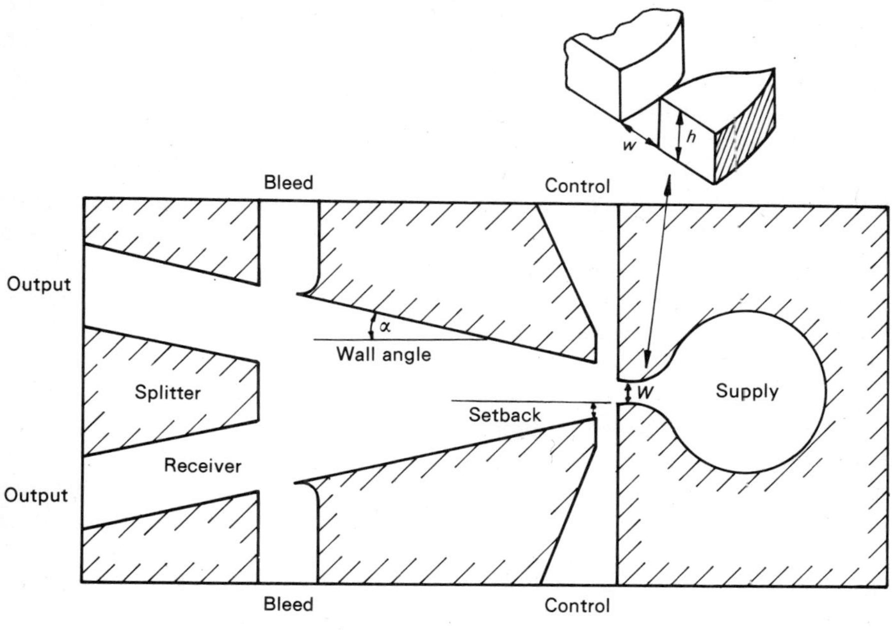

Fig. 13.4. A bistable amplifier with vents

and the controls are used to determine which wall the main jet will attach to. There are three main forms of control.

1. When the boundary walls are set far apart, a control flow into the separation bubble causes the reattachment point of the jet to move further and further downstream. The bubble becomes longer until eventually it reaches the end of the boundary wall, where it bursts and the jet detaches, switching to the opposite wall. Here once again a stable condition is achieved and the jet will remain attached even in the absence of any control flow. The maintenance of this stable condition after the removal of the control signal can be regarded as a memory function of the unit.

2. In the case of the boundary walls being set close together, say from zero set back, Fig. 13.4, from the orifice to a set back of $\frac{1}{4}$ power nozzle width, the injected control flow, while still moving the reattachment point downstream as in 1., also increases the radius of curvature of the jet so that the extreme edge of the jet may be brought into contact with the opposite wall before the reattachment point breaks away from the active wall. As soon as the extreme edge touches the far wall the jet reattaches there, another low pressure region is formed on the inactive wall and the flow switches across. This mode of operation is generally used in high gain, quick response units.

3. During operation this device tends to entrain or suck fluid in through the control ports. While the jet is attached to one boundary wall, a constriction of the opposite control port reduces entrainment through that control and the resulting suction draws the main jet across to the boundary on the constricted control side.

The effects [94] of load and of splitter position are examined more closely in Fig. 13.5. With the splitter very close to the nozzle the power stream divides and flows out of both outputs. A control pulse will drive it to one wall or the other but once this is removed it reverts to its original position of exit through both outlets. If the load on one output is increased, i.e. if an output is blocked, the jet will exit through the other side. When the splitter is greater than about five nozzle widths from the nozzle, the stream flows spontaneously to one outlet since the Coanda effect begins to show itself. When the splitter is more than ten nozzle widths away the device is very stable. A blocked load diverts the jet from the output it prefers to the opposite one, but when the block is removed it returns to its preferred outlet. When an output is blocked it is not always desirable that the stream issue from the other output and, for these cases, vents are used to give the fluid an escape path, Fig. 13.4.

Aspect ratio (h/w) is also an important parameter. If this is reduced too much, reattachment may not occur. At small aspect ratios the effects of the boundary layers on the top and bottom plates cannot be ignored. These provide leakage paths across the top and bottom of the jet into the reattachment bubble resulting in a lengthening of the bubble. As the aspect ratio is progressively decreased, a stage will be reached at which the bubble cannot be maintained and the device ceases to function properly. This critical aspect ratio is related to the Reynolds number. Parker [95] has argued that a Reynolds number of 4000 is the lowest

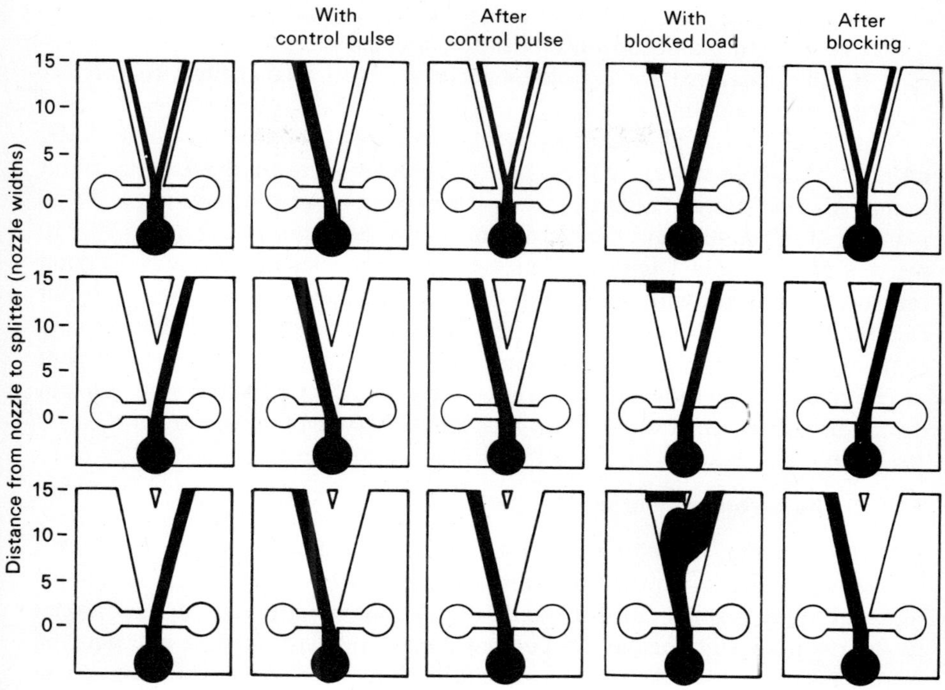

Fig. 13.5. The effects of load and of splitter position on bistable amplifier performance (after Kirshner [94])

practical limit for the operation of bistable amplifiers. Reynolds number here is defined as

$$R = \frac{Q}{wh}\frac{D_h}{v} = \frac{2Q}{v}\frac{1}{w(1+n)}$$

Q is the measured flow rate, w is width of the power nozzle, h is the depth of the power nozzle, D_h its hydraulic diameter and n is the aspect ratio h/w.

Since these elements are digital, the definition of gain used earlier for proportional amplifiers is not applicable. For bistable amplifiers the flow gain is defined as the ratio of the change of output flow to the change in control flow necessary *for switching to occur*. This has been referred to as an instantaneous gain since it only applies at the instant of switching, after which the control signal may be removed. Flow gain is usually in the region of 10. Pressure gain ($\simeq$ 10) and power gain ($\simeq$ 100) are defined in a similar manner as flow gain. It is important to note that gain and stability of attachment are a compromise in these devices —one can only be increased at the expense of the other.

Wall reattachment devices have switching times in the range of 0·5–2 ms, this consisting of three distinct phases

1. signal propagation time in control and output channels
2. time to disturb and to re-establish the flow pattern and
3. the time required for a fluid particle to travel from orifice to receiver (transport lag).

Commercially available elements work with supply pressures 10 kNm^{-2}–300 kNm^{-2}. Nozzle widths are in the range 0·25–0·50 mm and depth is usually about 1 mm. As with proportional elements, the maximum output pressure is 30–50 per cent of the supply and the maximum output flow can be as high as 90–110 per cent of the supply. Digital amplifiers can be connected in cascade and a good measure of the ease with which a device can be staged is given by its fan-out capability. The fan-out of an element specifies the number of downstream elements it can control. Similarly, fan-in specifies the number of elements that can be connected to the element's input. For wall attachment elements, fan-out cannot usually exceed four and fan-in usually is between two and four.

13.2.3 The turbulence amplifier [96]

In this device, Fig. 13.6(*a*), a laminar jet is directed at a collector tube in which a substantial proportion of the pressure driving the jet is recovered. If the conditions are right, a small control signal from a control jet placed at right angles to the main jet will change the laminar stream into a turbulent one and the pressure recovery in the collector tube will be reduced. A typical variation of output pressure with control pressure is shown in Fig. 13.6(*b*). Care must be taken in choosing a supply pressure, for if this is too large, Fig. 13.7, the jet will become turbulent without the use of controls. On the other hand if the supply pressure is too low the gain will be low. High pressure gains ($\simeq$ 1000) can be achieved if this amplifier is operated at a supply pressure slightly less than that which causes turbulence in the supply jet. Indeed the device can be so sensitive that a high pitched sound can cause a change in output.

The turbulence amplifier operates at very low pressure (about 1 $kN\ m^{-2}$ gauge), uses only a few cubic metres of air per hour and consumes about 0·1 watt of power per unit. Its main use is in logic circuitry, chiefly because it can drive up to ten or more secondary units and these are easy to interconnect. In addition a single amplifier may have a fan-in of 6. In commercially available devices, jet diameters are 0·75 mm and the distance between supply and output tubes ranges from 10 to 40 mm. Power gains of 40–80 and pressure gains of 1000 are possible. It is, however, a relatively slow acting device taking about 2 ms to turn off the output and about 5 ms to reset. The longer reset time is caused by the delay in the jet resuming a laminar condition after the control is removed.

The turbulence amplifier is not easy to mass produce. Difficulties can arise in producing consistently accurate long tubes and tiny burrs can produce unexpected results. Because of this, interest is now being shown in the planar turbulence amplifier. This is a two dimensional device in which the power and control lines are slots and not tubes and these can be cleaned and carefully aligned before the coverplate is put on. They can be stacked in very compact

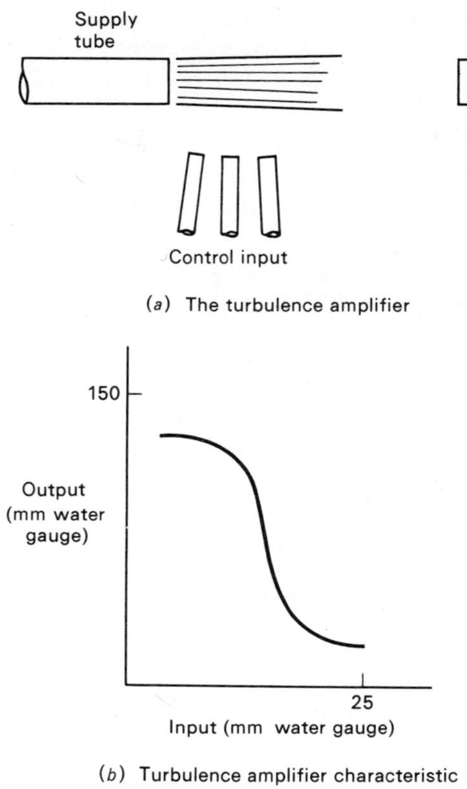

Fig. 13.6. (*a*) The turbulence amplifier (*b*) output pressure of a typical turbulence amplifier

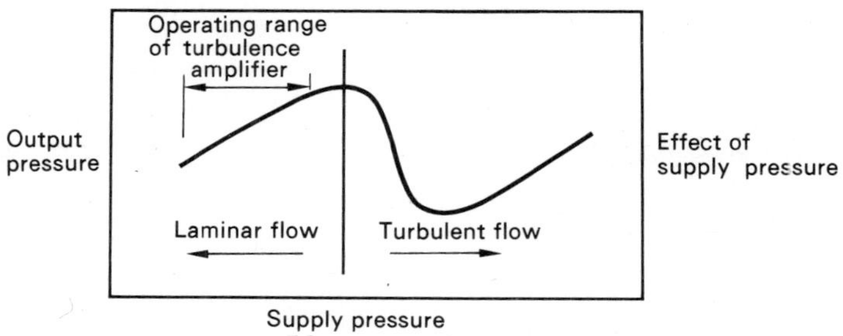

Fig. 13.7. Effect of supply pressure on recovered pressure in the turbulence amplifier

packages. The response time of the planar element is superior to that of the axisymmetric type, taking only 1·2 ms to switch on or off.

13.2.4 The vortex amplifier [97]

The vortex amplifier, Fig. 13.8, is essentially a variable restriction. The device includes a main power jet which enters a circular chamber and exhausts through

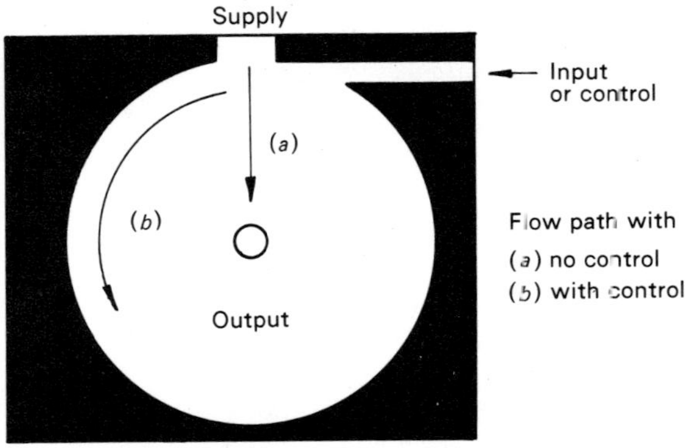

Fig. 13.8. The vortex amplifier

a hole in the centre of the chamber. (In some devices there are holes in both top and bottom plates.) A control jet is placed at right angles to the incoming power jet. With no control flow the amplifier presents simple orifice resistance, the outlet pressure being determined by the pressure losses in inlet and outlet orifices. When the control jet is present the fluid in the chamber moves in a spiral path. The centrifugal pressure drop across the vortex chamber reduces the pressure at the outlet orifice and hence the output flow rate decreases. Typical curves relating output flow to control pressure are shown in Fig. 13.9. Note that the vortex amplifier has a negative gain, an increase in control flow causing a reduction of output flow. Note also that the control pressure must exceed the supply pressure before it can cause throttling of the output flow. It is usually

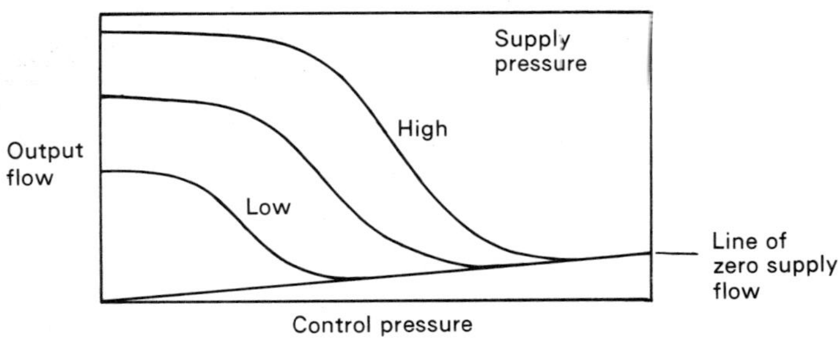

Fig. 13.9. Characteristics of a vortex amplifier

20–50 per cent higher than the supply pressure—if it is not so the control flow will not be able to force its way into the chamber. This is a disadvantage of vortex amplifiers. In Fig. 13.9, the line marked 'zero supply flow' indicates the lower limit on the output flow. At this condition the centrifugal pressure drop across the vortex is equal to the supply pressure and thus no flow is taken from the source and the outlet flow is the control flow. The ratio of the maximum to the minimum output flow rate at a given supply pressure is called the turn down ratio. This is typically 5–10 but can, by careful design and at the expense of stability, be raised as high as 25.

The fact that the output flow rate cannot be reduced to zero may be an embarrassment in some applications but this can be overcome by venting, Fig. 13.10. The receiver tube in this case is not attached directly to the body of the

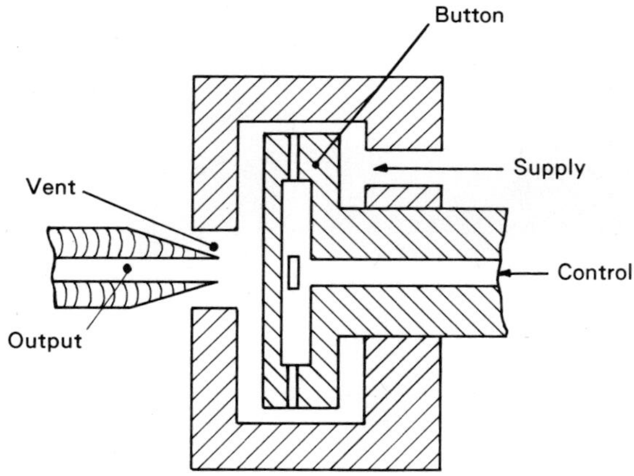

Fig. 13.10. A button vortex amplifier with an output pick off.

valve. It is separated from the valve body by a vent. As the control flow increases the amount of swirl in the outlet increases until ultimately the outgoing fluid spreads out in a hollow cone and does not enter the receiver tube. The figure also illustrates another design of vortex valve—the button vortex amplifier. In the simple vortex device of Fig. 13.8 the flow field is extremely non-uniform when there is no control flow. This is basically because there is only one supply port. As a result there is poor mixing of the supply and control flows and indeed under some circumstances two contra-rotating vortices may form. Mixing can be improved by providing several inputs and controls around the circumference. The button configuration is extremely effective in this respect. In this case, the supply enters the vortex chamber uniformly through an annular gap in which mixing takes place. The control flow may come from the button or the vortex chamber wall [156].

The vortex amplifier can be used as a high gain proportional amplifier. Flow gains of 200 are possible. Its response can be slow since it is essentially a fluid flywheel that takes some time to move from one speed to another. This, however, can be improved by raising supply pressure and reducing size. As the size is reduced, however, the effect of the boundary layers on top and bottom plates becomes important. The fluid velocity is zero at these boundaries; the tangential velocity component and hence the centrifugal force field is absent. Hence, near the boundaries, a radial flow can exist in spite of the presence of the vortex. This reduces turn-down ratio.

13.2.5 The impact modulator [98]

The impact modulator, Fig. 13.11, consists of two collinear supply tubes, referred to as the collector and the emitter. These two jets impact and form a

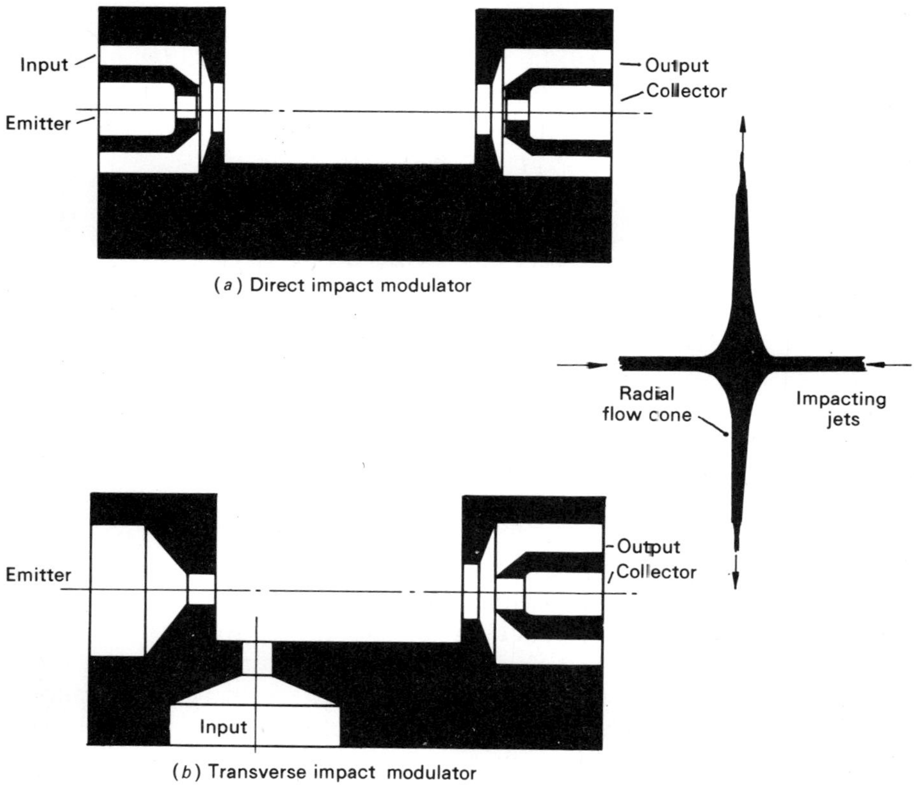

Fig. 13.11. (*a*) A direct impact modulator (*b*) a transverse impact modulator

radial flow cone which is collected at the output of the amplifier. If the momentum of either power jet is changed, the balance point will move and this in turn will cause a change in the output pressure. There are two ways of affecting the balance point.

1. The direct impact modulator uses an input or control signal applied around the emitter stream, Fig. 13.7(*a*). As the input increases the balance point moves

nearer the output and the output pressure increases. This is a positive gain proportional amplifier and pressure gains of 200:1 are possible. The flow gain is high also.

2. The transverse impact modulator, Fig. 13.11(*b*), uses a transverse jet for control purposes. As control flow increases, the balance position moves further from the output and the output pressure falls. This is a negative gain proportional amplifier. The no load pressure gain of these units is of the order −20 to −100.

Figure 13.12 shows the output pressure as a function of emitter pressure. The The data were taken with the collector pressure held constant and with no control flow. These curves are analogous to pressure recovery curves since they show the maximum output pressure range that can be achieved for a particular supply or emitter pressure.

The transverse impact modulator has a relatively fast response. Its frequency response shows no reduction in gain up until about 100 Hz.

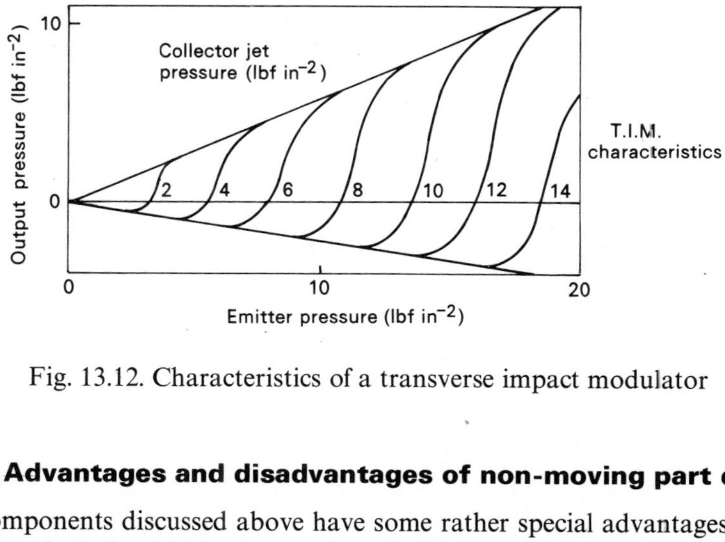

Fig. 13.12. Characteristics of a transverse impact modulator

13.3. Advantages and disadvantages of non-moving part devices

The components discussed above have some rather special advantages.

1. They have long life since there are no moving parts. For example it is reported [99] that a fluidic multivibrator has shown no deterioration in performance after $3{\cdot}5 \times 10^9$ cycles of operation on standard shop air. This also results in improved reliability and makes the use of these devices attractive in redundant aircraft control systems where reliability is at a premium.

2. They can tolerate extremes of temperature. Vortex amplifiers [100] have been operated at 1400°F and in liquid nitrogen.

3. They are insensitive to shock and vibration. Tests [101] run on digital devices indicate satisfactory operation at accelerations up to 500 ms^{-2}

4. They are unaffected by radiation and could be operated in the heart of a nuclear reactor. This is also important in space applications where radiation tolerance is highly desirable.

5. They are inexpensive when mass produced. Chemical machining, moulding or forming can be used to produce planar devices cheaply.

6. They are relatively small and can be joined together to form a compact package. Power nozzle widths are typically as small as 0·5–1·00 mm.

A big disadvantage is a relatively slow speed of response when compared with electronics. They are in the 10^{-3}–10^{-4} s, range whereas electronic circuits can operate in the 10^{-6}–10^{12} s, range. This is due to the fact that the speed of propagation of pressure waves in a fluidic system is determined by the speed of sound in the fluid medium. In electronics the speed of light is the limiting factor.

The power consumed by fluid amplifiers may prohibit their use in some cases. Gray and Stern [101] have plotted, Fig. 13.13, orifice power consumption against nozzle pressure drop, both for air flow and for water flow, and this shows that at the higher pressures, pneumatic elements need a relatively large amount of power. In calculating these values ideal adiabatic compression and cooling (constant pressure) to initial temperature of 61°F was assumed. The downstream nozzle pressure was one atmosphere. For example, an orifice

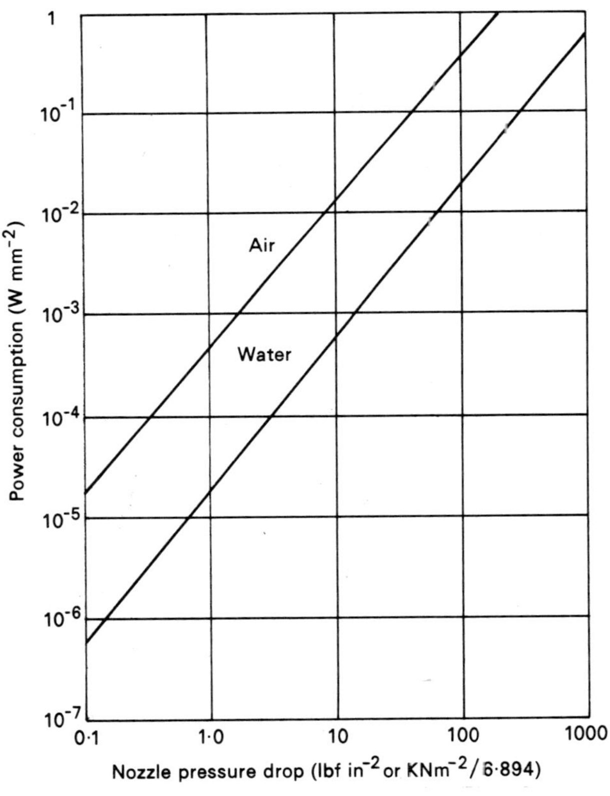

Fig. 13.13. Orifice power consumption (after Gray and Stern [101])

0·5 mm × 1 mm, operating at a pressure drop of 1 mN m^{-2}, consumes about $\frac{1}{2}$ hp. A circuit using several of these elements could be wasteful of power. However many existing devices operate at pressure drops nearer 5 kN m^{-2} and hence the power consumption can be small.

In the majority of cases air is the working fluid, although of recent years some attention has been given to hydraulic non-moving part elements. These are particularly useful in aircraft systems in which sensing, control and actuation could all be operated from hydraulic energy. The vortex valve holds most promise as an hydraulic device and has already been used in a non-moving part hydraulic servovalve. Beam deflection amplifiers can also be operated on oil but deaeration and cavitation can cause problems. Wall reattachment hydraulic elements are severely restricted in their range of operation. In order to exceed the minimum Reynolds number for successful operation, supply pressure or element size must be increased, leading to high powers and resultant high power wastage by the relatively inefficient devices. At the other extreme, if supply pressure is increased too much, cavitation may occur in the mixing zone of the jet or in the reattachment bubble, ultimately leading to failure of operation [157].

13.4. Auxiliary equipment

13.4.1 Inputs for fluidic circuits

Most means for supplying inputs to fluidic circuits rely either on the flapper-nozzle principle or the interruptible jet principle.

The flapper nozzle valve, Fig. 13.14(a), is a position to pressure transducer. If the control port of a fluidic amplifier is connected to chamber B, then movement of the flapper will cause the amplifier to respond. For instance, in thickness control, the surface of the material itself can be used as the flapper. If the nozzle traverses the material, then changes in thickness will cause pressure changes which can drive a proportional amplifier. The amplifier can then be used to start corrective action if the thickness is not right.

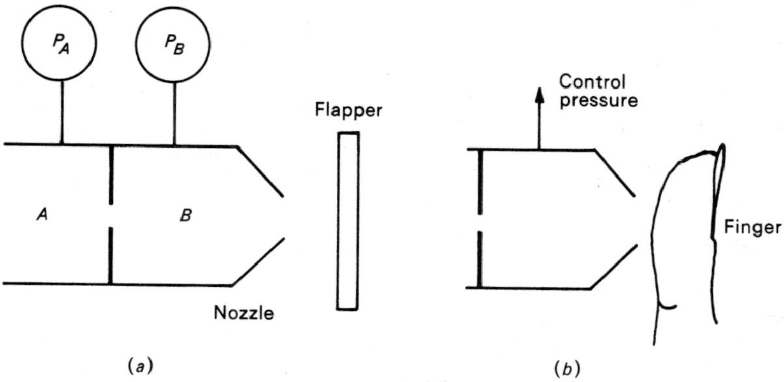

Fig. 13.14. (a) A flapper nozzle valve (b) using the finger as a flapper

The flapper nozzle has been used for many years in servo-mechanisms where it acted in a proportional manner, i.e. the output pressure proportional to the flapper displacement. However, in many cases, this property of proportionality is not required and the resulting flapper nozzles are very simple to construct. One such example is shown in Fig. 13.14(*b*). This is a manually operated touch sensor. The finger is the flapper in this case. Blockage of the nozzle with the finger sends a pressure signal to the fluidic element.

When an interruptible jet is used as an input, no physical contact with a component is necessary. It is the fluid equivalent of the photocell and is indeed sometimes referred to as the pneumatic eye. An air jet is transmitted from a transmitter orifice to a receiver orifice and some pressure is recovered in the receiver. If the jet is interrupted, the pressure in the receiver drops and hence a signal is generated. There are several possible configurations of jet and receiver. In Fig. 13.15(*a*), the jet entrains ambient air and the mixture enters the receiver. Unfortunately, if the device is operating in dusty conditions, such as flour mills or cement works, the receiver and the equipment it feeds are likely to become clogged. This can be overcome by pressurising the receiver as shown in Fig. 13.15(*b*). This is a type of direct impact modulator. Normally the impact of the transmitter jet on the receiver jet gives rise to a pressure build up in the receiver. If the transmitter jet is interrupted the receiver pressure falls. Both of these methods have been successfully applied with transmitter receiver gaps up to 30 mm. Greater separation requires higher pressure and hence greater power loss. One way round this is to use a type of transverse impact modulator as shown in Fig. 13.15(*c*). Here the distance from the control nozzle to the transmitter nozzle can be as much as 300 mm without power losses being of concern.

13.4.2 Output devices for fluidic circuits

The output pressure of most fluidic devices is too low to operate conventional pneumatic valves of cylinders. This is particularly true of turbulence amplifiers whose output pressures are only about 1 kN m^{-2}. Step up relays or boosters are necessary and these are commercially available with a capability of amplifying 80 mm water gauge to 800 kN m^{-2}. Most of these devices rely on a diaphragm first stage to provide sufficient force to move a spool or flapper second stage. Simple diaphragms can also be used to convert a pressure signal to an electric signal. Pressure on the diaphragm actuates an electric limit switch, this signal in turn being used to drive solenoid operated hydraulic or pneumatic valves. Hence means are available for providing both pneumatic—hydraulic and pneumatic—electric interfaces.

13.4.3 Interconnection of elements

Fluidic elements can be simply connected by pipe or tube. This method is flexible, allowing elements to be replaced or discarded with a minimum of inconvenience and is thus invaluable for laboratory work, research or develop-

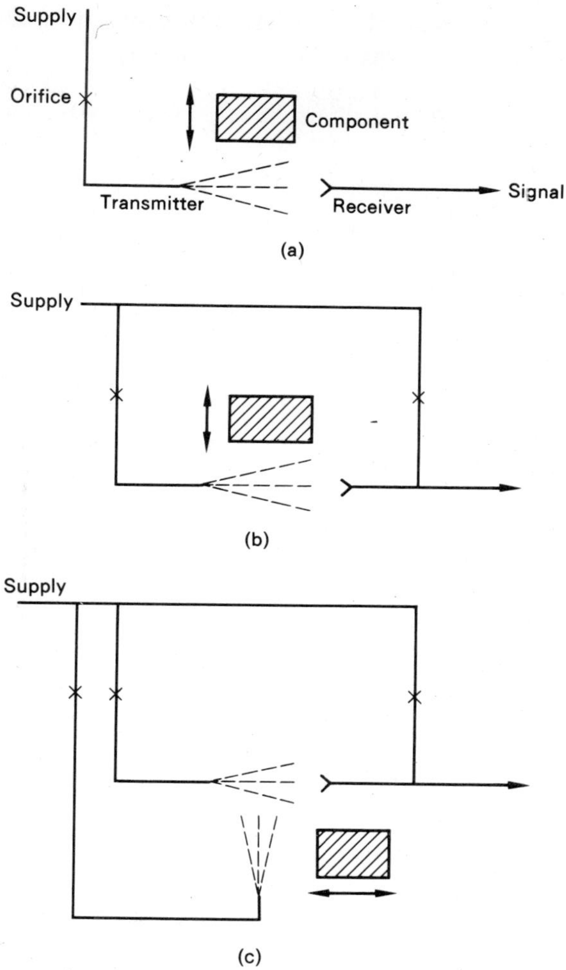

Fig. 13.15. Use of interruptible jets for component detection

ment. The reliability of such systems however, with their many fittings and interconnections can be low.

The use of the integrated circuit, in which the elements and their interconnections are contained within a sheet of material improves reliability. However, unless very large quantities of the same system are to be produced this method can be relatively expensive. At present, integrated circuits are rare but the increasing popularity of fluidics indicates that greater use of these and perhaps greater standardisation of circuits will occur in the future.

A discussion of the dynamics of component interaction is beyond the scope of this book. The reader should be warned however that, since most non-moving part fluidic elements are sensitive to both upstream and downstream conditions, unexpected things can occur when one element is loaded by another. The use of vents to reduce load sensitivity is one way of curing such problems when they arise.

Chapter 14

Boolean algebra and fluid logic

14.1. Introduction

The word fluidic was originally derived from the two words fluid logic. Although, as stated in chapter 13, the fluid carries information as well as power in a fluidic circuit, it is essential that the information be used in a logical fashion if a reliable and efficient system is the aim. In this context logic is a means of achieving the desired system behaviour by means of a set of unambiguous decisions.

Boolean algebra is the algebra of logic and although it was originally an abstract mathematical form, including a switching algebra, an algebra of sets and a propositional calculus, it can be applied with relative ease to the design of fluidic systems. George Boole, its inventor, published the important paper *Mathematical Analysis of Logic* in 1847. It was not until 1938 however, that Shannon, an electrical engineer, blew the dust off Boole's work and showed how it could be used in the design of switching circuits in digital computers and telephone exchanges.

Modern computing and logic devices use binary arithmetic. With such a system any desired function can be generated using a simple two-state device, such as a switch which can be either on or off. In addition, manipulation of binary numbers is simple. The addition tables are very easy to remember, e.g. $0 + 0 = 0$, $0 + 1 = 1$, $1 + 1 = 0$ and carry 1. As a memory refresher a few examples of conversion from binary to decimal are listed below.

1. The binary number 10101

$$\begin{aligned} 10101 &= (1 \times 2^4) + (0 \times 2^3) + (1 \times 2^2) + (0 \times 2^1) + (1 \times 2^0) \\ &= 16 + 4 + 1 = 21 \end{aligned}$$

2. The binary number 1·0101

$$\begin{aligned} 1{\cdot}0101 &= (1 \times 2^0) + (0 \times 2^{-1}) + (1 \times 2^{-2}) + (0 \times 2^{-3}) + (1 \times 2^{-4}) \\ &= 1 + 1/4 + 1/16 = 1\,5/16 \end{aligned}$$

14.2. The language of Boolean algebra

In all that follows it is assumed that a variable can only have two values, 1 or 0.

This does not mean, for example, that the output of a fluidic element is either 1 kN m^2 or 0, but rather that the output can have two states, on (1) or off (0). The variables can then be written as

$$A = 1 \text{ or } B = 0$$

The three basic operations in Boolean algebra are (1) NOT, (2) AND, (3) OR.

14.2.1 The NOT operation

NOT implies inversion and is written as

$$\bar{A} = \text{NOT } A$$

Thus if at any particular time

$$A = 1 \text{ then } \bar{A} = 0$$

A variable without a bar over it is said to be in its true form. The inverted form of the variable has a bar over it.

The symbol for the NOT operation is shown in Fig. 14.1. This can be accomplished in practice by many fluid elements and Fig. 14.1 also shows how the turbulence amplifier, the jet deflection element and the spool valve can act as invertors. The spring logic element, Fig. 14.1(*d*), uses a close coiled spring as a variable restrictor. The signal A makes the diaphragm move, deflecting the spring and allowing air to escape through the open coils. In all cases shown in the figure, the output S of the device is off when the control A is on, and vice versa. This is written as

$$S = \bar{A}$$

14.2.2 The AND operation

Before discussing the AND operation it is necessary to introduce the logic gate. This is a device having several inputs and designed so that there is an output when and only when a certain definite set of input conditions are met. The AND gate only has an output when all of its inputs are present. This is described by

$$S = A \wedge B \wedge C \wedge D \cdots$$

or

$$S = ABCD \cdots$$

meaning that the output S is one, only if the inputs A *and* B *and* C *and* D, etc., are each one. Figure 14.2 shows the symbol for the AND gate and a few fluidic components that are capable of generating this function. The jet interaction device is passive, in that no power is supplied to it other than by the control signals. The fluid will only flow out of the output exit if both controls A *and* B are on. The spool valve element can handle much more power and again only

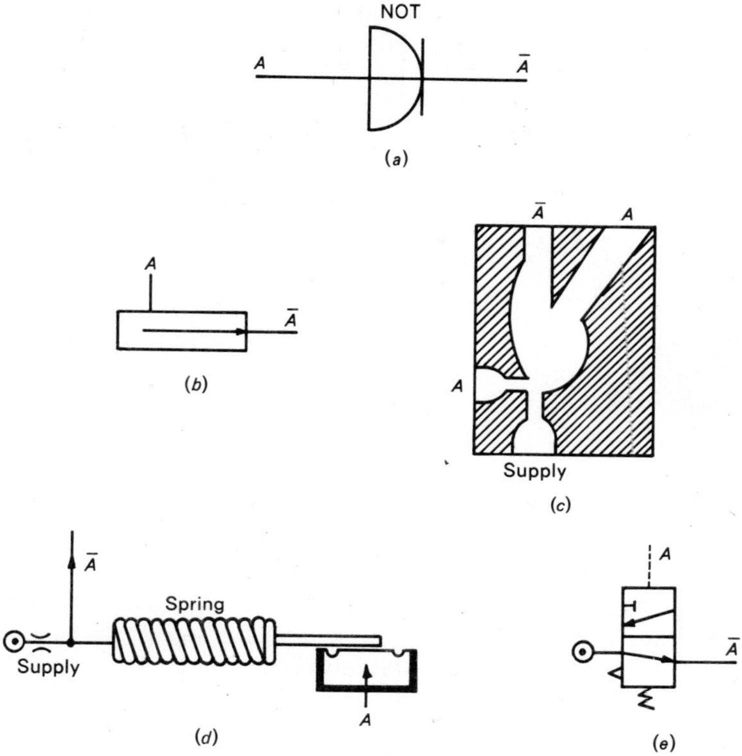

Fig. 14.1. (*a*) symbol for inversion (*b*) the turbulence amplifier inverter (*c*) a NOT gate relying on jet deflection (*d*) a spring element and diaphragm (*e*) a three-way spool valve used as a NOT gate

gives an output if both signals *A* and *B* are present. The spool valve gate shown in Fig. 14.2(*c*) is also a passive device. Combination of an active and a passive element gives the gate of Fig. 14.2(*d*).

It is sometimes useful to construct a table showing the value of the output of a gate for all possible combinations of the inputs. This is known as a truth table, so called after its original connection with the truth and falseness of statements in symbolic logic. The truth table for an AND gate is shown below

A	*B*	*S*
0	0	0
0	1	0
1	1	1
1	0	0

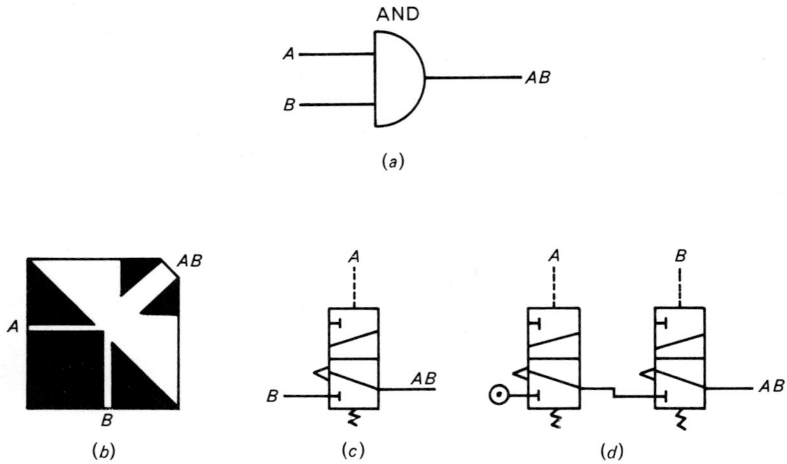

Fig. 14.2. (*a*) AND gate symbol (*b*) a passive jet interaction device (*c*) a three-way spool valve (*d*) two three-way valves in series

Note that in going from one line on the table to the next only one variable is changed in value. The number of lines in the table is 2^n where n is the number of variables. It will be shown later that this table is useful when proving if Boolean expressions are equal or not.

14.2.3 The OR operation

The OR gate gives an output if any one of the inputs is present. This is written

$$S = A + B + C + D + \cdots$$

meaning that the output S is one only if the input A *or* B *or* D, etc., is one. Some fluidic OR gates are illustrated in Fig. 14.3. In the wall attachment element the flow is normally attached to the upper wall. It is a monostable device. Controls A or B or C will switch the jet to the lower wall. Spool valves can again be used to implement this function and Fig. 14.3(*c*) shows how two active three-way valves can be used. The simple shuttle valve is basically an OR gate and Fig. 14.3(*d*) shows how two of these can act as a three input OR gate.

The truth table for a two input OR gate is shown below

A	B	S
0	0	0
0	1	1
1	1	1
1	0	1

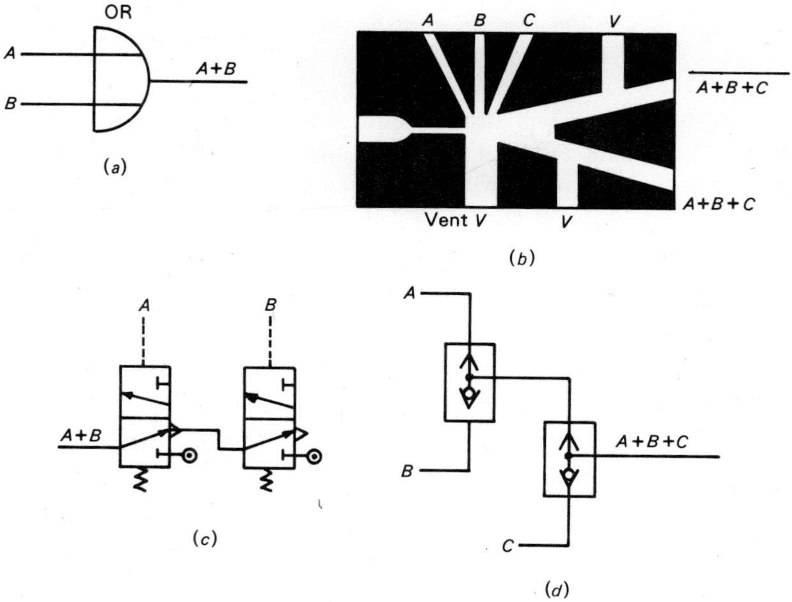

Fig. 14.3. (*a*) OR gate symbol (*b*) a wall attachment OR gate (*c*) OR gate using two active three-way valves (*d*) use of shuttle valves

14.3. Some further logic gates and devices

There are other logic operations in addition to the above three, but it is important to note that any logic can be carried out using only the three basic gates NOT,

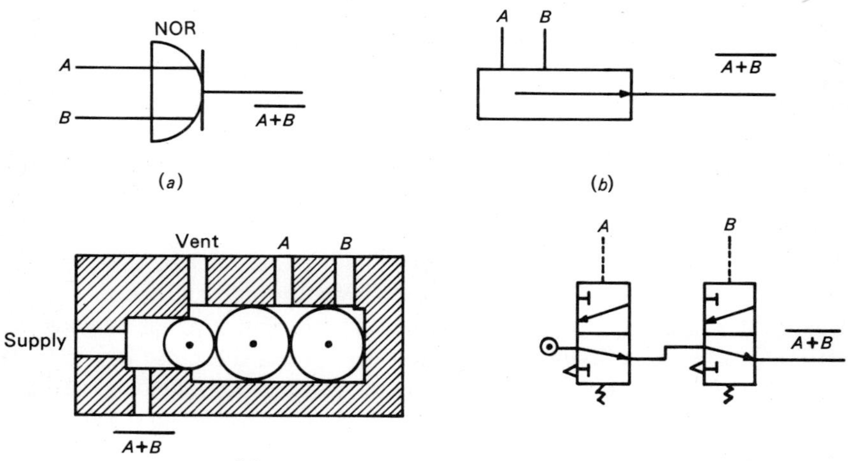

Fig. 14.4. (*a*) NOR gate symbol (*b*) the turbulence amplifier as a NOR (*c*) the Stanford ball element (*d*) series three-way valves

AND and OR. The NOR operation is the inversion of OR and is written

$$S = \overline{A + B + C + \cdots}$$

The output S is one, only if all inputs are zero. The turbulence amplifier is a very simple NOR gate, Fig. 14.4(*b*). The Stanford ball element, which is capable of working at high pressures with either liquids or gases, is illustrated in Fig. 14.4(*c*). Imbalance of forces ensures that the small ball will be moved to the left, closing off the supply, when either A or B is activated. The spool valve NOR gate is very similar to the AND gate of Fig. 14.2(*d*), the interconnections being slightly changed. Reference again to Fig. 14.3(*b*) will now show that the wall attachment OR element is in fact an OR-NOR gate, depending on which output is considered. NOR can also be obtained by cascading NOT and OR gates.

The truth table for NOR is

A	B	S
0	0	1
0	1	0
1	1	0
1	0	0

The NAND operation is the inverse of AND and is commonly used in logic circuits. It is described by the equation

$$S = \overline{ABCD \cdots}$$

The flip-flop or bistable, is another important device. Its symbol is shown in Fig. 14.5(*a*). If the last input was A, output S_2 will be on. If it was B, then output S_1 will be on. The memory inherent in these devices results in the state of the output being affected by not only its present inputs but also its past inputs. The bistable amplifier, Fig. 14.5(*b*), relying on reattachment for this its operation, is perhaps the simplest fluidic bistable. Having been switched to one side by a signal A it will remain there even when signal A is removed. Further signals A will not change the state of the output. This can only be changed by a signal B.

Bistable action can also be achieved by the use of two NOR gates interconnected as in Fig. 14.5(*c*). Turbulence amplifier NOR gates can be used as in Fig. 14.5(*d*). If an input pulse is supplied at B, the lower amplifier is turned off, no signal exists at S_2 and an output S_1 is obtained. This output supplies a signal which continues to keep the lower amplifier off even when the pulse B is removed. This is the memory. A pulse at A turns off the top amplifier and the lower amplifier comes on giving an output S_2.

The element in Fig. 14.5(*e*), the Kearfott ball element, is a moving part bistable. As shown there is no input and the ball is held in the upper position by a force imbalance achieved by a careful sizing of the orifices. Output S_2 is then pressurised and S_1 sits at a much lower pressure level. A pressure pulse at B will

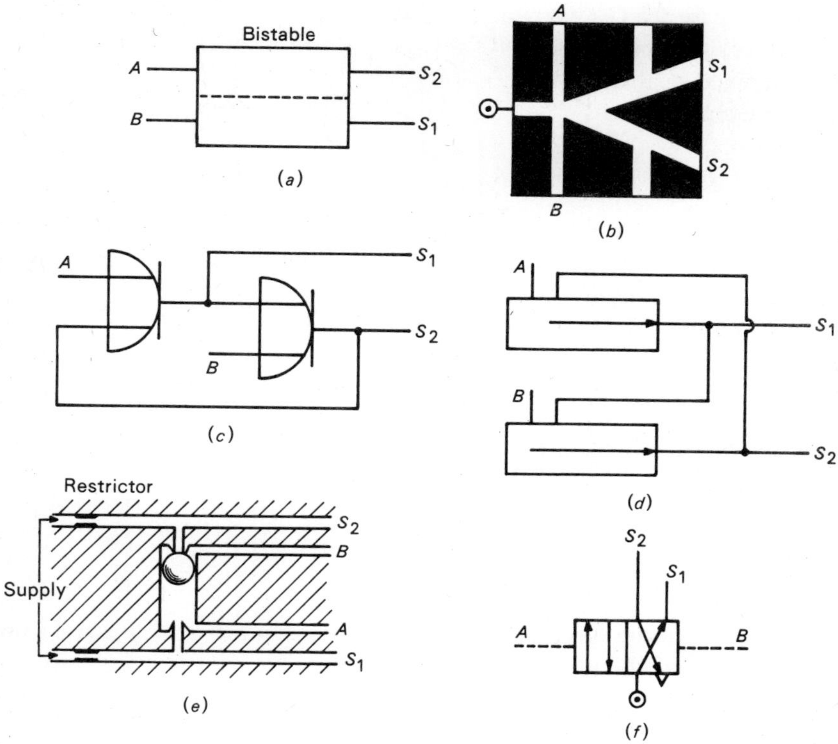

Fig. 14.5. (*a*) Symbol for bistable element (*b*) reattachment device (*c*) two NORs can make a bistable (*d*) use of turbulence amplifiers (*e*) the Kearfott ball element (*f*) the double-pilot operated four-way valve

disturb the equilibrium and the ball will fall to its lower position, in which output S_1 is pressurised and S_2 relieved. The ball will remain there after the removal of signal B.

In conventional pneumatic circuits the pilot operated, two-position, four-way valve is commonly used as a memory, Fig. 14.5(*f*). Here, friction between spool and valve body provides the memory mechanism. This of course, is unreliable if used under conditions of severe vibration or impact where the spool may move without the application of a pneumatic signal.

14.4. Manipulation of logic expressions [102]

There are several rules and theorems involved with Boolean algebra and these must be appreciated before an attempt can be made to design logic circuits with the basic fluidic elements.

1. Identities

$$A + 0 = A$$
$$A \cdot 1 = A$$

In the first case the OR function is solely dependent on the value of A. In the second, since one input to the AND always exists, then the output depends only on the presence or absence of A.

2. The idempotent laws

$$A + A = A$$
$$A \cdot A = A$$

The output of the OR is solely dependent on the value of A, since both inputs are A. The reader should verify that this is equivalent to saying that a switch in parallel with itself is itself.

3. The commutative laws

$$A + B = B + A$$
$$AB = BA$$

Both of these laws indicate that the order in which signals are considered is of no significance. Their truth is evident from the truth tables.

4. The associative laws

$$(A + B) + C = A + (B + C)$$
$$(AB)C = A(BC)$$

These show that the way in which functions are grouped is of no significance. It does not matter where the parentheses are placed.

5. The distributive law

$$A(B + C) = AB + AC$$

This can easily be verified from a truth table.

A	B	C	AB	AC	$AB + AC$	$B + C$	$A(B + C)$
0	0	0	0	0	0	0	0
0	0	1	0	0	0	1	0
0	1	0	0	0	0	1	0
0	1	1	0	0	0	1	0
1	0	0	0	0	0	0	0
1	0	1	0	1	1	1	1
1	1	0	1	0	1	1	1
1	1	1	1	1	1	1	1

6. $A + AB = A$

The truth of this equality can again be easily checked by means of a truth table. However it can also be checked by deduction. The equality states that the value of the left-hand side is independent of B. If A is one, then regardless of the value of B, the OR function will have a value of one. If A is zero then AB is zero and the OR function has value zero also.

7. $A + \bar{A} = 1$

A or $\bar{A}$ has at any instant the value of one. The OR value is then always one.

8. $A\bar{A} = 0 \quad A \cdot 0 = 0 \quad A + 1 = 1$

9. De Morgan's theorems

$$\overline{ABC} = \bar{A} + \bar{B} + \bar{C}$$
$$\overline{A + B + C} = \bar{A}\bar{B}\bar{C}$$

These expressions are very important and show that the inverse of an expression is obtained by inverting all the variables and then changing the AND to OR or the OR to AND.

The use of these theorems is illustrated in the following example. Simplify the expression

$$(XY + ABC)(XY + \bar{A} + \bar{B} + \bar{C})$$

This expands to

$$XYXY + \bar{A}XY + \bar{B}XY + \bar{C}XY + ABCXY + A\bar{A}BC + AB\bar{B}C + ABC\bar{C}$$

Now from theorem 2. $(XY)(XY) = XY$

from theorem 8. $(A\bar{A})BC = (0)BC = 0$
from theorem 8. $A(B\bar{B})C = AC(0) = 0$
from theorem 8. $AB(C\bar{C}) = AB(0) = 0$

The expression then reduces to

$$XY + \bar{A}XY + \bar{B}XY + \bar{C}XY + ABCXY = XY(1 + \bar{A} + \bar{B} + \bar{C} + ABC)$$

From theorem 8. $1 + (\bar{A} + \bar{B} + \bar{C} + ABC) = 1$

The expression now becomes simply $(XY)1 = XY$. De Morgan's theorem is extremely useful for determining how one logic function can be derived from the others, e.g. how can AND logic be obtained using NOT and NOR logic? This is very important if the basic elements at hand are NOT and NOR units and AND is desired. Consider

$$S = AB$$

De Morgan's theorem gives

$$\bar{S} = \bar{A} + \bar{B}$$

To put this in NOR form, invert again

$$S = \bar{\bar{S}} = \overline{\bar{A} + \bar{B}}$$

Hence S = neither (NOT A) *nor* (NOT B)

This is illustrated in Fig. 14.6 in which the turbulence amplifier, a basic NOR element, has been used. Note that the number of bars in the logic expression is equal to the number of turbulence amplifiers.

Use of De Morgan's theorum can again be illustrated by considering how OR logic can be derived from AND logic.

$$S = A + B$$

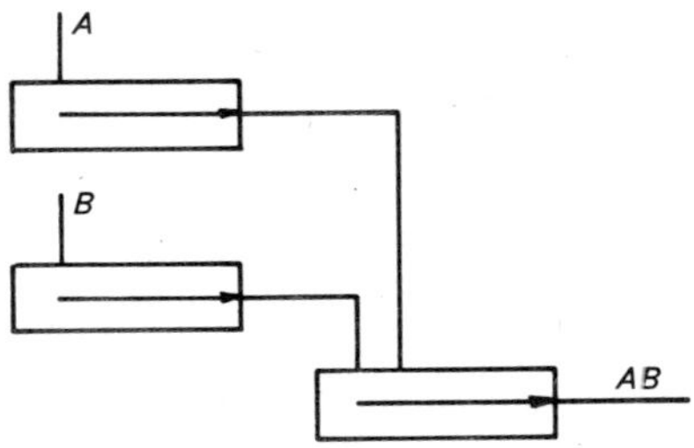

Fig. 14.6. AND logic using NOR elements

Inverting,

$$\bar{S} = \bar{A} \cdot \bar{B}$$

Inverting again,

$$\bar{S} = \overline{\bar{A} \cdot \bar{B}}$$

A practical fluidic OR gate, consisting of NOT and AND elements, is shown in Fig. 14.7.

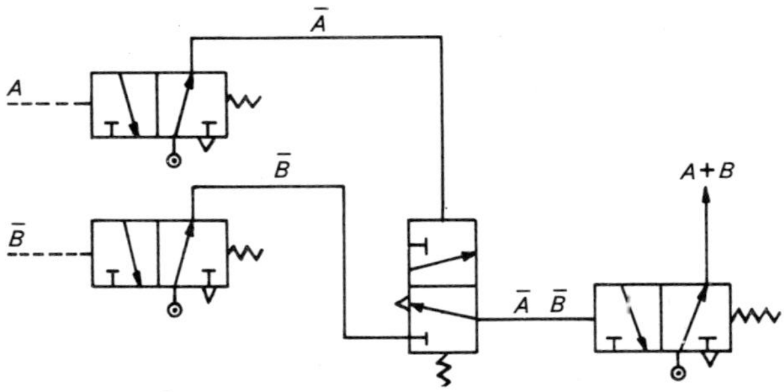

Fig. 14.7. OR logic using NOT and AND gates

14.5. A circuit design

Consider the following problem. Design the logic machine for controlling an automatic drill operation. The drill should only operate if the start button is pressed, if the workpiece is in position, and if the clamp is operated. The drill should not operate if the tool is broken or if the number of operations with the tool has exceeded a certain limit. In addition the operator should be able to start the operation from a remote position.

Firstly, it is assumed that transducers are available to provide signals indicating the state of the machine.

Signal A = start button pressed
B = workpiece in position
C = guard in position
D = tool is broken
E = tool life is exceeded
F = start from remote position

If S is the drill actuating signal then

$$S = (A + F)BC\overline{(D + E)}$$

The logic circuit is shown in Fig. 14.8 and this can be implemented in practice by using the OR, NOR and AND gates described earlier. Sometimes, however, the

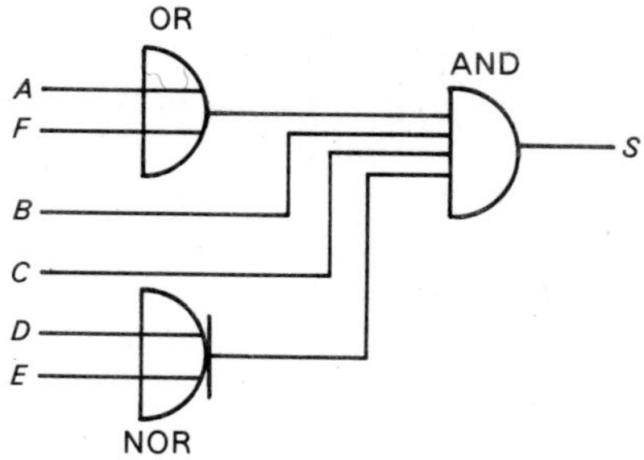

Fig. 14.8. Basic logic circuit for design example

number of available different gates is limited and the logic circuit must be changed. If, for example, only turbulence amplifiers are available, the NOR logic must be used. Inversion gives

$$\bar{S} = \overline{(A + F)} + \bar{B} + \bar{C} + (D + E)$$

and a further inversion gives

$$S = \bar{\bar{S}} = \overline{\overline{(A + F)} + \bar{B} + \bar{C} + D + E}$$

Four turbulence amplifiers (the number of bars in the equation) are necessary. The circuit is shown in Fig. 14.9. Note that the NOR gate has five inputs and this

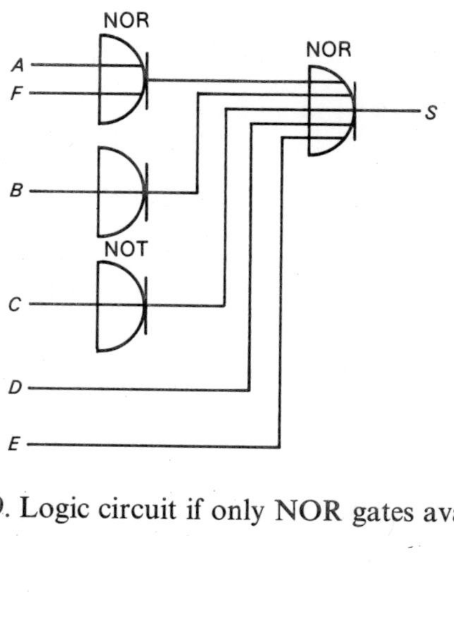

Fig. 14.9. Logic circuit if only NOR gates available

may be difficult to realise in practice. If only four inputs can be handled then the logic expression can be rewritten as

$$S = \overline{\overline{A + F} + \bar{B} + \bar{C} + \overline{\overline{D + E}}}$$

Six turbulence amplifiers are now needed.

If AND logic is preferred, the equation is manipulated as follows

$$S = (A + F)BC(\overline{D + E})$$
$$\bar{S} = (\overline{A + F}) + \bar{B} + \bar{C} + (D + E)$$

or

$$\bar{S} = (\bar{A} \cdot \bar{F}) + \bar{B} + \bar{C} + D + E$$

$$S = \bar{\bar{S}} = (A \cdot F)BC\bar{D}\bar{E}$$

In this case, two AND gates and five invertors are needed, Fig. 14.10.

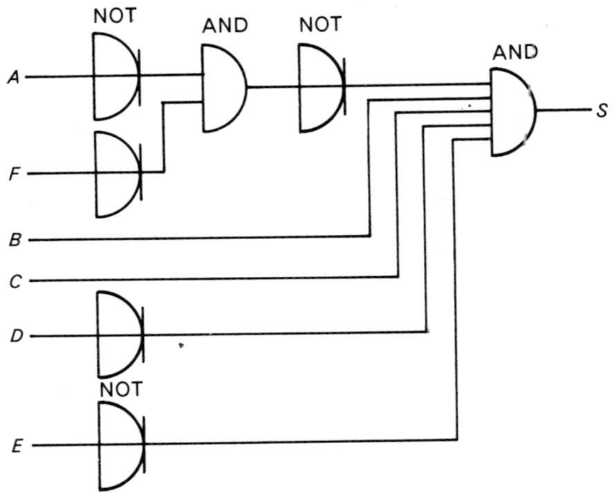

Fig. 14.10. Logic circuit if only NOT and AND gates available

This system can be made semi-automatic by using a bistable or memory at the output. In Fig. 14.11, output S operates the drill which, on completion of its

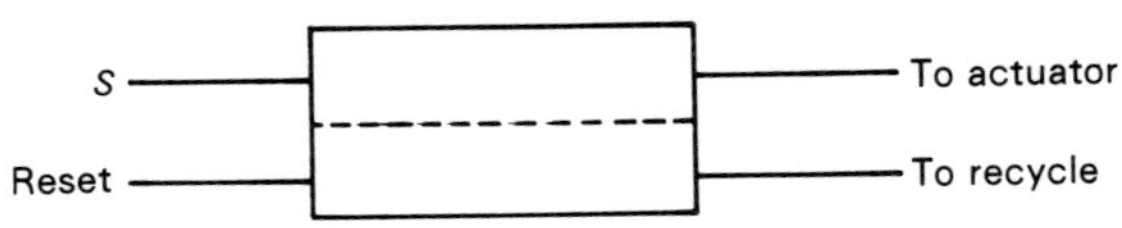

Fig. 14.11. Use of bistable to make sequence semi-automatic

task, sends a reset signal to the bistable. The bistable, when reset, sends a recycle signal which can be used to initiate another cycle of operation. A practical circuit, using turbulence amplifiers with a fan-in of four is shown in Fig. 14.12.

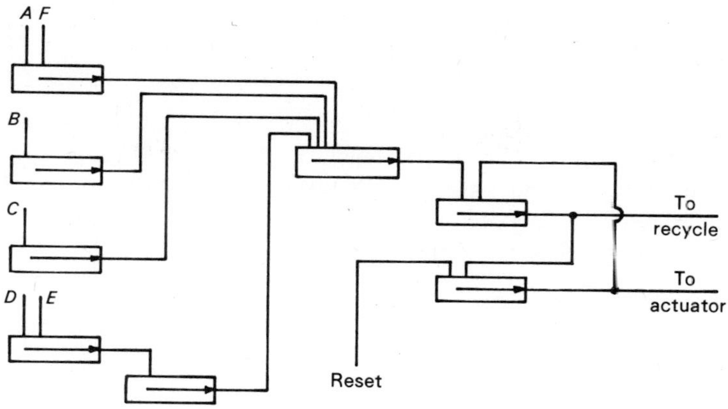

Fig. 14.12. A practical circuit using turbulence amplifiers

14.6. Some special circuits

The half adder is an important device and is used extensively in fluidic shift registers. It is closely related to the OR gate and is sometimes called an exclusive OR element. In an earlier discussion of OR elements, it was shown that an output exists if either *A* or *B* is present *or* if both *A* and *B* are present. The exclusive OR elements only gives an output if *A or B* is present; not when *A and B* are present.

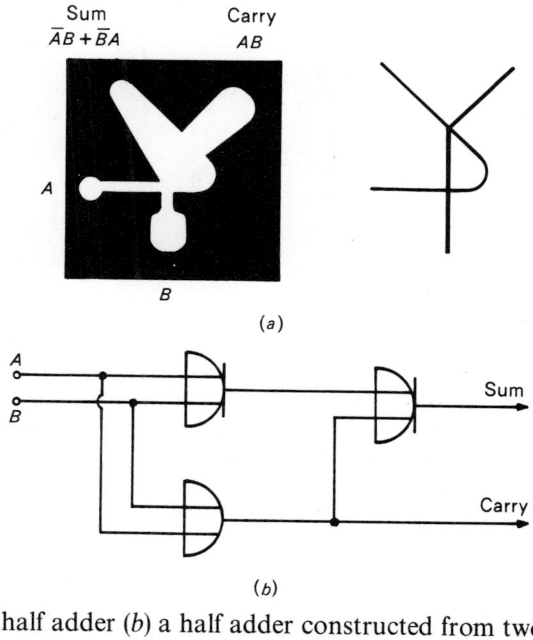

Fig. 14.13. (*a*) a half adder (*b*) a half adder constructed from two NOR and one AND gate

The output can then be written as

$$S = \bar{A}B + \bar{B}A$$

The device shown in Fig. 14.13 is so shaped that if jet A is on by itself it will be deflected and will enter the left-hand output. If jet B alone is present it attaches to the right-hand wall and goes out by the left-hand output. If both A and B are present the jet exits by the right-hand output. Operation as an adder is made clearer by reference to the truth table

A	B	SUM (S)	CARRY (C)
0	0	0	0
0	1	1	0
1	1	0	1
1	0	1	0

A half adder can also be made from two NOR gates and an AND gate, Fig. 14.13(*b*).

A full adder can be constructed from two half adders and an OR gate. Figure 14.14 shows a two-stage full adder for adding the binary numbers A_2A_1 and

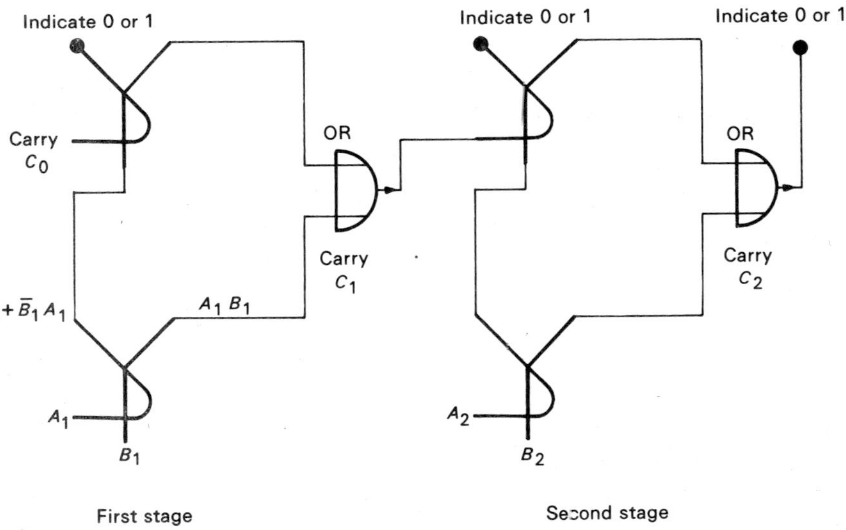

Fig. 14.14. Binary adder using two full adders

B_2B_1. For example, if $A_2 = A_1 = B_2 = B_1 = 1$, the sum is $11 + 11 = 110$. In the figure, there is assumed to be no carry signal C_o. In the first stage, there is a signal from the right-hand output of the lower half adder, since both A_1 and B_1 are present. The first stage indicator registers 0. The input to the OR gate generates a carry signal C_1. In the second stage, there is also a signal from the right-hand output of the lower half adder, since A_2 and B_2 are both present. The second stage indicator registers 1. The input to the OR gate generates a carry signal C_2. Reading from right to left the indicated sum is 110.

While on the subject of adding, it is appropriate to show how fluidics can count. A rather special flip-flop, known as the wall attachment binary flip-flop, Fig. 14.15, can be used for this purpose. This consists of a power inlet joined by a

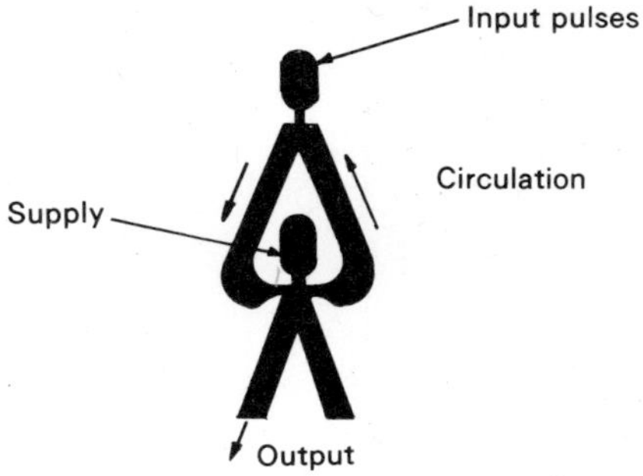

Fig. 14.15. The wall attachment binary flip flop

heart shaped loop to a *single* control or input. There are two outputs and each input pulse switches the main jet from one output to the other. In the figure, the jet exits from the left-hand output and entrainment into the reattachment bubble sets up an anticlockwise circulation of flow in the feedback loop. If a control pulse is added, it follows the loop circulation and makes the jet switch to the other side. The circulation then changes direction. As illustrated in Fig. 14.16 the binary counter is in the reset position and indicates zero. Zero output from a flip-flop implies the absence of a signal at the output. When a pulse input is applied to flip-flop A, its output switches to the right-hand exit, registering 1 there. The controls to flip-flops B and C are cut off and their outputs still read

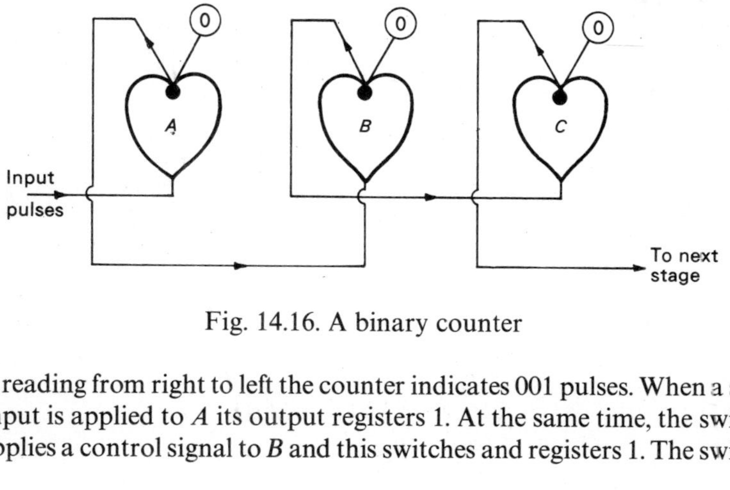

Fig. 14.16. A binary counter

0. Thus reading from right to left the counter indicates 001 pulses. When a second pulse input is applied to A its output registers 1. At the same time, the switching of A supplies a control signal to B and this switches and registers 1. The switching

of *B* removes the control signal from *C* and so its output is unchanged and still reads 0. The counter will count up to seven pulses (111). The eighth pulse resets the counter to zero. A possible application of such a counter will be discussed in chapter 17.

Binary flip-flop action can also be achieved by using spool valves. Two four-way double pilot operated valves and four shuttle valves are needed, Fig. 14.17.

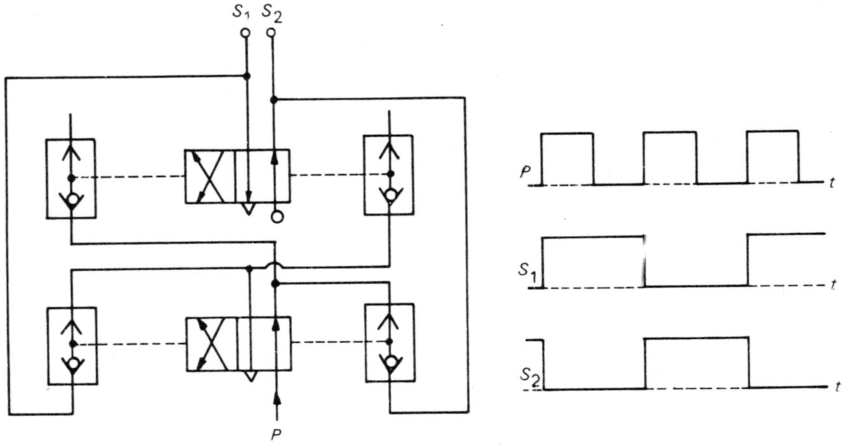

Fig. 14.17. Binary flip-flop using spool valves and shuttle valves

Here again an output S_1 or S_2 changes after every second input impulse *P*. The operation of the circuit will not be described but will be left as an exercise for the reader. This binary stage can be built from conventional pneumatic valves, but since it is essentially a pure information circuit, small spools with low output can be used. Thus a binary counter, like that shown in Fig. 14.17, can be compact and efficient.

Chapter 15

Some applications of non-moving part fluidic elements

15.1. Introduction

When the first fluidic elements were introduced about 1958, they inspired a tremendous wave of interest and within a short period many people began to think that here was a new technique that would solve all control problems at very low cost. Components would last forever and maintenance would be unnecessary. Electronics was a thing of the past. Within a few years, however, it was evident that things were not going as well as expected. There were very few reported successful applications of fluidics and doubts about the future of the subject began to grow. This was largely due to the fact that there was a great temptation to use fluidic components simply because they were new and in many cases no attempt was made to compare the cost of the fluidic system with a conventional system. Unproven components were used and matching of components presented a problem. As a result interest began to wane. Fortunately, at that period, there were successful applications but these for security reasons and proprietary reasons were kept from the general public. By the time that these came to light, several industrial concerns had started to manufacture reliable non-moving part elements and the combined effect resulted in a new wave of enthusiasm. People began to realise that in certain areas the use of fluidics could pay off. One of the most fruitful areas is in non-contact sensing where fluidics offers advantages of cheapness and simplicity over other methods, such as the photoelectric cell. There is also advantage to be gained by using fluidics in fluid power systems. Interface devices are reduced to a minimum, both control and actuation using fluid power, and the high reliability of the fluidic element can result in a trouble free system. Applications of this type include sequence controlled machine tools and automatic handling and inspection systems. Other factors, such as environmental tolerance, have made fluidics an attraction. Elements can be immune to extremes of temperature, illumination, radiation, corrosion, mechanical shock, electrical shock and vibration and this has resulted in successful application in the aerospace field, in nuclear reactors, in steel mills and in the control of cryogenic pumps.

Several possible uses for and applications of fluidics are described in this chapter, and it is hoped that these will give the reader an idea of the potential of fluidics. There are many excellent publications covering applications [103–109] and these are recommended for further study.

15.2. Gauging and positioning

The interruptible jet and the back pressure sensor have been described in an earlier chapter. Neither of these methods need contact with a component and, in general, it is not necessary for a part's motion to be stopped for inspection or detection by them. Fluidic sensing can thus be used to advantage.

The measurement of position is required in many systems. In systems employing hydraulic or pneumatic cylinders, it is often necessary to know when the cylinder rod is fully extended or retracted. In the newspaper and clothing industries, many cases arise where the relative position of the edge of a web or sheet of material must be known. In machining processes, it is often necessary to know if the work-piece is properly positioned. Many other examples could be quoted in which fluidic sensing can be used. One is illustrated in Fig. 15.1

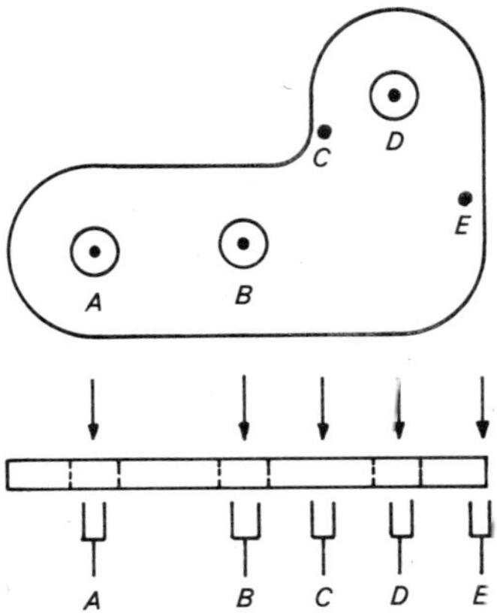

Fig. 15.1. Use of interruptible jets for part inspection

which shows how air jets can be used to inspect critical hole locations in a part. In Boolean algebra terms, the part should be rejected if $\bar{C}\bar{E}(\bar{A} + \bar{B} + \bar{D}) = 1$.

Gauging or inspection is another possible application of fluidic sensing. Indeed, air gauging, with the use of the back pressure sensor, has long been established. The dimensional tolerance of components can be checked by back pressure sensors, as shown in Fig. 15.2. The desired thickness is 50 mm ± 0·10 mm. It is assumed that the sensing heads develop enough pressure to operate the succeeding logic elements when the distance between component and head is less than 2·0 mm. If head *A* is then positioned 52·1 mm above the reference plane, it will pass a signal if the part thickness exceeds 50·1 mm. If head *B* is positioned 51·9 mm above the reference plane, it will *not* pass a signal if the part thickness is less than 49·9 mm. The logic is then

$$\text{Accept if } \bar{A}B = 1$$
$$\text{Reject if } A + \bar{B} = 1$$

The use of OR-NOR gates for generating the accept signal is shown in Fig. 15.2(*b*).

There are other applications for these sensors. For example, liquid level can

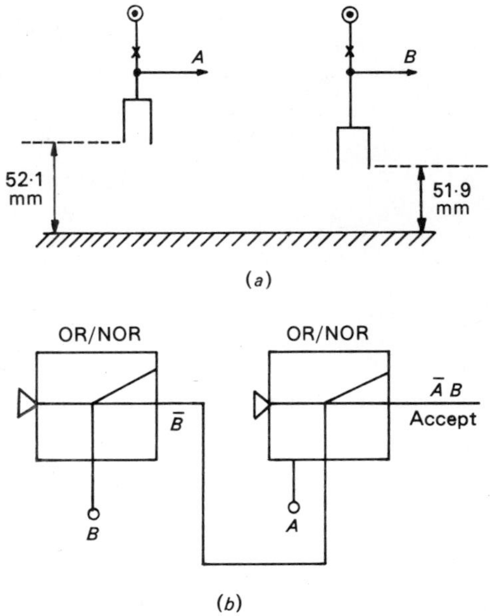

Fig. 15.2. Gauging with back pressure sensors (*a*) the mechanical arrangement (*b*) the fluidic circuit

be measured by using the liquid itself as the 'flapper' in a back pressure sensor. Other variables such as force, torque and temperature can be measured, provided that there is some means of converting changes in the variable to changes in displacement. Measurement of force is particularly simple, since the force can be applied directly to a cantilever flapper. A swash plate can be attached to the shaft when torque is to be measured. As the torque twists the shaft, a point on the swash plate has a component of movement parallel to the shaft axis. This can be used to signal a back pressure sensor. For temperature measurement, the flapper may be bimetallic.

Both interruptible jet and back pressure sensors are analogue devices, their outputs being proportional to their inputs. Their range of operation is small, however, and creates problems when trying to use them in some situations. For example, they would be very difficult to use for measuring the output position of a pneumatic servo. In these cases, it is better to use digital data than analogue data, so that each time the output travels a fixed small increment a pulse is generated giving information about the position. This information is derived from an encoder [110] which subdivides the physical motion into incremental motions and converts it to digital form by determining the number of increments contained in the given motion. Encoders are usually electronic but attention is now being paid to the use of fluid controls. Interruptible jets can be used as shown in Fig. 15.3. In the incremental pattern encoder, a pulse is generated in the jet

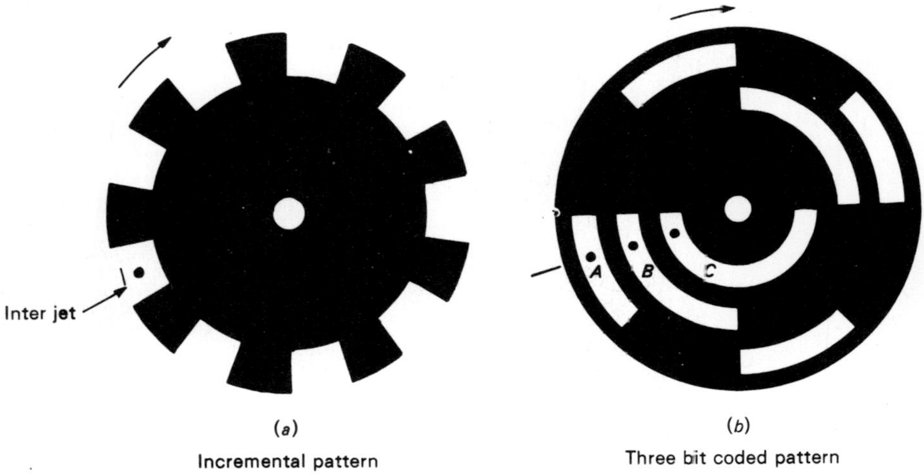

Fig. 15.3. (*a*) The incremental pattern encoder (*b*) A 3-bit coded pattern

receiver line each time the jet is uncovered. These pulses can be counted by a binary counter like that shown in Fig. 14.16. Since each pulse represents a rotational increment, the total movement relative to a datum can be deduced. This encoder needs to be checked against a reference at intervals in order to check that it has not missed a pulse. The coded pattern encoder, Fig. 15.3(*b*), does not suffer from this disadvantage since the number corresponding to any position can be determined directly from the encoder output and there is no need to refer relative displacements to a datum position. In Fig. 15.3(*b*), the revolution is divided into eight increments, each position being indicated by a three bit binary number *CBA*. This system has another advantage over the incremental pattern encoder in that the direction of rotation is evident from the succession of the output numbers.

15.3. Oscillators and their uses

The oscillator provides a timing unit or clock which is an essential part of many sequential control systems in which an ordered series of events must occur at fixed times. In addition, the frequency of a fluidic oscillator is usually sensitive to changes in fluid properties and this raises the possibility of using the oscillator as a sensor. A simple oscillator using an OR–NOR gate is shown in Fig. 15.4(*a*). When the supply is turned on, the NOR output is active, but after a time delay, determined by the magnitude of the orifice resistance *R* and the chamber and line capacitance *C*, the pressure at control port *X* builds up sufficiently to switch the output to the OR side. The time of dwell on this output is determined by the magnitude of resistance R_1. In addition, the oscillation can be stopped by supplying a signal at R_1.

Better control over the mark space ratio can be obtained if a bistable amplifier is used; Fig. 15.4(*b*). The length of dwell at any output depends on the feedback time constants $T = RC$. The output pressure of such a bistable is load dependent, with the result that the frequency of this type of oscillator is also load dependent. It is useful for generating frequencies in excess of 50 Hz, but for lower frequencies the size of the feedback chambers becomes unwieldy.

A bistable amplifier will also oscillate if its control ports are joined as in Fig. 15.4(*c*). This is sometimes known as the sonic or acoustic oscillator, since its

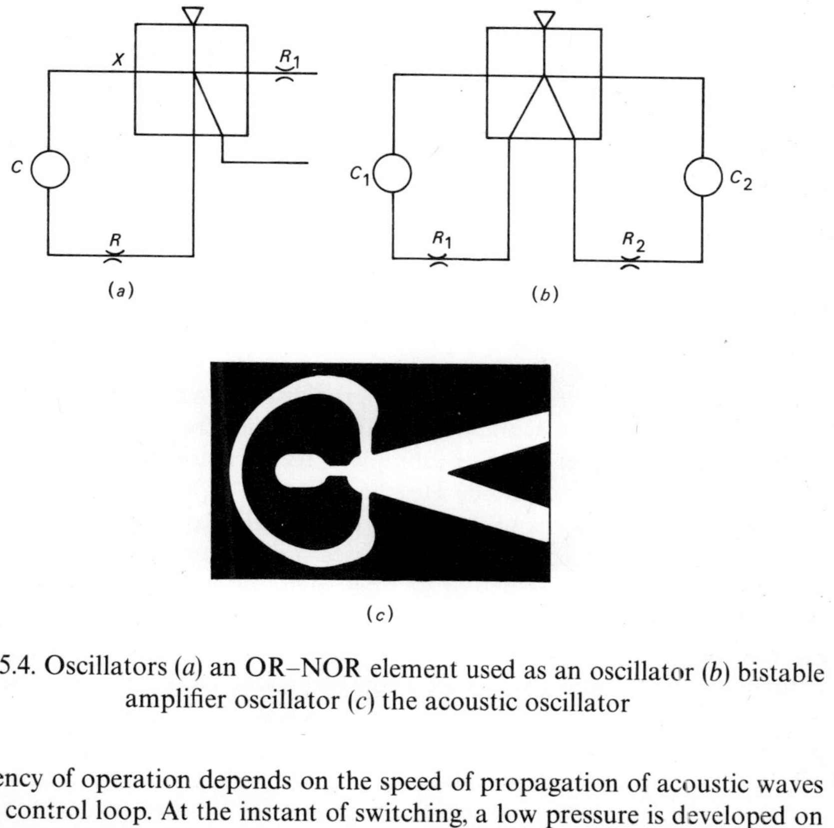

Fig. 15.4. Oscillators (*a*) an OR–NOR element used as an oscillator (*b*) bistable amplifier oscillator (*c*) the acoustic oscillator

frequency of operation depends on the speed of propagation of acoustic waves in the control loop. At the instant of switching, a low pressure is developed on one side of the amplifier and a high pressure in the other. Thus a low and a high pressure wave are propagated in different directions along the control loop. When these waves reach the control ports, they induce switching and another cycle commences. The velocity of propagation is proportional to the square root of the fluid temperature and hence much attention has been given to the use of the sonic oscillator as a temperature sensor. As such it can have better response and higher reliability then the thermocouple. It has been used in gas turbine research [111] and, made from tungsten, it could be used for measuring the temperature of molten metal.

The control circuitry [112] necessary for converting the oscillator output to an analogue quantity is shown in Fig. 15.5. The oscillator output is compared

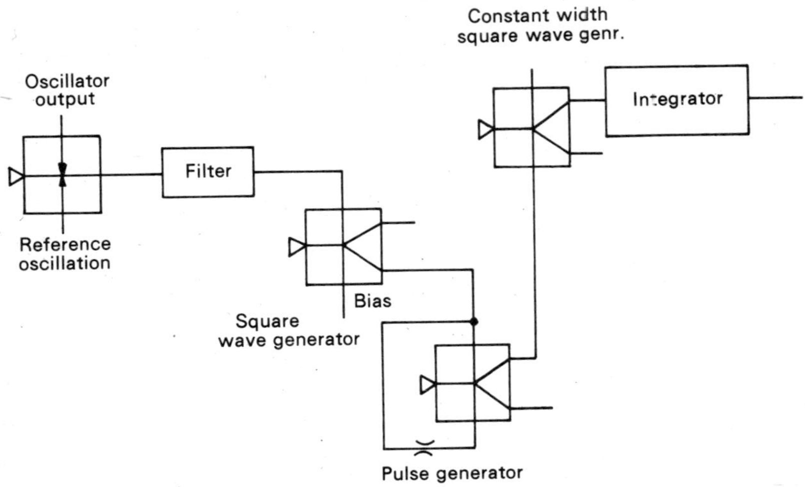

Fig. 15.5. Circuit for converting a variable frequency oscillator output to an analogue signal

to a reference frequency (which could be provided by a tuning fork heterodyne) by means of a proportional amplifier. Beat frequencies are developed in addition to the two basic frequencies. For example, if the oscillator output frequency is 520 Hz and the reference is 500 Hz, then the beat or difference frequency is 20 Hz. A low pass filter is used to remove the high frequency signals and the remaining signal at beat frequency is passed to a square wave generator. The resultant square waves are passed to a differentiator which produces pulses or spikes at its output. The method of operation of this differentiator is left as an exercise for the reader. These pulses are employed in a third digital amplifier, with fixed time constant, to generate a constant width square wave. An analogue output pressure proportional to temperature is then obtained by integrating the constant width square waves in an output chamber and resistor.

In the bio-medical field, the fluid oscillator forms the basic building block for a whole range of new devices including heart pumps, external cardiac compressors and artificial respirators. One such respirator [113] is illustrated in Fig. 15.6. During the inhalation phase, the supply is attached to the right-hand wall and air is directed to the lungs. The feedback path from the lung to control port *B* causes the supply to switch to the left-hand exit when a predetermined pressure has been reached. As the supply leaves through port *C*, air is entrained from the lung and exhalation is induced. In practice, the device is about 5 cm square and is easily fitted to a face mask.

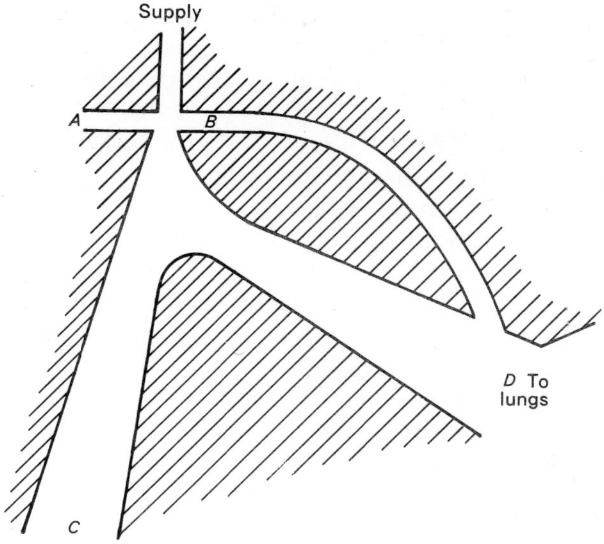

Fig. 15.6. A fluidic respirator

15.4. Rate measurement

Speed of rotation of a shaft can be measured by the beat frequency method described earlier for temperature measurement. In this case, a reference frequency is compared to the frequency of shaft rotation and an analogue output is generated proportional to shaft speed. Back pressure sensors or interruptible jets can easily be used to generate the signals at shaft rotation frequency.

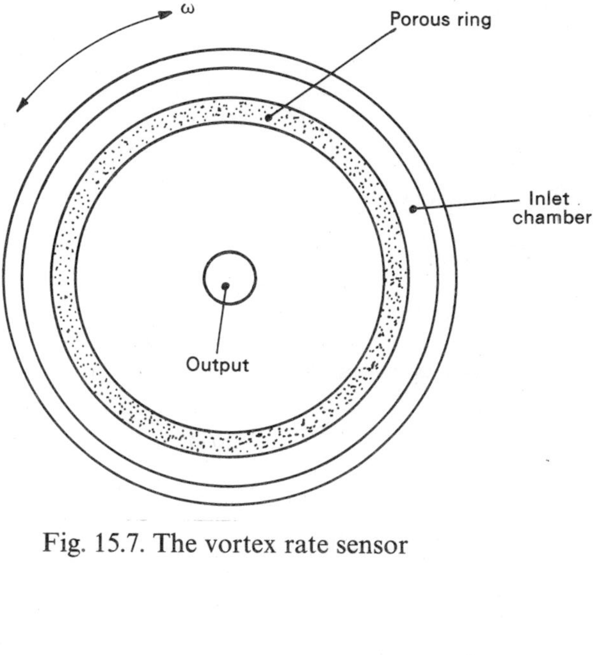

Fig. 15.7. The vortex rate sensor

There are however several simpler analogue rate sensors which depend for their operation on the jet deflection caused by linear or rotational velocity. The Coriolis effect has been used to advantage, as has the deflection of a jet issuing from a moving element.

The vortex rate sensor, Fig. 15.7, uses a different principle. It can provide a differential pressure signal proportional to the rate of turn about its axis of symmetry. Without rotation the fluid flows radially to the output port. If the sensor is rotated about its axis, a tangential velocity component is imparted to the fluid as it passes through the porous ring. As the fluid flows towards the outlet, its tangential velocity increases (because of the conservation of angular momentum) and the resultant vortex motion at this output can be detected and used to drive other fluidic elements. This device is used in systems for missile and space vehicle stability augmentation, for aircraft autopilots and for torpedo guidance. The vortex rate sensor output signals can be integrated to provide attitude information.

15.5. Control systems

15.5.1 On-off control of cloth alignment

When winding cloth or paper on to rolls, it is important that the winding be uniform, i.e. that all edges are in the same plane at right angles to the roll axis. If this is not so, then subsequent trim operations may cause excessive waste. An on-off operated controller can be constructed using interruptible jets and OR–NOR gates, Fig. 15.8. When the cloth is too far left, it interrupts both jets 1 and 2. In this condition signal *A* is amplified in the step up relay and drives the pneumatic cylinder and guide to the right. Similarly, the guide is driven to the left when neither of the jets are interrupted. When only jet 2 is interrupted, no

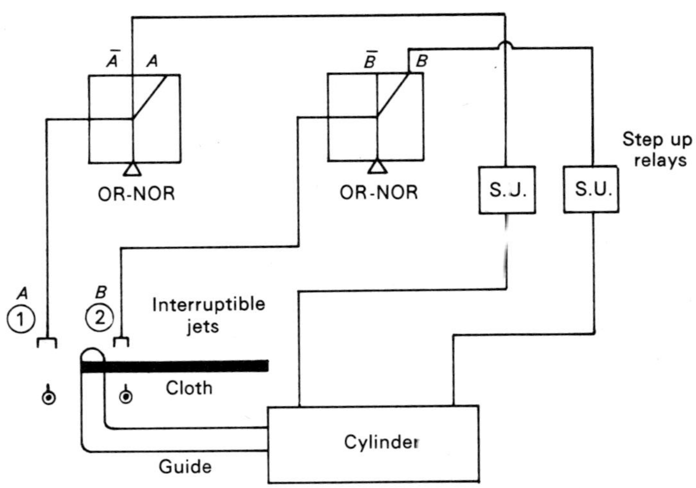

Fig. 15.8. On-off control of cloth alignment

signal goes to the cylinder, the cloth being in the desired position. Cloth centring accuracy is determined by the separation of the two jets and an accuracy of ± 0·5 mm has been reported [114]. The guide should be heavily damped in order to avoid limit cycle oscillations. A pilot operated closed centre three-position four-way valve driven by the step up relays may also be useful for reducing limit cycle oscillation amplitudes.

15.5.2 Automatic yarn tensioning [115]

In the spinning process, it is very important that the yarn be fed to the machine at the correct tension. This example, in contrast to the above mentioned digital edge control, shows how fluidic elements can be used in a proportional control systems, in this case proportional control of tension. Figure 15.9 shows that the

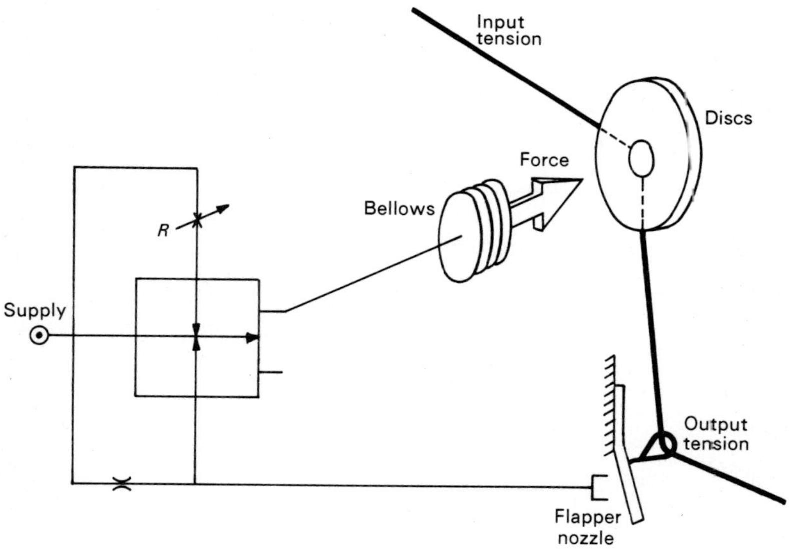

Fig. 15.9. Automatic yarn tensioning

tension is measured by a cantilever spring which acts as the flapper of a flapper nozzle valve. When the output tension is too low, the flapper gap is small and the pressure developed in the sensing head directs the output of a proportional amplifier into a bellows or capsule. The yarn is fed between two discs and the capsule forces these together, causing the output tension to increase. The required tension is adjusted by varying the signal to the upper control port of the amplifier.

15.5.3 Gas servo for engine actuation

When a control system has to function in extreme environmental conditions, fluidic non-moving part devices offer an attractive solution. Such extremes are encountered by rocket engine gimballing systems, where severe vibration and extreme heat or cold may be encountered. In a reported application [116], an engine actuator was designed to operate on hydrogen gas at a temperature of

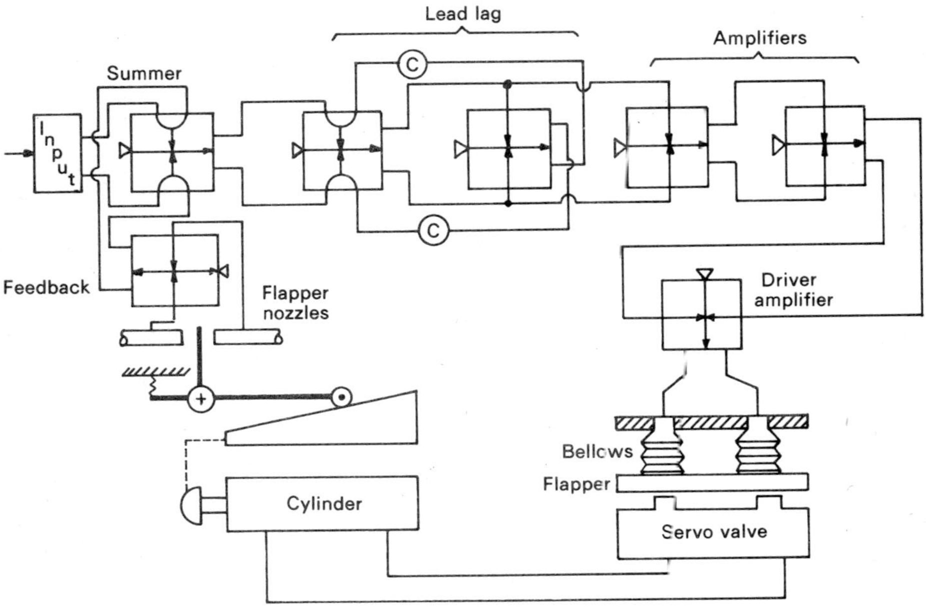

Fig. 15.10. A gas servo for engine actuation

$-250°F$ and under severe vibration. The system is shown in Fig. 15.10. It is an all fluid system and had to be capable of performing the following

Sense the engine position.
Determine the error in position.
Provide a lead lag characteristic in the forward path of the servo.
Provide the necessary loop gain.

The figure shows that all of these functions were accomplished by using in all seven proportional amplifiers. The position was sensed by making a roller follow a cam attached to the cylinder rod. Movement of the roller changes the output pressure of the flapper nozzle bridge circuit, thus generating a pressure signal proportional to actuator position. An error signal is then generated by the summer amplifier. The lead lag circuit shown is analogous to a d.c amplifier with resistance and capacitance in series in the feedback line. The driver amplifier is a proportional amplifier for raising the signal level at the compensation networks to a value sufficient to operate the servovalve. Its output ports are connected to bellows which make a flapper rotate providing a differential pressure across the servovalve spool.

15.5.4 Vortex servo valve

The vented vortex amplifier has the highest power recovery of all fluidic amplifiers and hence has shown promise as a non-moving part servovalve. The use of a vortex valve in place of the conventional spool valve offers several advantages [117, 156].

1. There are no sliding spools and hence threshold and contamination effects are minimal.
2. Environmental tolerance is excellent. Hot gases or liquid gases can be used.
3. Erosion of valve lands and metering edges is no problem.
4. Because there are no moving parts the bandwidth of the fluidic amplifier is wider and can exceed 100 Hz.
5. Fewer precision machined parts are needed and tolerances are not so tight.

In the arrangement shown in Fig. 15.11, two vented amplifiers are driven in a push pull fashion by a proportional amplifier. Increase of control signal C_1

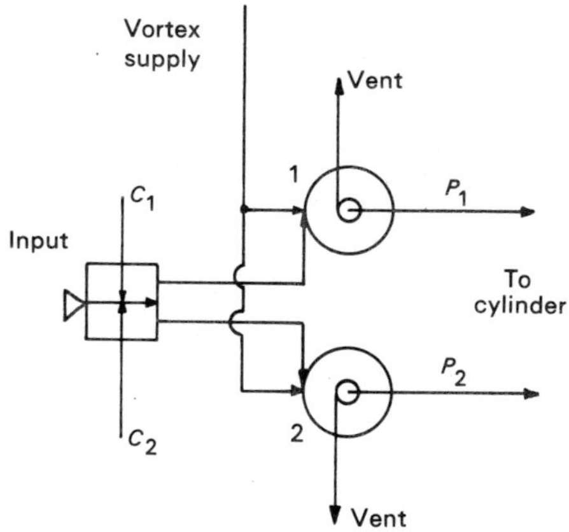

Fig. 15.11. The vortex servovalve

results in a reduction of control pressure at amplifier 1 and an increase of control pressure at amplifier 2. As a result, pressure P_1 increases and P_2 decreases. These devices have already found successful application in helicopter control. A case, described in [117], uses MIL-H-5606 oil at a pressure of 10·34 MN m^{-2}.

15.6. Sequences and interlocking logic systems

The examples of control systems discussed above make it clear that, under certain circumstance, non-moving part elements can be used with advantage in closed loop servo systems. But many applications are not concerned with closed loops and feedback. A large area in which fluidics is making a contribution is that of the open loop system, in which there is no feedback. These systems, in

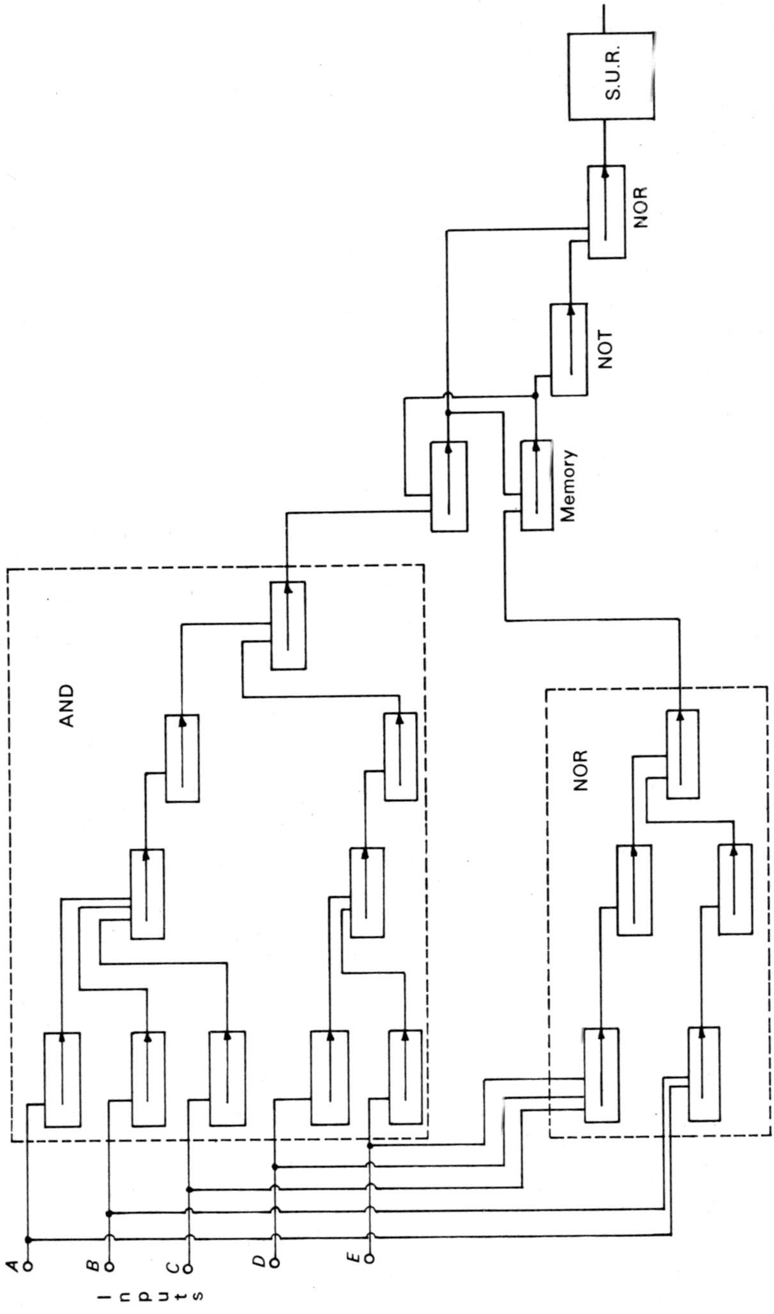

Fig. 15.12. Turbulence amplifier circuit for detonation crimping

general, use digital elements. For example, in the majority of industrial systems involving hydraulic or pneumatic cylinders, the cylinder rod only has two rest positions—fully retracted or fully extended. There is thus no need to continually monitor output position and to feed this back for comparison with an input signal. All that is needed is a two-position valve for driving the rod to one extreme or the other. There is, of course, usually the proviso that the rod should not move unless a specified set of conditions are fulfilled and this is where Boolean algebra comes into its own.

For example, a successful application [118] was concerned with the crimping of detonators—a dangerous task. The system was developed to improve the safety of the process. The explosive nature of the product plus the fact that the girl operators were on piece rate had increased the possibility of an accident above the tolerable level. Originally, the operator removed the detonators from a cylindrical safety holder at her right side and positioned one detonator for crimping with her left hand. The danger arose that she might hold other detonators in her right hand to speed up the operation and any explosion during the crimping operation could cause a serious explosion in her right hand. Her left hand was normally well protected by the built in safety features of the crimping machine.

A logic circuit was developed which required that each finger of the operator's right hand covered an orifice (touch sensors) before the machine would work. Figure 15.12. shows how nineteen turbulence amplifiers were used to construct the circuit. When all five touch sensors are operated there is an output from the AND gate. This puts the upper element of the memory off and the lower on. The NOT is then off and the NOR on, sending a signal to the pilot valve on the crimping machine. If one or more of the fingers is raised the machine will not operate again until the memory is switched and this does not occur until all five fingers are released and the crimping action is released.

Some further examples of fluidic applications will be discussed in the following chapters.

Chapter 16

Sequential circuits

16.1. Introduction

In earlier chapters, servovalves were shown to be essential for accurate proportional control of the position of hydraulic and pneumatic cylinder rods. These servosystems are expensive but, fortunately, in the majority of applications servovalves are an unnecessary luxury and adequate control can be achieved by means of two- or three-position directional control valves [92, 119, 120, 121, 122]. These give simple 'bang bang' or on-off control, with the result that usually a cylinder rod only has two rest positions—fully extended and fully retracted. (There is, of course, the facility for stopping the rod at an intermediate position if a closed centre valve is used.)

In many applications, more than one working cylinder is needed and the movements of these are required to follow a fixed pattern or sequence. For example, an automatic drilling machine may require two cylinders; one to deliver and to withdraw components from the machine and one to power the drill movement. If the component handling cylinder is labelled A and the drill feed cylinder B, the sequence can be described

1. Cylinder rod A feeds component.
2. Cylinder rod B feeds drill.
3. Cylinder rod B withdraws drill.
4. Cylinder rod A withdraws component.

If the extended position of rod A is called the A_e position and the retracted position is the A_r position, this particular sequence can be described as $A_eB_eB_rA_r$ if it is assumed that both rods are retracted at the start. It is important to note that although the notation $A_eB_eB_rA_r$ indicates a certain sequence of actions, it does not imply that one action must be completed before the next commences. In the chosen example, a fast cycle could be achieved if the drill feed was arranged to approach the drilling position while the component was being fed to the machine. This, however, would involve consideration of timing which depends on the distance of the drill from the component, the feed rate, etc. It is generally safer in this case to wait until rod A has completed its stroke before initiating the movement of rod B. Interlocking is required to ensure that rod B does not move until rod A is fully extended. The same argument can be applied to the retraction strokes. Clearly the component cannot be withdrawn until the drill is clear and this can be assured if interlocks do not allow rod *A* to retract until rod *B* has retracted. This chapter consists mainly of a discussion of means of providing interlocks.

16.2. Unequal loading

If two nominally identical cylinders are driven by a common valve, Fig. 16.1, it is unlikely that both rods will travel at the same rate. Variations in piston leakage

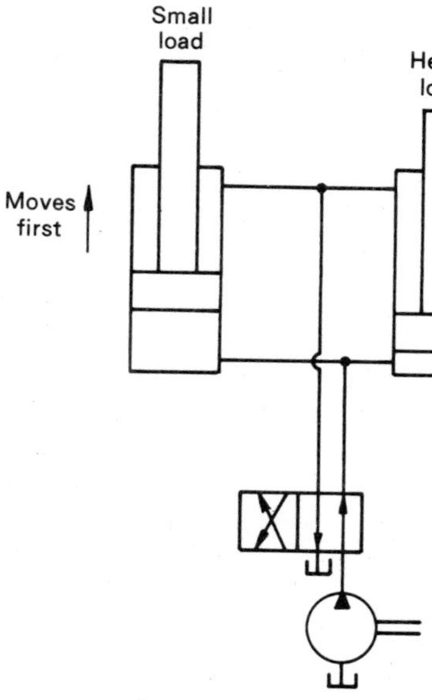

Fig. 16.1. Cylinder sequencing relying on difference of loads

or seal friction can lead to small differences in rate, but the most pronounced effects are due to load differences. The cylinder with the smaller load will move first and although, at first sight, this may appear to be an undesirable state of affairs, there are in practice applications where this can be used to advantage. For example, in the automatic stoking of a furnace, the lightly loaded cylinder might be the door driving cylinder and the heavily loaded cylinder might be the coal handling cylinder. If these cylinders are driven off a common valve, the door will open before the fuel is fed. This system is not interlocked and is hence only used in the simplest of circuits. It could, for example, lead to damage of the door of the feeding mechanism if the door were to stick. In this case, the feed cylinder would extend and strike the door.

16.3. Use of sequence valves

These valves have been described earlier in chapter 11. They allow fluid to pass

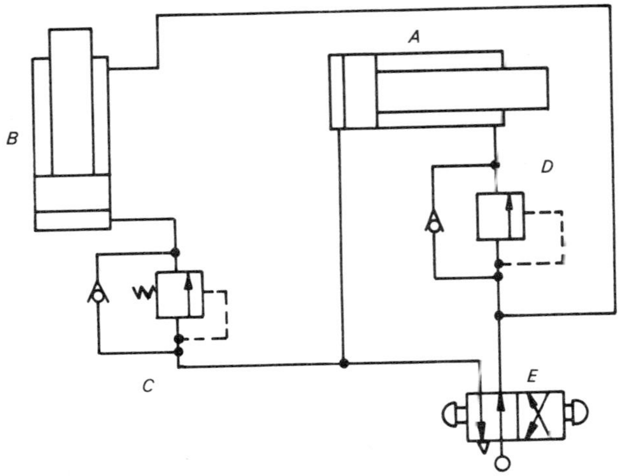

Fig. 16.2. A 'clamp and drill' circuit using sequence valves

only when its pressure exceeds a preset value and can be used in both hydraulic and pneumatic circuits. A typical 'clamp and drill' circuit is shown in Fig. 16.2. Cylinder A is the clamp and B operates the drill. When the four-way directional control valve E is actuated, pressure is applied to the blank end of rod A which extends clamping the workpiece. Pressure build up in the blank end of A then causes the sequence valve C to operate, connecting the supply to the blank end of cylinder B and making its rod extend. The sequence A_eB_e is now completed. If the valve E is reset, rod B will retract, sequence valve D will operate and rod A will retract. The full sequence $A_eB_eB_rA_r$ is complete.

Although this is a simple circuit it has several disadvantages:

1. There is no positive interlock. If, for example, some 'stickiness' in cylinder A causes premature pressure build up, sequence valve C may operate, moving rod B before A has completed its stroke. The drill would operate before the component was clamped.
2. Flow control valves cannot be used to control rod speeds. If meter out speed control were attempted for cylinder B, the pressure in the rod end of B may build up and cause premature operation of sequence valve D.
3. In hydraulic systems especially, care must be taken to ensure that premature pressure surges do not occur when the main directional control valve is switched.

On the whole, these valves must be used with extreme caution and should be reserved for the simplest of applications, such as that described—they are in fact commonly used in the woodworking industry for clamping operations using pneumatics.

In addition to this type of sequence valve which detects pressure rise, there are also those which detect pressure decay. Normally pressure in an exhaust line is higher than atmospheric due to line losses. When the rod completes its stroke the pressure falls to atmospheric and this can be used to actuate a valve which is normally held closed by pressure. This type of valve is also an indirect signalling device. It does not directly indicate rod position.

16.3. Programmers

A camshaft drive can be used to sequence any number of cylinders, either hydraulic or pneumatic. The camshaft lobes, Fig. 16.3, may contact electrical

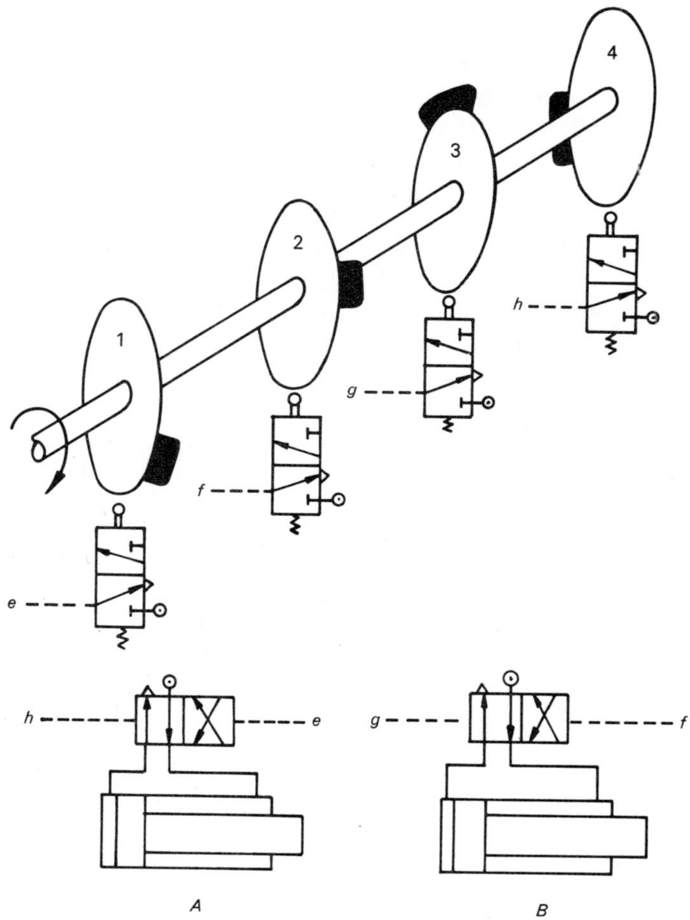

Fig. 16.3. Order of sequence controlled by a cam shaft programmer

limit switches, hence operating solenoid operated valves, or they may actuate three-way spring return pilot valves which in turn signal pilot operated main valves. In other cases, it may be convenient to use the cam shaft for direct operation of the main valves. The cams may consist of several lamina of a basic camshaft lying on top of each other and different sequences are obtained by rotating the lamina relative to each other.

The circuit for the sequence $A_e B_e B_r A_r$ is shown in Fig. 16.3. As shown, the first spring return pilot valve operated sends a signal to actuate the double pilot operated four-way directional control valve driving cylinder *A*. Rod *A* extends. The next pilot valve is tripped by cam 2 and causes rod B to extend and so on.

Although the camshaft programmer provides a flexible means for controlling sequences, its use must be treated with some caution. Timing is important and if in one part of a sequence it is essential that one rod is fully extended before another moves, then the camshaft speed and the relative angular position of the lobes must be carefully adjusted. In addition, this form of programmer is not positively interlocked and hence is not intrinsically safe. If one part of the sequence is not completed, perhaps because of a jammed cylinder, the camshaft will move relentlessly onwards to initiate the next movement. This risk of malfunction may be tolerated in some applications where it does not lead to danger to personnel or damage to the machine, e.g. in packaging of domestic foodstuffs. When it cannot be tolerated, some form of interlocking is required and pilot valves are necessary for detecting when the cylinder rods are fully retracted or extended, Fig. 16.4. In addition to the four cam operated valves, the sequence $A_e B_e B_r A_r$ now requires an additional four pilot operated, spring return three-way valves. The circuit in Fig. 16.4 is interlocked and the sequence will stop if some movement is incomplete. For example, when the camshaft actuates number 3 pilot valve, it in turn initiates movement B_r by passing a signal g to the main valve controlling cylinder *B*. The signal g will however only

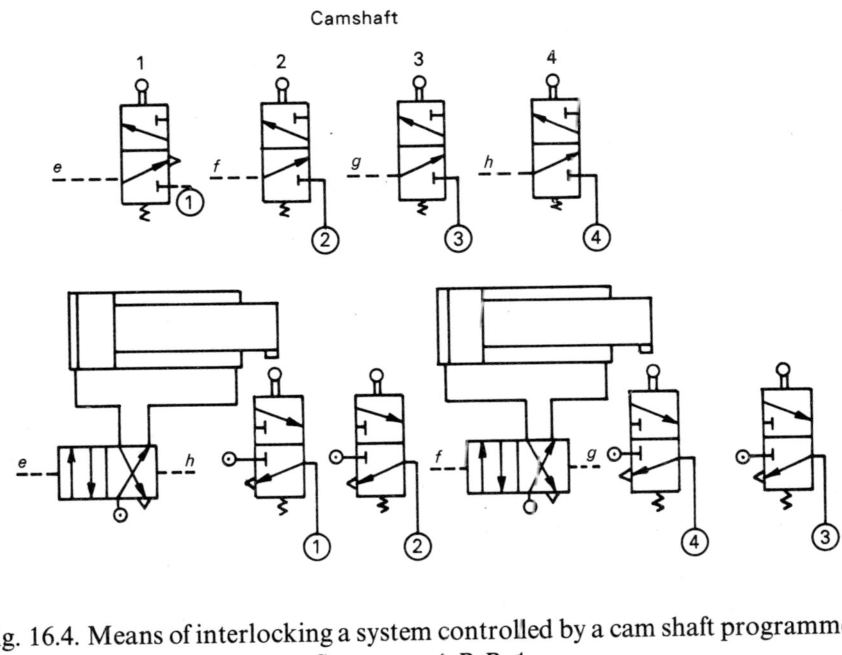

Fig. 16.4. Means of interlocking a system controlled by a cam shaft programmer. Sequence $A_e B_e B_r A_r$

exist if the previous movement B_r has been completed, this providing the supply to pilot valve 3. (Note that pilot valves are drawn in their normal position, even although those at the A_r and B_r positions in Fig. 16.4 are in practice actuated.)

This is the first example of a direct signalled sequence; one in which information concerning the position of the rods is directly obtained from the rods.

16.4. Direct signal interlocks

The sequence $A_eB_eA_rB_r$ is not so commonly found in practice as the sequence $A_eB_eB_rA_r$. A possible application could be found in a wire stripping machine in which cylinder A feeds the wire to the clamp B, which grips it while rod A retracts pulling the wire from its sheath. Clamp B can then retract.

Before studying this sequence in detail, it is necessary to introduce more notation. In most of the circuits to follow, each cylinder will need two three-way limit valves; one to detect the extended rod position and one for the retracted position. For cylinder A, the limit valve at A_e is referred to as a_e and the limit valve at A_r is referred to as a_r. In addition, the four-way double pilot operated directional control valve controlling cylinder A is called main valve A or MVA. The circuit for the sequence $A_eB_eA_rB_r$ is shown in Fig. 16.5. The cycle is initiated

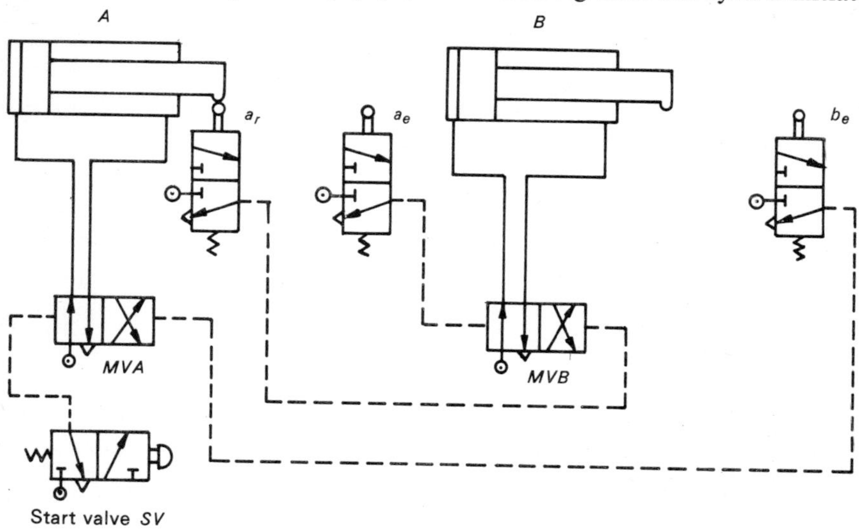

Fig. 16.5. Circuit for the sequence $A_eB_eA_rB_r$ single shot

by pressing the button on the start valve SV. This switches MVA and rod A extends to A_e. The signal from pilot valve a_e then switches MVB sending rod B to B_e. Pilot valve b_e resets MVA and rod A retracts actuating a_r and hence resetting MVB. Rod B then retracts and the cycle is complete. Instead of one shot operation the sequence can be made continuous if start valve SV is removed and the signal from a b_r pilot valve is used to reset MVA.

This circuit is easily constructed by using the signal generated at the end of each rod movement to initiate the next movement. It is important to note that there are, on no occasions, pilot signals applied simultaneously to both sides of a main valve. There are no *locked on* signals. A double pilot operated valve will only switch if (1) a signal is applied and (2) the opposing signal on the other side of the valve is removed. Care must be taken, when designing circuits, to ensure that clashing signals are avoided and this is done by arranging that when a pilot signal is applied to one side of a main valve, the other side is connected to exhaust or tank.

Although the limit valves in this example are pneumatic or hydraulic, they could just as well have been electric, in which case the main valves would be double solenoid operated valves.

16.4.1 The sequence $A_e B_e B_r A_r$

This sequence arises often in practice and has been discussed earlier· It can suffer from locked on signals and needs careful attention. Figure 16.6 shows how this can arise. The *SV* will switch *MVA* causing rod *A* to extend to A_e. This actuates a_e, switching *MVB* and driving rod *B* to B_e. The sequence now demands that rod *B* retract to B_r and the apparently obvious thing to do is to connect b_e to the left-hand side of *MVB*, hoping that this will reset *MVB* and make rod *B* retract. The pilot signal from a_e is however locked on the other side of *MVB* and *MVB* will not reset while this signal exists. Cylinder rod *A* holds valve a_e in the actuated position and so the only way of removing the signal from a_e is

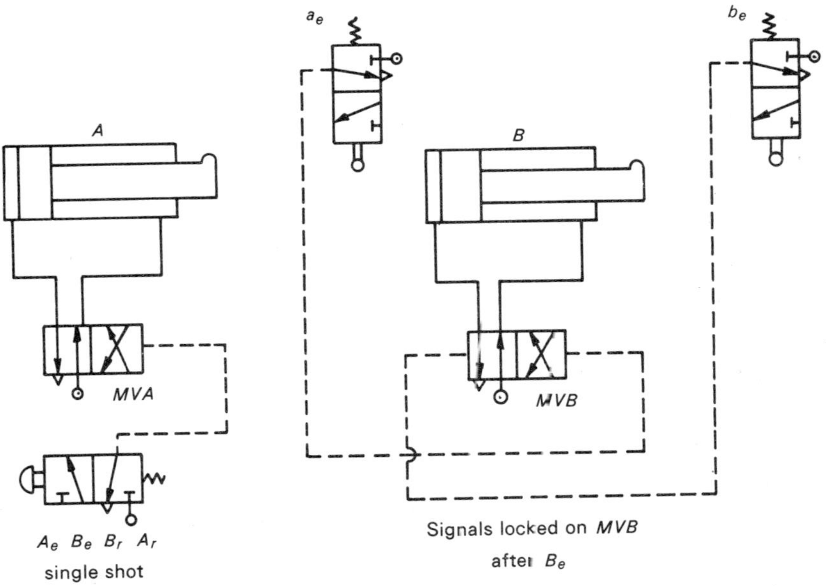

Fig. 16.6. Part of the circuit for the sequence $A_eB_eB_rA_r$ single shot, illustrating signal locked on *MVB* after movement B_e

to remove the supply from a_e. An additional four-way valve, the selector valve, Fig. 16.7, is used for this purpose. When b_e is actuated it cuts off the supply to a_e and is hence able to reset MVB. Cylinder rod B can now retract and valve b_r is used to reset MVA making rod A retract. The sequence is complete. The circuit is shown in Fig. 16.8. The selector valve is also used to control the supply to b_r since this can oppose the starting signal. Again the sequence can be made continuous if an a_r valve is used in place of the start valve.

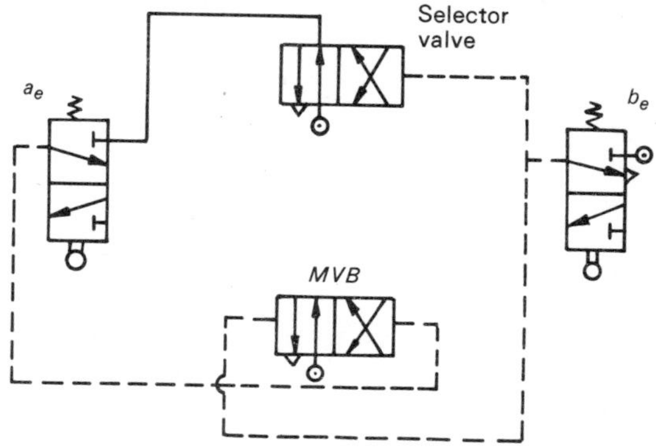

Fig. 16.7. Use of a selector valve to remove the locked on signal

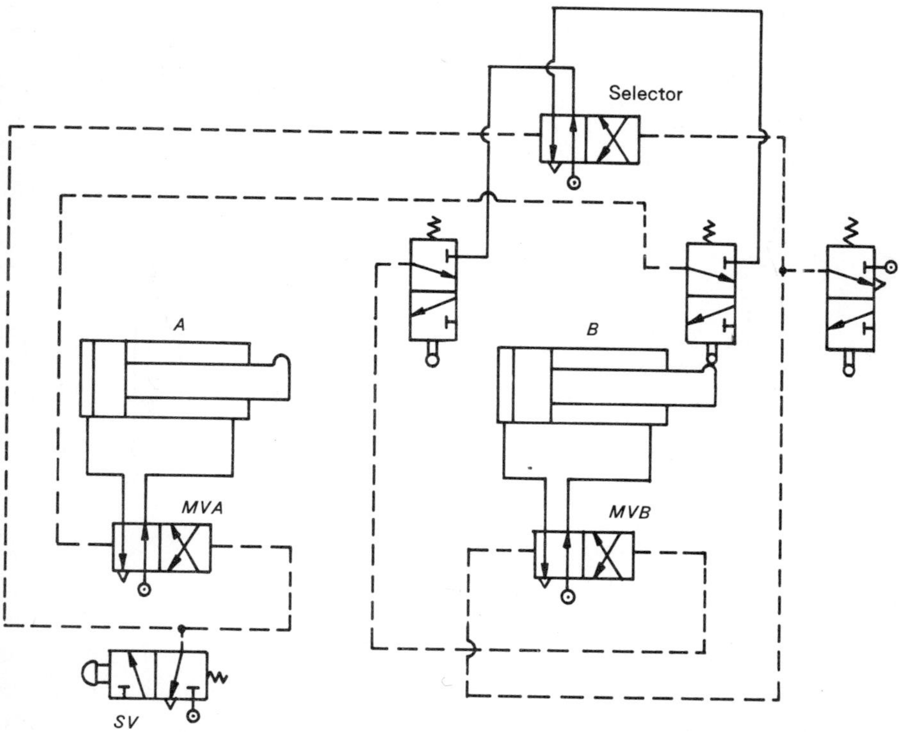

Fig. 16.8. Complete circuit for sequence $A_eB_eB_rA_r$ single shot

16.5. Other means of eliminating locked signals

The locked on signal condition arises in the last example because rod A in its extended position holds a_e actuated whilst rod B extends and retracts. The selector valve was used to limit the period during which the signal from a_e exists. There are other means of limiting the period of a signal from a valve. One method uses a reservoir, Fig. 16.9. Here the three-way valve, a_e in Figs.

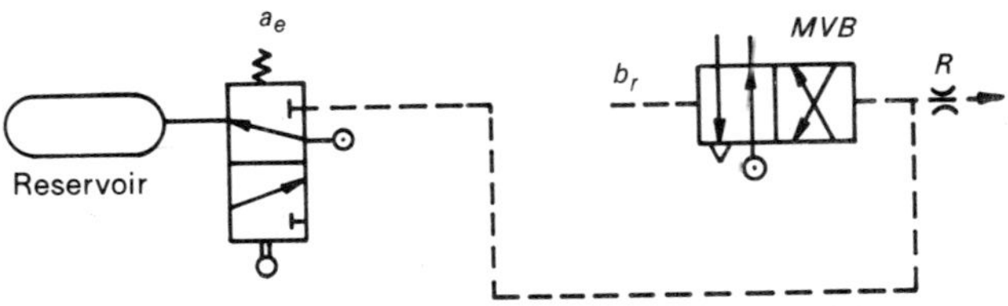

Fig. 16.9. A means of avoiding locked on signals

16.7 and 16.8, has the normal output port connected to a small air reservoir of 2–3 in^3 capacity which is normally connected to supply. When the valve is actuated by rod A, the supply is cut off and the air reservoir is connected to the main valve B. The signal operates MVB and then disperses through the bleed R, the magnitude of which affects the period of application of the signal from a_e. When the signal from a_e has dispersed there is no pressure to oppose the signal from b_r making the rod B retract.

Another device, the trip valve, for providing limited period signals, is shown in Fig. 16.10. This is so designed that it is only tripped as a rod passes in one

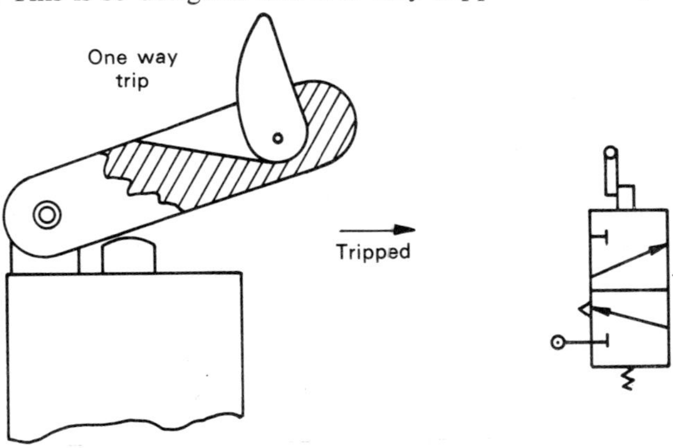

Fig. 16.10. The one-way trip valve

direction, in this case to the right. Unfortunately, its action requires that it be placed *near* the end of the stroke and not at the end. This allows the striker on the rod to trip the valve sending an impulse to the main valve and then to ride over the trip so removing the signal to the main valve. When the rod returns in the other direction it does not actuate the valve. The application of trip valves in the circuit for the sequence $A_eB_eB_rA_r$ is shown in Fig. 16.11.

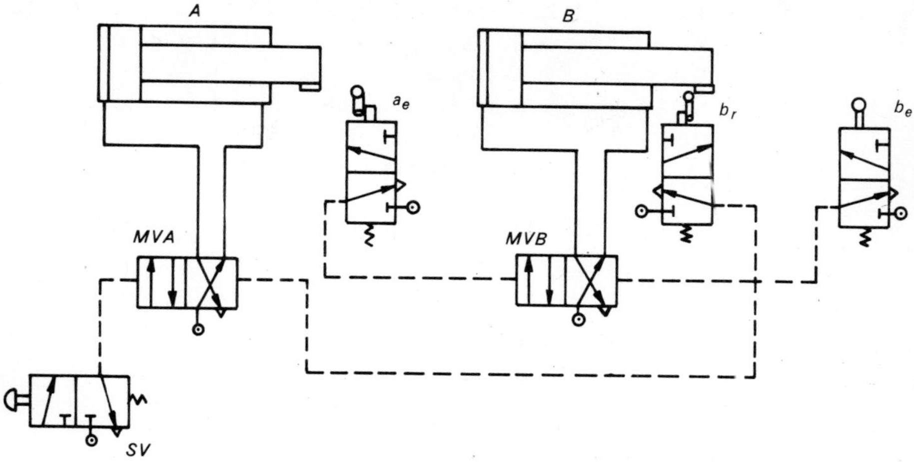

Fig. 16.11. The circuit for the sequence $A_eB_eB_rA_r$ using one-way trip valves

When limited period signals are used, there is always a chance that on some occasion the pulse signal may not be sufficient to move the main valve, because of seal friction or slight dirt build up. Greater reliability is achieved by using unlimited period signals.

16.6. Use of electrics

The above mentioned techniques for avoiding locked on signals are popular in pneumatic circuits but, in hydraulics, the use of electric limit switches and solenoid operated valves is common practice. In Fig. 16.12, the electrical cir-

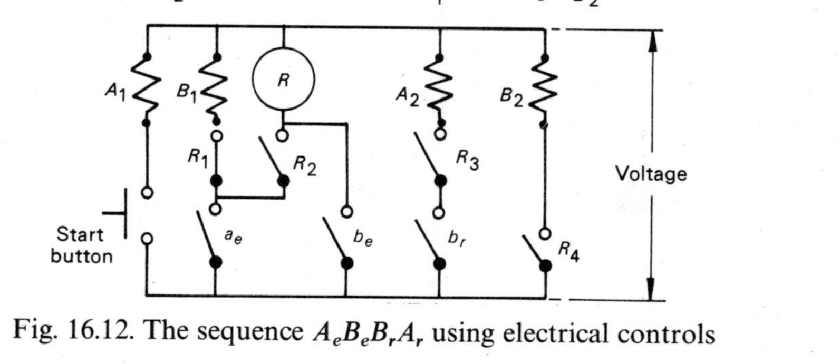

Fig. 16.12. The sequence $A_eB_eB_rA_r$ using electrical controls

cuitry for the sequence $A_eB_eB_rA_r$ is shown. There are four main types of component; a start button, solenoids A_1, A_2, B_1 and B_2, micro switches a_e, b_e and b_r, and a latching relay with four contacts R_1, R_2, R_3 and R_4. When the start button is depressed solenoid A_1 is actuated causing rod A to extend. Microswitch a_e then activates solenoid B, through contacts R_1, of the relay R. The rod B extends and b_e activates the relay, breaking contacts R_1 and making contacts R_2, R_3 and R_4. Solenoid B_2 is now activated through R_4 and the rod B retracts. The microswitch b_r then closes and activates solenoid A_2 through contacts R_3. Rod A then retracts and as microswitch a_e breaks, the current to relay R is interrupted and its contacts revert to their original position. The relay fulfils the same function as the selector valve of the earlier example.

16.7. A simple design technique for sequence circuits

This technique, proposed in [119], will be demonstrated by applying it to the sequence $A_eB_eB_rA_r$.

1. Draw the cylinder, main valves and pilot valves. Note, Fig. 16.13, the simplified symbol for a pilot valve. The left-hand connection is the supply to the valve and the right-hand connection is the output.

2. Determine the pilot signals acting on the main valves. In this case, the signal a_r must initiate the movement A_e and hence must act on the left-hand side of MVA. After A_e is completed, the signal from a_e must initiate B_e and hence must act on the left-hand side of MVB. The positions of b_r and b_e are similarly determined.

3. Draw the cylinder rods in their different sequential positions and hence determine if there are locked on signals on the main valves. In this example, pilot signals a_r and b_r must not exist simultaneously since they act on opposite sides of MVA. Similarly, the simultaneous existence of signals a_e and b_e must be avoided. The first step of the chosen sequence shows a_r and b_r actuated. If both these valves are powered from the supply then there is a locked on signal on MVA. Similarly, in the third step a_e and b_e are actuated and there is the possibility of locked on signals on MVB. Consider the step 1 where a_r and b_r are both actuated. One of these signals must be eliminated—which one? Signal a_r initiates the next step in the sequence to A_e and so must be retained; b_r *must then be eliminated at this stage*. This is indicated on the figure by putting an x at the b_r signal and a $\surd$ at the a_r signal. Similarly at step 3, a_e must be removed and b_e must be retained to initiate the next part of the sequence to B_r. A selector valve a_e. At the end of step 1, a_r switches the selector valve to remove the supply a_e is removed by b_e switching the selector valve. This removes the supply from valve a_e. At the end of step 1 a_r switches the selector valve to remove the supply from b_r. Since valves a_r and b_e control the selector valve, they must be supplied from the main supply. The two other pilot valves are supplied by the selector valve.

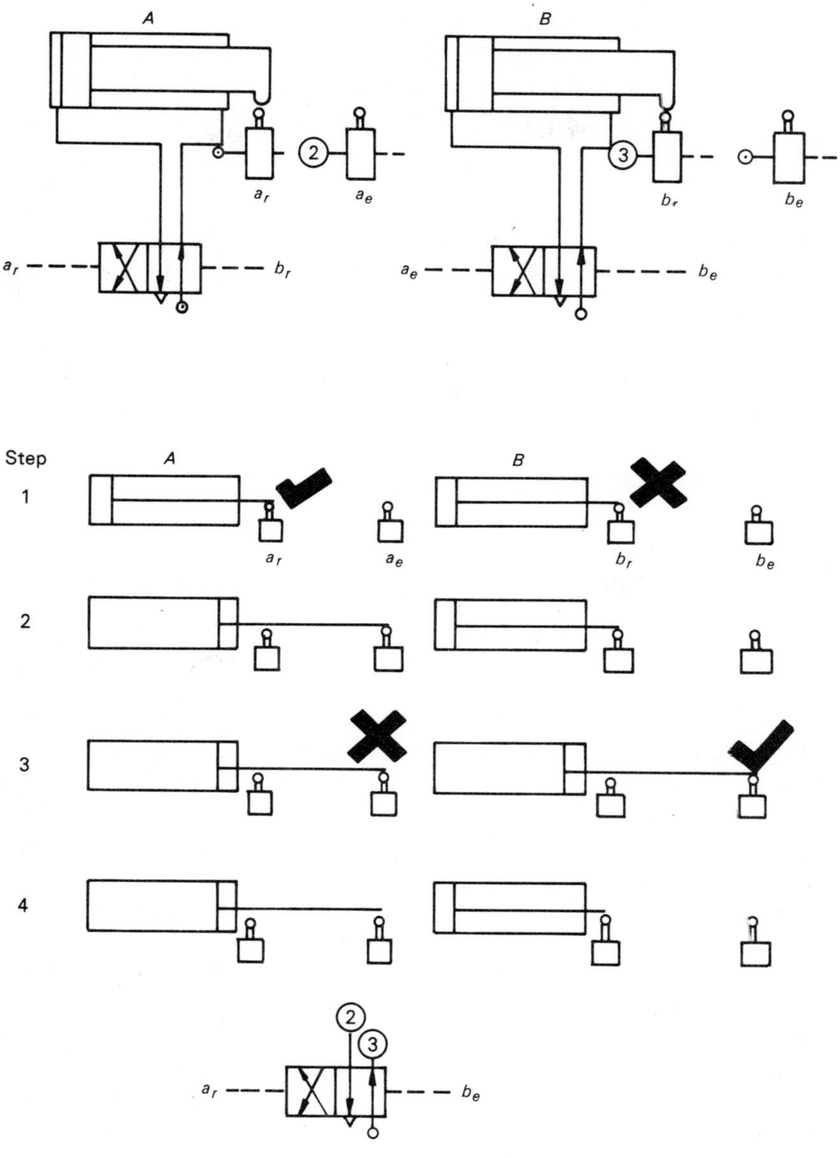

Fig. 16.13. A breakdown of the design technique for the sequence $A_eB_eB_rA_r$

16.7.1 The sequence $A_eA_rB_eB_r$

This sequence is unusual in that conditions at the start of step 1 and step 3 are identical, i.e. both rods retracted. The design method described above will be adopted, but in future problems the method will be reduced to tabular form. This particular sequence can be described in the following way:

Table 16.1

Output	A_e	A_r	B_e	B_r
Input	b_r	a_e	a_r	b_e

The input on the lower row is the signal needed to produce the movement on the row above.

Table 16.2

Step	Output	a_r	a_e	b_r	b_e
1	A_e	1	0	1	0
2	A_r	0	1	1	0
3	B_e	1	0	1	0
4	B_r	1	0	0	1

Use of this table avoids the need to draw the cylinder rods in their various positions throughout the sequence. The table relates the inputs $a_r a_e b_r b_e$ at the start of any step in the sequence to the movement required in that step. 1 indicates that a limit valve is actuated and an 0 indicates that the valve is not actuated, e.g. at the start of step 1 both rods are retracted, both valves a_r and a_e are actuated giving $a_r = b_r = 1$. The other two valves a_e and b_e are not operated. Hence when $a_r = b_r = 1$ and $a_e = b_e = 0$, the sequence demands that rod A extends to A_e.

Table 16.1 shows that main valve A is operated in different directions by inputs a_e and b_r. Similarly, main valve B is operated by inputs a_r and b_e. Table 16.2 is now inspected to see if locked on signals occur. At the start of step 2 there is a possibility of a locked on condition on main valve A. Input a_e initiates the next part of the sequence and must be powered, but b_r can be eliminated by removing its supply. Again in step 4 the supply must be removed from limit a_r to avoid a locked signal on MVB. The circuit is drawn in Fig. 16.14.

This design method does not always produce the circuit with the minimum components. Here, four pilot valves and three four-way valves are needed. Figure 16.15 shows a single cycle circuit which will give the same sequence of operations without the use of a selector valve. This circuit distinguishes between the similar conditions at the start of steps 1 and 3 by using the A_e movement to preset an air return, roller operated pilot valve a_e. This is reset later in the sequence. The valve a_e acts as a memory; it remembers that the last movement was A_e and only allows B_e while it is set.

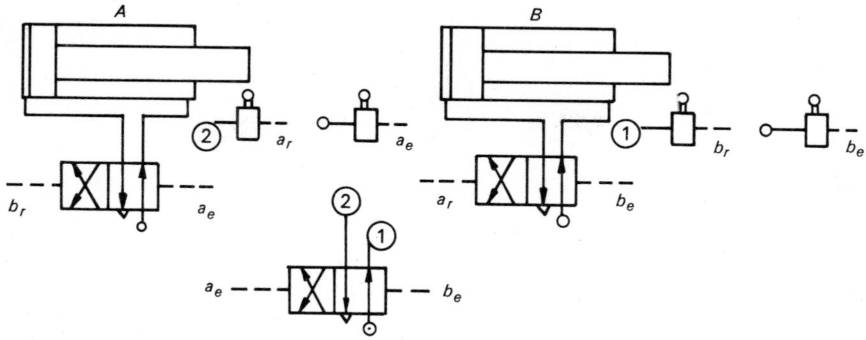

Fig. 16.14. Circuit for the sequence $A_eA_rB_eB_r$

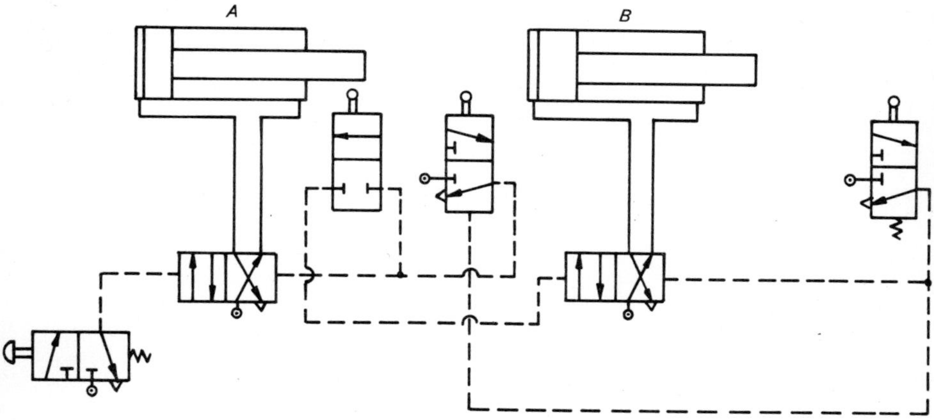

Fig. 16.15. Circuit for the sequence $A_eA_rB_eB_r$ single shot, using an air return pilot valve

16.7.2 The sequence $A_e B_e B_r C_e C_r A_r$

This design method can be extended to any number of cylinders. This example shows how it can be applied to a three cylinder sequence of the 'clamp and operate' type. In practice, cylinder *A* might clamp while cylinders *B* and *C* fed drills and taps respectively. Again the input output relationships are presented in tabular form.

Table 16.3

Output	A_e	B_e	B_r	C_e	C_r	A_r
Input	a_r	a_e	b_e	b_r	c_e	c_r

Table 16.4

Step	Output	a_r	a_e	b_r	b_e	c_r	c_e
1	A_e	1	0	1	0	1	0
2	B_e	0	1	1	0	1	0
3	B_r	0	1	0	1	1	0
4	C_e	0	1	1	0	1	0
5	C_r	0	1	1	0	0	1
6	A_r	0	1	1	0	1	0

Table 16.3 shows that simultaneous existence of signals a_r and c_r, a_e and b_e, b_r and c_e must be avoided. Table 16.4 shows that there is a possibility of locked on signals at steps 1, 3 and 5 and at these steps limits c_r, a_e and b_r respectively must have their supply removed. This is achieved by the use of two selector valves, Fig. 16.16.

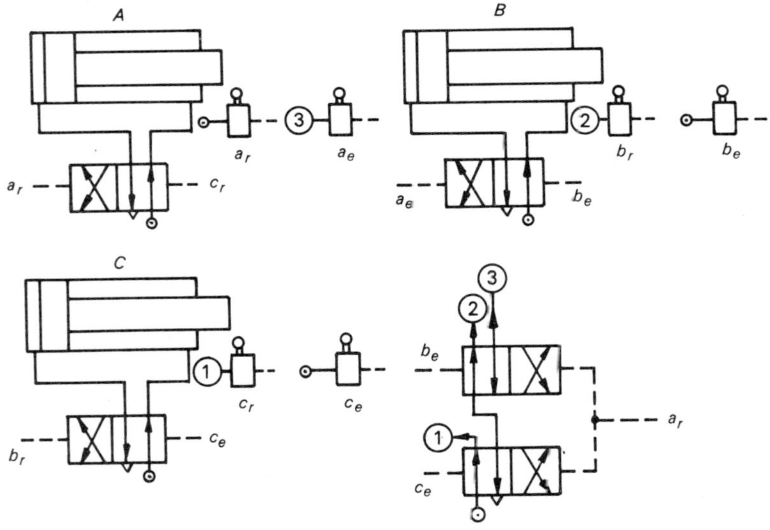

Fig. 16.16. Circuit for the sequence $A_eB_eB_rC_eC_rA_r$

16.8. The Cascade system [121]

This provides a straightforward method of designing sequential circuits and always gives a workable circuit which can rarely be improved upon. A clear set of rules has been laid down for this method with the result that a circuit can be designed on a step by step basis without a lot of mental effort. In this respect it is superior to the above mentioned technique. Consider, for example, $A_eB_eB_rC_eC_rA_r$. The first task is to split this sequence into groups so that no letter appears twice in any group. In this case

Group I is A_eB_e
Group II is B_rC_e
Group III is C_rA_r

The importance of correct sequence splitting cannot be overemphasised, since it determines the number of components in the circuit. Splitting should be applied both in the forward and reverse directions as a check. For example, consider the sequence $A_eB_eA_rC_eC_rD_eD_rC_eC_rB_r$. Forward splitting gives apparently five groups

$$\underbrace{A_eB_e}_{1}/\underbrace{A_rC_e}_{2}/\underbrace{C_rD_e}_{3}/\underbrace{D_rC_e}_{4}/\underbrace{C_rB_r}_{5}$$

Backward splitting makes it clearer that there are only four groups

$$\underbrace{A_e/\underbrace{B_eA_rC_e}_{2}/\underbrace{C_rD_e}_{3}/\underbrace{D_rC_e}_{4}/C_rB_r}_{1}$$

Having established the number of groups, the next job is to ensure that power supplies for all the pilot valves in each group are removed as soon as the group is completed, and at the same time that the supplies to the pilot valves of the next group are turned on. Figure 16.17 shows how two selector valves can provide three supply lines, one to each group. In this example, the first group is A_eB_e and the pilot valves a_e and b_e must be supplied. At the start of this group both selector valves are moved to the left by signal a_r, the last signal from the preceding group C_rA_r. This powers line I, switches *MVA* and sends rod *A* to A_e. The signal from valve a_e, which is also supplied from line I, switches *MVB* and rod *B* goes to B_e. It trips valve b_e, which is supplied from line *I*. This is the last signal from group I. It is used to remove all supplies from group I pilot valves and to supply the group II valves.

It is easy to work out the number of components required for a cascade circuit. If there are n cylinders, n main valves and $2n$ pilot valves are needed. In addition, $m - 1$ selector valves are needed, where m is the number of groups. It should be noted, however, that as the group number increases the method of grouping selector valves shown in Fig. 16.17 has a disadvantage in that a control signal may have to traverse many valves before arriving at the required point. The pressure loss may be intolerable. This can be overcome by the methods shown in Figs. 16.18 and 16.19. Here a signal has at most two valves to traverse.

The cascade system has some notable advantages:

1. A minimum number of different valve types is needed.
2. Circuits can be easily and quickly prepared.
3. Bad practice of using limited period signals is avoided.
4. Fault tracing is simplified.
5. It is easy to arrange a sequence to stop at any desired point. In Fig. 16.17, for example, the sequence can be stopped at C_e by breaking the line from c_e to the selector valve.

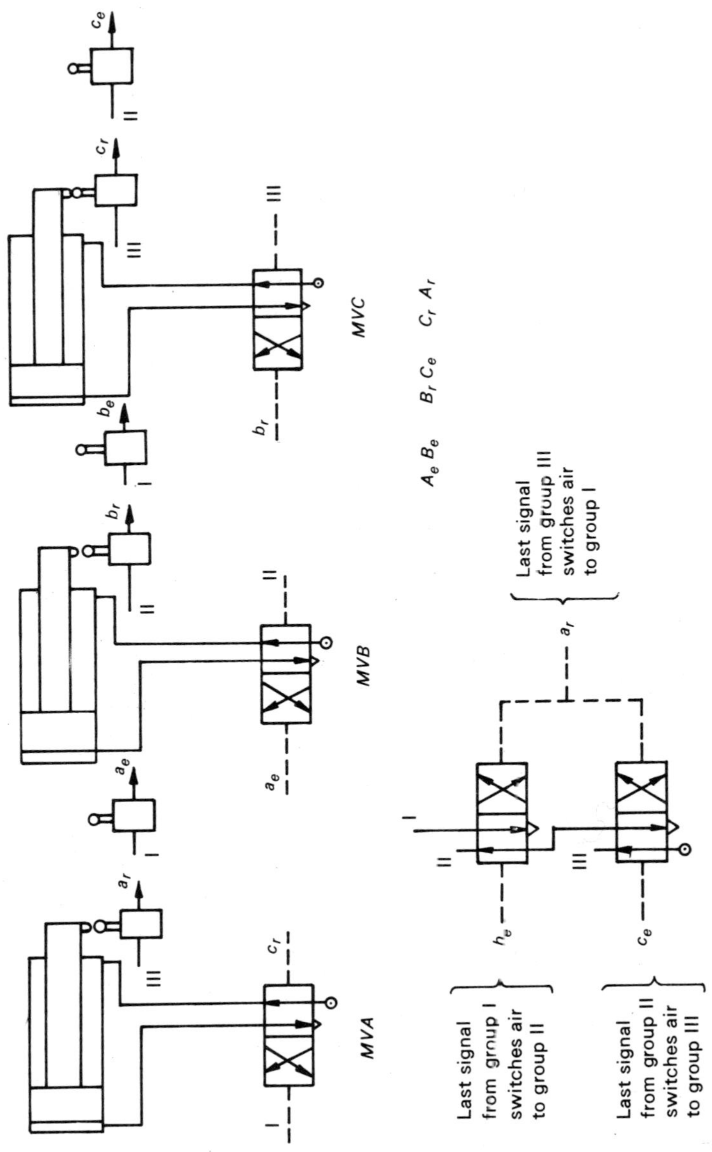

Fig. 16.17. Use of the Cascade method for the sequence $A_eB_eB_rC_eC_rA_r$

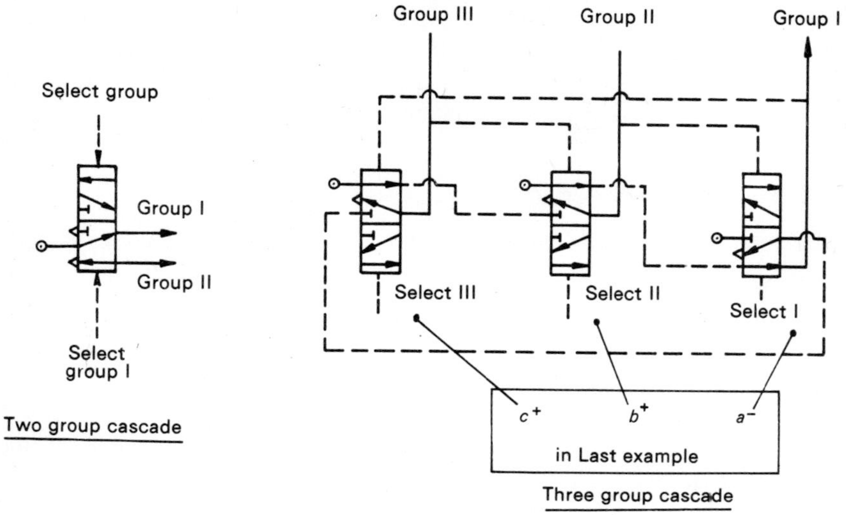

Fig. 16.18. Ways of grouping selector valves for 2- and 3-group cascades

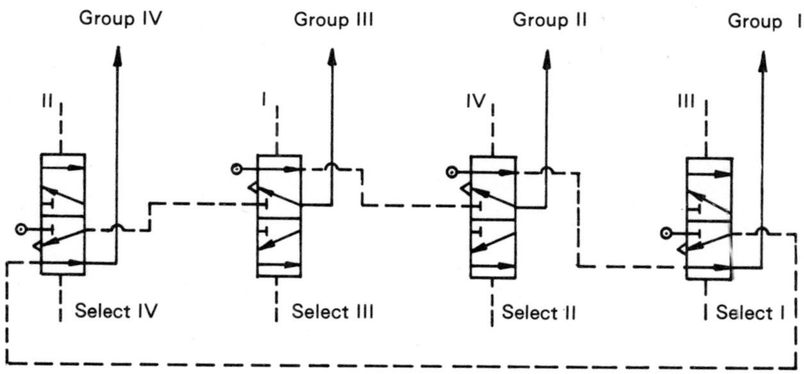

Fig. 16.19. Selector valve grouping for a 4-group cascade

The main disadvantage is that the method does not always result in the circuit with the minimum number of components. As seen earlier (Fig. 16.5) the repetitive sequence $A_eB_eA_rB_r$ in which extensions and retractions are in the same order, does not need a selector valve. The cascade method would include one. In addition, oscillating sequences such as $A_eA_rB_eB_r$ can be built more cheaply in other ways, see Figs. 16.14 and 16.15. The cascade method is commonly used in the design of pneumatic circuits and with a few modifications it can be used in hydraulic circuits too. The cascade method, for example, recommends that the last limit valve in any group be used to remove the supply from all valves in that group, including the last limit valve itself. In a pneumatic system, the compressibility of the air makes it possible for a valve to be used in this way to remove its own supply, e.g. valve b_e in Fig. 16.17. Hydraulic fluids are practically incompressible in comparison, however, and if this technique were used trouble might arise. This problem, in the circuit in Fig. 16.17, can be avoided if the last limit valve is permanently connected to supply.

16.8.1 The sequence $A_e B_e A_r A_e B_r A_r$

This sequence is unusual in that cylinder rod A makes two strokes. Fortunately this is easily dealt with using the cascade method if the second stroke of cylinder A is treated as if it were by a second cylinder A_2. The sequence is then written

$$\underbrace{A_{1e}B_e}_{1}/\underbrace{A_{1r}A_{2e}}_{2}/\underbrace{B_rA_{2e}}_{3}$$

Cylinder A has two limit valves a_{1e} and a_{1r} and cylinder A_2 has two limits a_{2e} and a_{2r}, Fig. 16.20. The construction of the circuit in Fig. 16.20 is left as an

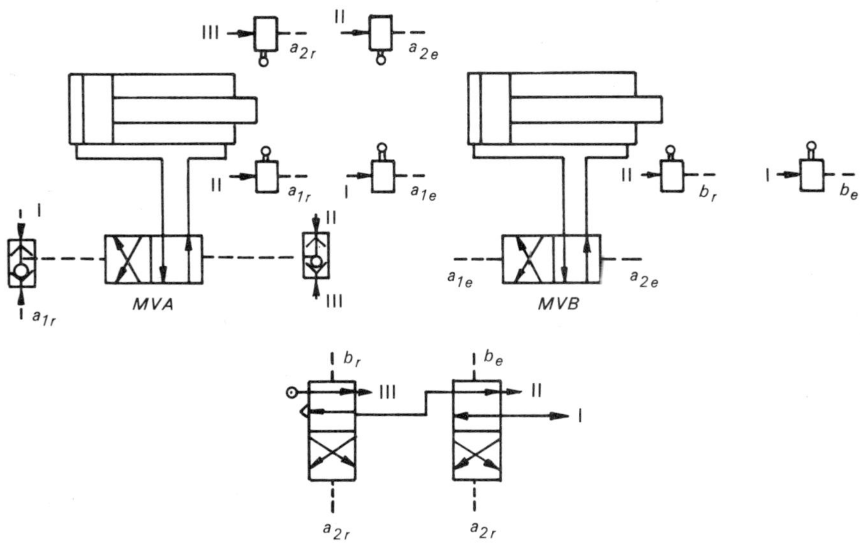

Fig. 16.20. Circuit for the sequence $A_eB_eA_rA_eB_rA_r$ using the cascade method

exercise for the reader. The only new point is the use of shuttle valves to isolate the double signal sources to MVA.

16.9. The logic approach to circuit design

Speed of response of a circuit can be greatly improved if fluidic elements are used for control instead of conventional valves. Interruptible jets can be used to detect the position of the cylinder rod. For example, Fig. 16.21 shows how two jets are used for this purpose. When the rod is retracted both jets are picked

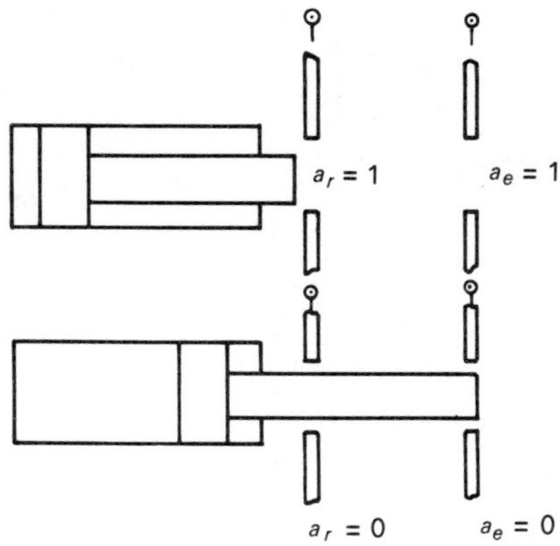

Fig. 16.21. Use of interruptible jets as limits

up by their receivers, $a_r = 1$, $a_e = 1$, and when the rod is extended both jets are interrupted $a_r = a_e = 0$. Consider again the repetitive sequence $A_eB_eA_rB_r$. A truth table, relating inputs and outputs can be constructed.

Table 16.5

	Inputs			
Outputs	a_r	a_e	b_r	b_e
A_e	1	1	1	1
B_e	0	0	1	1
A_r	0	0	0	0
B_r	1	1	0	0

This tells us, for example, that when both jets are interrupted on the A cylinder, and the B jets are not interrupted then the next movement is B_e. The logic circuit is shown in Fig. 16.22, in which the NOT and AND gates could in practice be any of the appropriate fluidic elements discussed in chapter 13. The output from an AND gate is amplified by a step up relay (S.U.R.) which enables it to actuate a double pilot operated main valve or memory valve. When this circuit is compared with that using conventional pneumatic components, it is seen that instead of four three-way roller operated valves, the system now requires at least eight fluidic elements capable of dealing with four inputs, there may be a greater number of elements required. Thus it is clear that the improved speed of response is at a price.

However, closer examination of this sequence reveals that the solution shown in Fig. 16.22 has several redundant elements. Consider again the truth table,

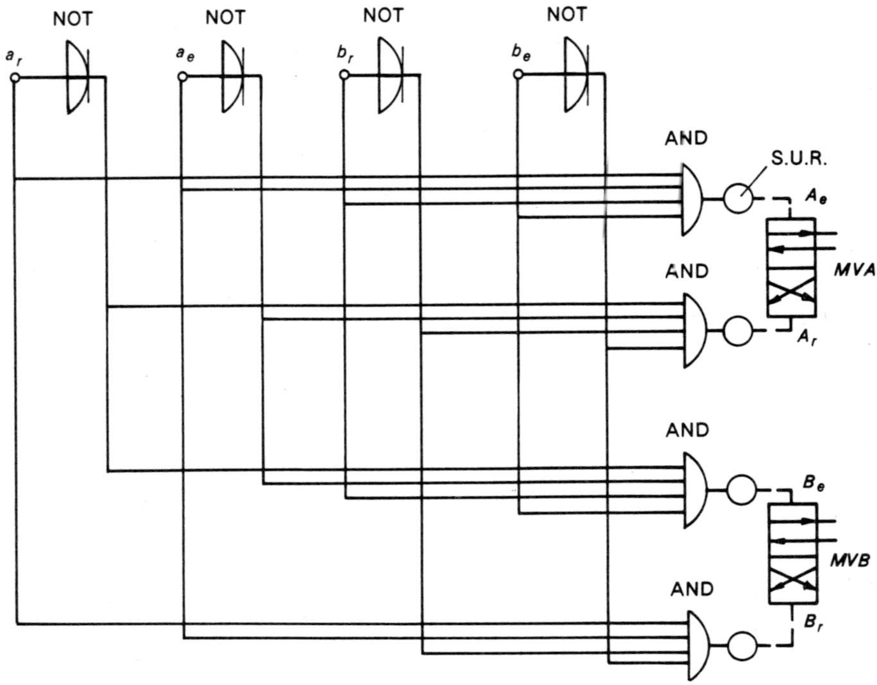

Fig. 16.22. Logic circuit for the sequence $A_eB_eA_rB_r$

(Table 16.5), which shows for example that

$$A_e = a_r \, a_e \, b_r \, b_e$$

The physical construction of the cylinder however ensures that signal a_e exists if a_r exists. Similarly b_e is on when b_r is on. Hence the logic simplifies to

$$A_e = a_r \, b_r$$

Further inspection of the truth table shows that before the command A_e is given, the cylinder B is in the retracted position and hence signal b_r is on. The logic is thus further reduced to

$$A_e = a_r \cdot 1 = a_r$$

If the other outputs are treated similarly the logic is:

$$B_e = \bar{a}_e$$

$$A_r = \bar{b}_e$$

$$B_r = a_r$$

The circuit is shown in Fig. 16.23 and this is seen to give a considerable reduction in components over that of Fig. 16.22.

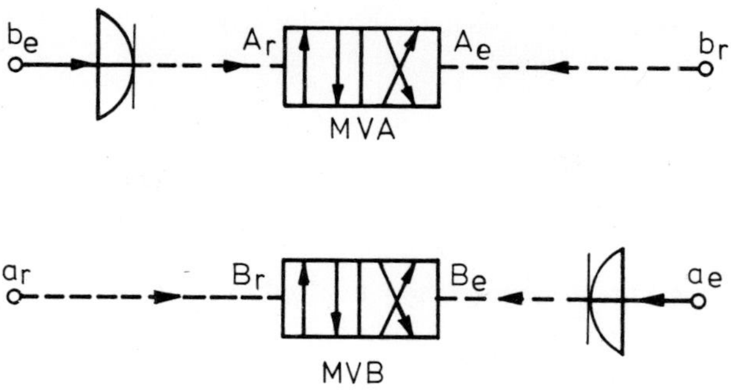

Fig. 16.23. Reduced circuit for the sequence $A_eB_eA_rB_r$

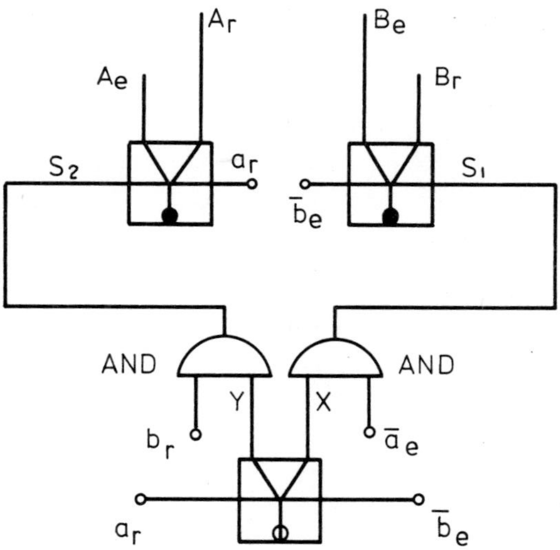

Fig. 16.24. Logic circuit for the sequence $A_eB_eB_rA_r$

The sequence A_e B_e B_r A_r suffered from locked on signals and it is of interest to see what this leads to if the logic approach is used. The truth table in this case is:

	Inputs					
Outputs	a_r	a_e	b_r	b_e	X	Y
A_e	1	1	1	1	1	0
B_e	0	0	1	1	1	0
B_r	0	0	0	0	0	1
A_r	0	0	1	1	0	1

If one follows the procedure described above for sequence A_e B_e A_r B_r the following logic results (ignoring infants X and Y for the moment)

$$\begin{aligned} A_e &= a_r \\ B_e &= \bar{a}_e \\ B_r &= \bar{b}_e \\ A_r &= b_r \end{aligned}$$

However, if these signals are applied to double pilot operated main valves (bistables), the problem of locked on signals arises and the system does not work. In this case, as seen earlier when discussing the use of valves in this sequence, the $\bar{b}_e$ and the $\bar{a}_e$ signals interfere, as do the a_r and b_r signals. The existence of such conditions can also be gathered from the truth table. Their presence is indicated by different output conditions being demanded from similar inputs. For example, Table 16.6 shows that outputs B_e and A_r both require the same input combination. Some means is required to differentiate between these two states and this is usually achieved by using a bistable whose output is different at each of the two states. A suitable circuit is shown in Fig. 16.24. A bistable and two AND gates are needed, in addition to the components of Fig. 16.23.

Chapter 17

Discontinuously controlled and digital servos

17.1. Introduction

The familiar electrohydraulic servo is an analogue device and much effort has been directed at determining its stability and performance on a theoretical and practical basis. Nowadays, the digital computer is commonly used in systems, either for presenting input data to the control system or as part of the control loop and, as a result, fluid power engineers have been trying to devise means of using conventional fluid power systems in conjunction with digital data. Digital systems have several advantages over their analogue counterparts.

1. Digital systems can be very accurate. The main limit on accuracy is the amount of data or bits of information that have to be handled.
2. Digital data is not impaired by manipulation, two state devices being relatively insensitive to noise.
3. Systems are generally more reliable because of the simplicity of the basic two state element.

Their main disadvantages are:

1. Complexity of controls.
2. High cost of digital to analogue conversion, where necessary.

This chapter presents an introduction to methods of digital control, on-off control and to pulse width modulation. It includes an analysis of cavity formation in on-off controlled hydraulic servos.

17.2. Digital to analogue converters

Since the familiar analogue hydraulic servo is well established and is used on many machine tools, the most direct approach to making hydraulics and digital data compatible is to use digital to analogue conversion. In some systems the converter is placed between the digital data source and the input to the control system. In others, higher accuracy is achieved by placing two d.a. converters in the loop, one for the error signal and one for the feedback signal. This necessitates the use of a digital comparator.

The fluid power engineer is well aware that bistable fluidic elements are ideal for digital data handling. Interface problems between the fluid power system and the data processor are reduced and in some cases eliminated. In addition, the added advantages of environmental tolerance and reliability make their use attractive. An example, proposed by Parker [123], of a pneumatic digital to analogue converter is shown in Fig. 17.1. This could be interposed between

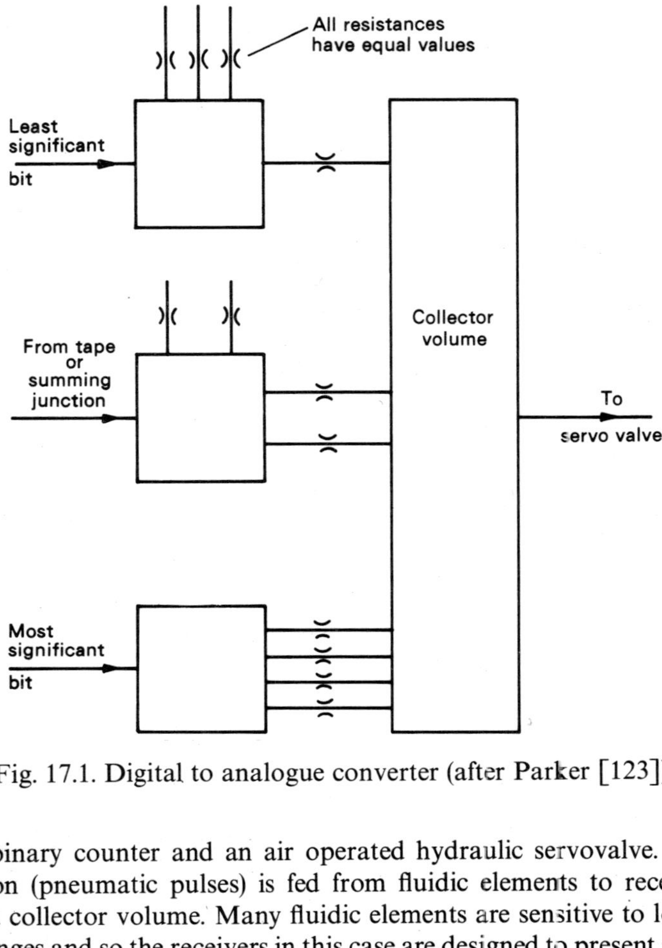

Fig. 17.1. Digital to analogue converter (after Parker [123])

a fluidic binary counter and an air operated hydraulic servovalve. Basically, information (pneumatic pulses) is fed from fluidic elements to receivers and hence to a collector volume. Many fluidic elements are sensitive to load impedance changes and so the receivers in this case are designed to present a constant load impedance. This is achieved by ensuring that each input 'sees' the same orifice impedance—four similar orifices. The most significant bit will have the most significant effect on the pressure rise in the collector volume, since it supplies the volume through four orifices.

Another more direct approach is to use the valve itself as the converter so that the valve can accept digital signals directly without the use of prior decoding devices. Two possible arrangements are shown in Fig. 17.2. In the first [124], the first stage of the valve consists of four arms of a hydraulic bridge, in which the controlling resistances are two sets of normally closed flapper valves in

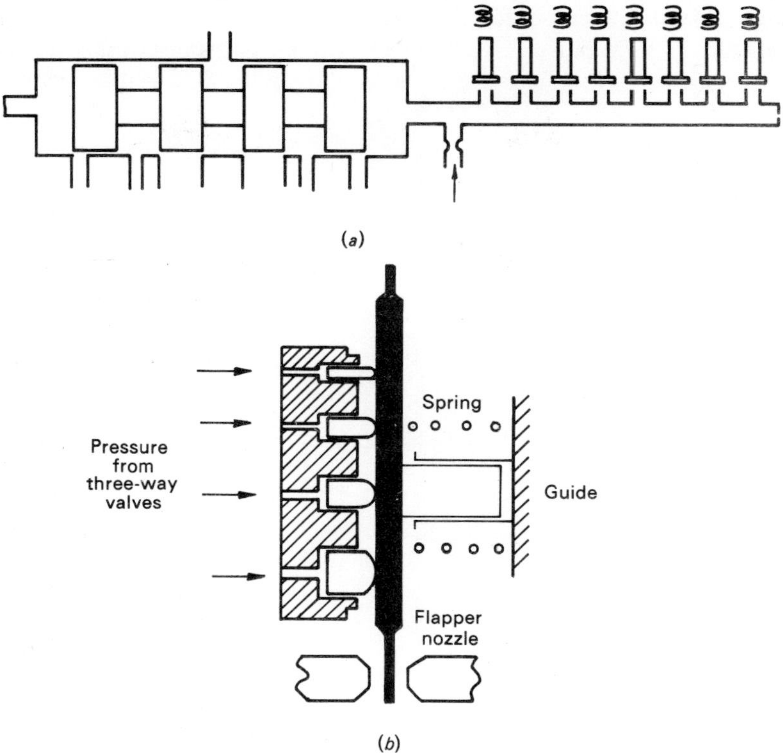

Fig. 17.2. (*a*) Digital to canalogue valve using binary orifices (after Wells [124]) (*b*) Digital to analogue valve using binary cylinders (after Himmler [125])

parallel. The orifices are sized in binary progression so that opening the first gives a pressure drop of x lbf in^{-2} across the second stage spool and opening the second gives $2x$ lbf in^{-2} and so on. The solenoids are driven by a parallel digital system. The second example uses binary forces [125], acting on a spring loaded flapper first stage. Both systems can accept a tape input, the magnitude of the signal not being critical so that in the first case voltage stabilisation is unnecessary. The main disadvantage is that much data must be handled if high accuracy is desired.

17.3. Pulse controls

In the systems described above, parallel digital information is used, i.e. all bits of information can be supplied simultaneously to the actuator. Another approach is to use sequential control, in which the number of control pulses arriving in sequence determines the output of the device. This method has been used successfully with fluid systems. The main element is the stepping motor, which rotates through a fixed movement for each applied pulse. Electrical stepping motors can operate at 1000 pulses/s, each of 1·5°. Figure 17.3 shows the use of such a

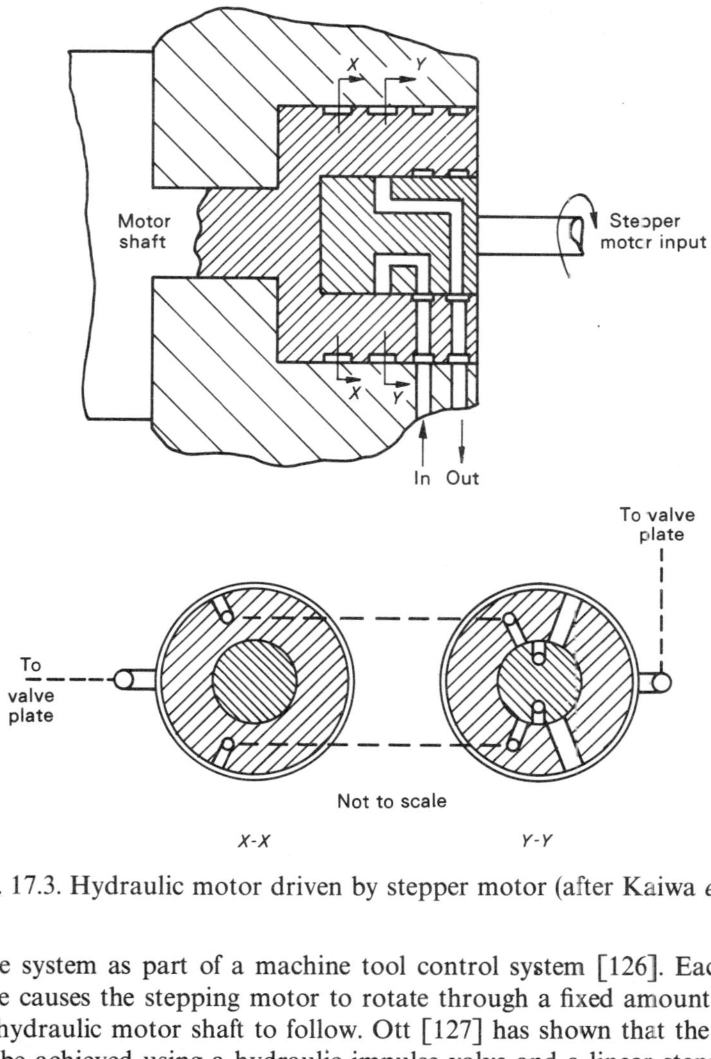

Fig. 17.3. Hydraulic motor driven by stepper motor (after Kaiwa *et al.* [126])

pulse system as part of a machine tool control system [126]. Each electrical pulse causes the stepping motor to rotate through a fixed amount and causes the hydraulic motor shaft to follow. Ott [127] has shown that the same thing can be achieved using a hydraulic impulse valve and a linear stepping motor. The impulse valve accepts low power positive and negative pulses (from perhaps a diaphragm driven by a bistable fluidic element) and sends high power pulses to a stepping motor. Figure 17.4 shows that, for a rightward movement of the direction control valve, the input pulses $\pm\Delta$ will pressurise alternatively ports 1 and 2, while simultaneously exhausting ports 3 and 4. These ports 1–4 lead to the inlet and outlet passages of the fixed central element of the stepping motor. As shown, a pulse at 2 will send fluid to the right-hand side of the main piston, causing it to move to the left. As it moves it closes off the connection to passage

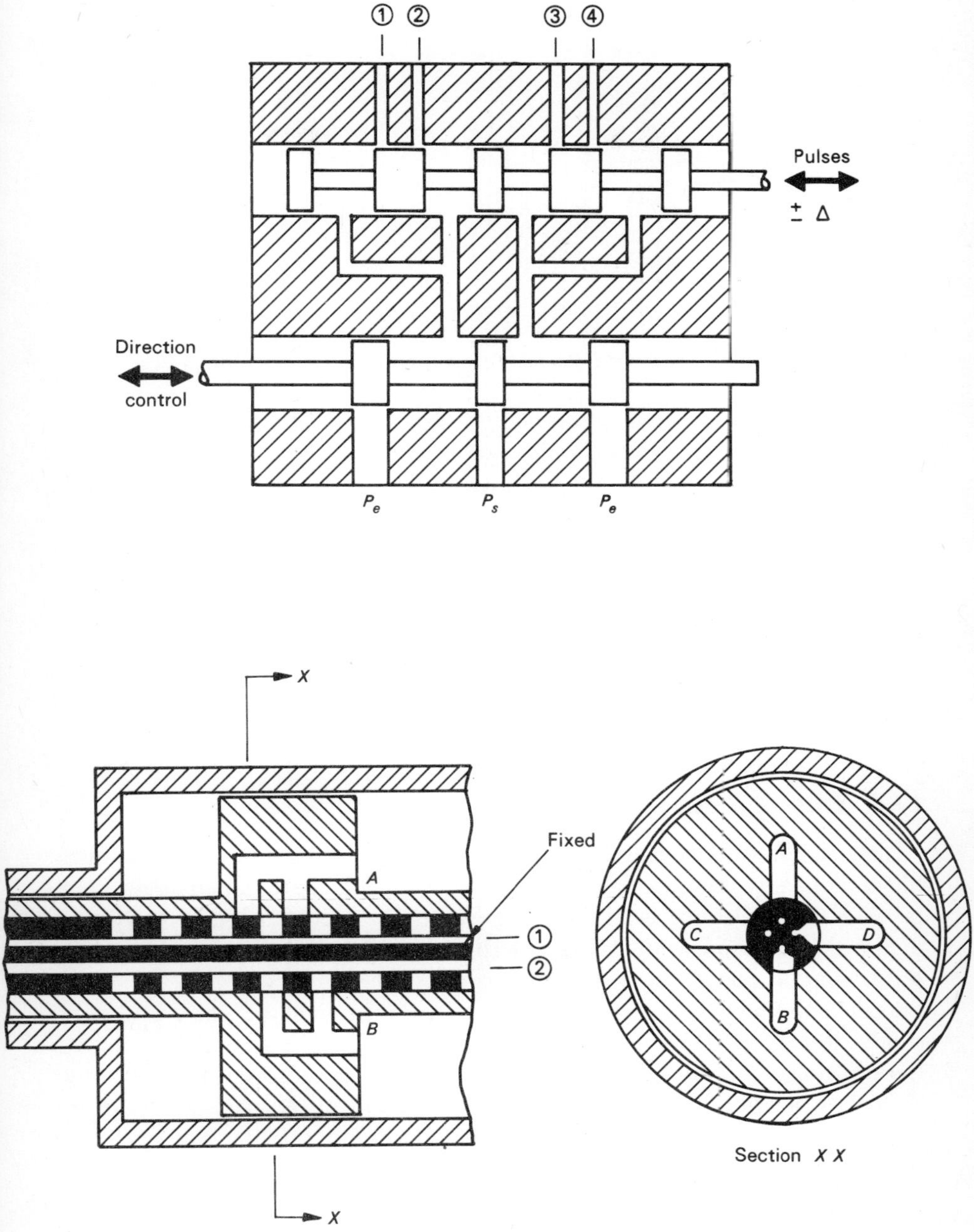

Fig. 17.4. (*a*) hydraulic impulse valve (*b*) linear stepping motor driven by the impulse valve (after Ott [127])

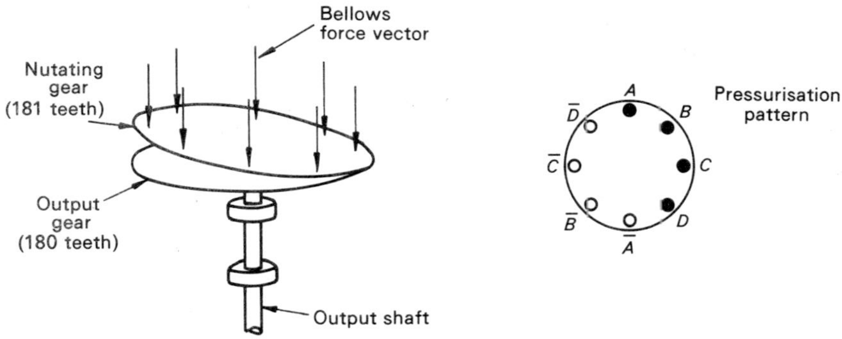

Fig. 17.5. The nutating gear stepping motor (after Griffin [128])

2 and when it has moved one port width it will stop. Other ports allow the fluid on the left-hand side of the piston to exhaust.

Another example, Fig. 17.5, is the nutating gear stepping motor [128], which has been designed for use with fluidic circuits. There are only two moving parts—a driving gear which is free to wobble but not to rotate, and an output gear which can rotate but cannot nutate. Unequal pressurisation of the driving bellows causes the driving gear to tilt and contact the output gear. Since the output gear has 180 teeth and the input gear 181 teeth, every complete revolution of the point of contact will make the output gear advance by 1 tooth or rotate 2°. Fluid logic circuitry has been developed to provide the correct sequence of bellow pressurisation.

17.4. Digital actuators

Nearly all of the above mentioned techniques have used in one form or another the analogue hydraulic servo. Another approach eliminates all analogue components and hence the need for digital to analogue conversion by relying on digital signals throughout; this uses a digital actuator which is capable only of positioning at one or more discrete positions. The simplest digital actuator is the conventional cylinder, positioned by a two-position valve. The cylinder rod can be considered as having only two discrete positions if such a valve is used, the positions being fully extended or fully retracted and clearly the accuracy of such a positioning device depends on the accuracy of manufacture of the cylinder. An important feature with this digital actuator is that it is an open loop system—positioning the rod at either of its two positions does not require the use of feedback and, as a result, stability problems should not exist. Digital actuators are used for machine tool control, programmed handling, indexing systems, gear changing and other applications where precise location at discrete positions is required. Figure 17.6 shows a means of providing more discrete positions [129]. The stops are driven either mechanically or by fluid power and setting the actuator in the desired position requires the movement of the appropriate stop and the pressurisation of the cylinder.

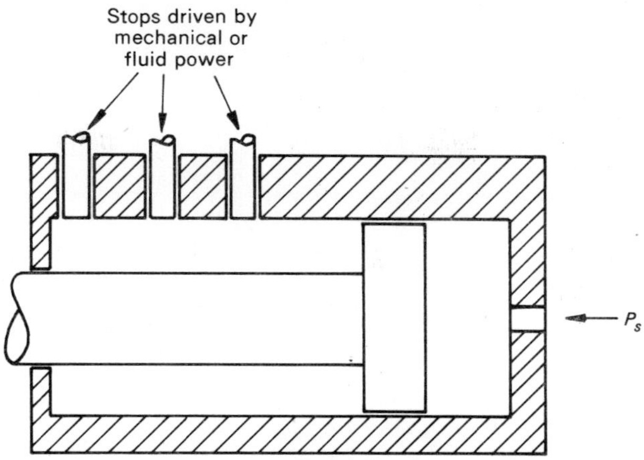

Fig. 17.6. A discrete position cylinder using mechanical stops (after Miles [129])

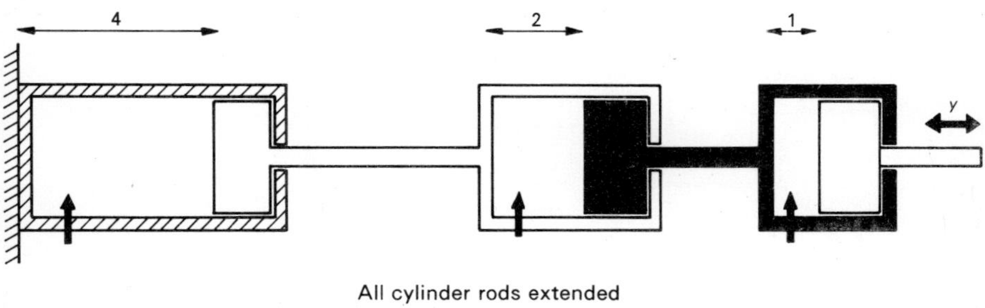

Fig. 17.7. A digital positioner using series cylinders

Another form of digital positioner uses cylinders connected in series, Fig. 17.7. With this device 2^n discrete positions are available if n cylinders are used. In addition, the cylinder strokes can be arranged in binary progression so that parallel binary information can be used to directly operate valves controlling each cylinder. One disadvantage of this simple system is its bulk and Fig. 17.8 shows a more compact design [130]. The Production Engineering Research Association have developed such a digital actuator with a stroke of approximately 32 in, smallest increment of 1/64 in, an accuracy of 0·0015 in and a speed of 50 ft min^{-1}. Others of shorter stroke and greater accuracy have been developed [131] for aircraft and aerospace applications where on-board digital computers are used to operate the actuators.

Binary information can be supplied from a fluidic counter using for example the binary flip-flop which changes its output condition on the reception of each pulse. The outputs from the counter can be amplified by step up relays to drive the appropriate cylinders. In some applications, conventional pneumatic valves are sufficiently fast in operation.

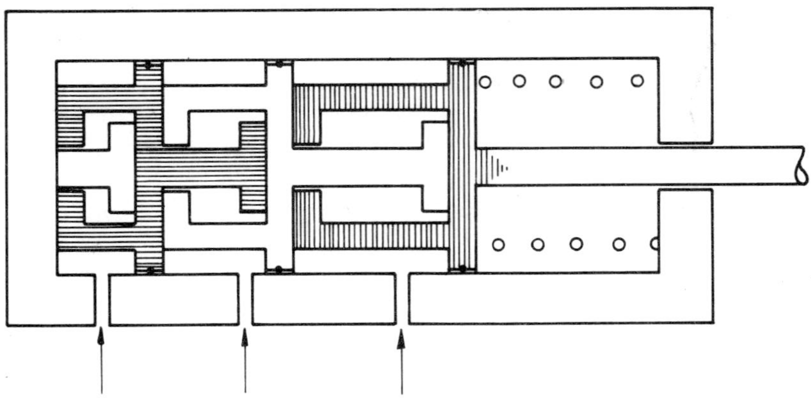

Fig. 17.8. A compact digital actuator

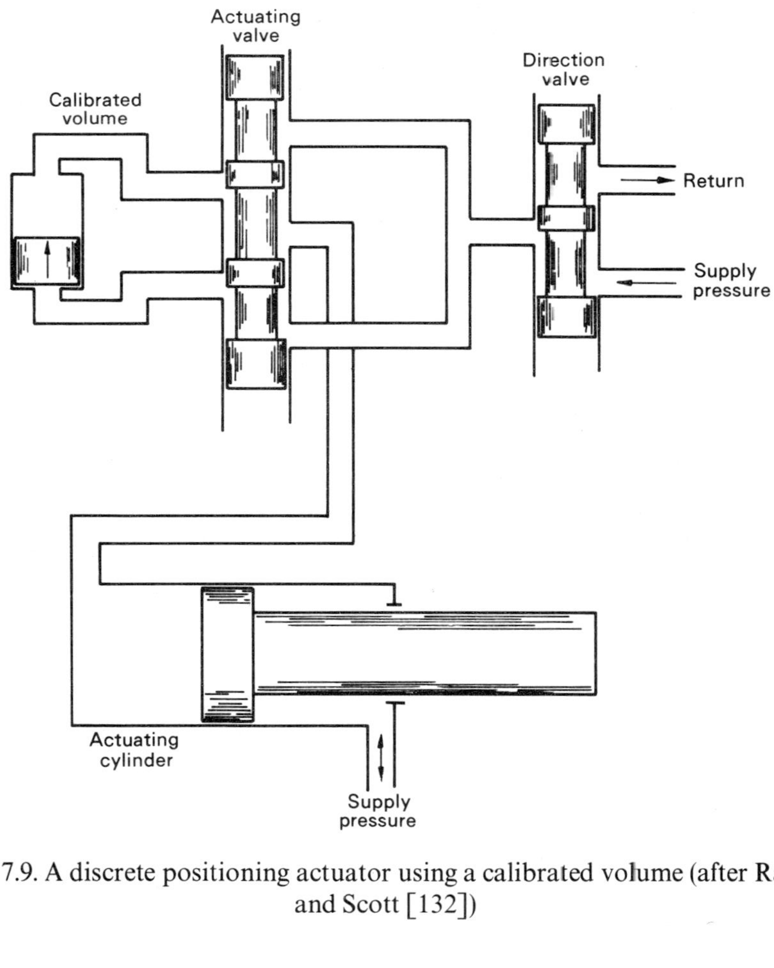

Fig. 17.9. A discrete positioning actuator using a calibrated volume (after Raider and Scott [132])

The main advantages of digital actuation can be summarised as:

1. They are basically open loop systems. Compressible fluids, which can cause stability problems in analogue systems, give no trouble.
2. The two-position valves associated with the device are cheap, reliable and not prone to contamination. A reported application uses hot gas at 2000°F and 3000 lbf in^{-2}.
3. The lack of stability problems allows tighter accuracy and faster response than analogue systems.
4. Digital to analogue conversion is unnecessary.
5. They can be operated directly from tape parallel information or from sequential data from counters.

A last example, Fig. 17.9, of a discrete positioning actuator uses the calibrated volume method [132]. Here two valves are used, a directional control and an actuating valve. The directional control determines which way the output will move. Each time the actuating valve changes position (up to down or vice versa), the piston displaces a calibrated volume into the actuating cylinder and the output steps a fixed increment. Leakage from the calibrated volume can cause problems. This is an incremental digital actuator and can only be commanded to a position adjacent to its current position. Full value digital actuators, such as that shown in Fig. 17.8, can be commanded to any position at any time.

17.5. On-off controlled systems

Comment has been passed earlier on some advantages of two-state devices. The simple two- and three-position control valves also have several advantages over the sophisticated servovalve. They are much cheaper. They require less elaborate driving equipment, e.g. the torque motor is designed for maximum torque and not for maximum linearity. They are less sensitive to the environment, and dirty oil or hot gas can be handled readily. The price to be paid for these advantages is the reduced accuracy of a simple on-off controlled servo. The major problem of on-off control is the limit cycle [142] and attempts to remove this by using relay dead zone or valve overlap tend to aggravate the problem of inaccuracy. Nevertheless, there are applications, such as rudder control [133], where slight inaccuracy can be tolerated. In on-off controlled hydraulic servos, cavitation is likely to occur, and analysis at the end of this chapter shows how this affects response.

17.6. Oscillating control systems—pulse width modulation

In order to retain the advantages of the two-position valve and to obtain better performance, the pulse width modulated valve was developed. The feature which distinguishes this form of control from the conventional proportional control is that, instead of generating a signal proportional to the error, the amplifier generates a series of pulses which alternately drive the valve to its two extreme positions. Pulse length modulation of an input signal θ_i can be

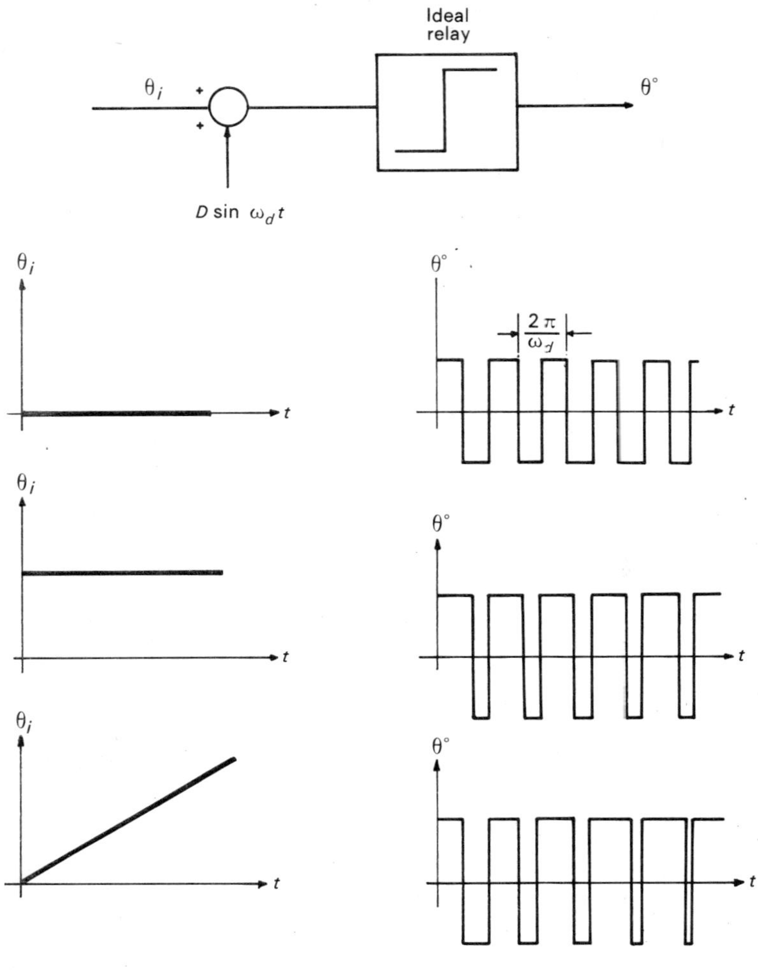

Fig. 17.10. Pulse width modulation using a bistable, and response to different inputs

achieved by adding a dither signal $D \sin \omega_d t$ to the input and passing the combined signal through a bistable element. Figure 17.10 shows the bistable output for various inputs. A step input for example, makes the valve dwell longer on one side than the other. In the acceleration switching valve [134], the output from the bistable drives a flapper nozzle valve, whose output is amplified by a second stage spool. Because of the double integration inherent in this two-stage valve, the no load cylinder rod acceleration is proportional to flapper position—hence the name 'acceleration switching valve'.

The addition of the dither signal to the input signal and the use of the bistable provides another means of processing analogue data in digital fashion. As discussed earlier, the use of the bistable in a hydraulic servo can lead to stability problems but, fortunately, this is not so in this case. The dither signal effectively linearises the characteristic of the non-linear bistable, the modified characteristic being referred to by Taft *et al.* [135] as the altered characteristic of the element. This is the plot of the average value of the output of the bistable over one cycle of the dither versus the input signal magnitude. For example, for sinusoidal dither, input to relay $= \theta_i + D \sin \omega_d t$. Then the average value $\bar{\theta}_o$ of output over one cycle can be shown to be

$$\bar{\theta}_o = \frac{2}{\pi} \sin^{-1} \frac{\theta_i}{D} \tag{17.1}$$

Figure 17.11 shows the effect of dither amplitude on the altered characteristic. As this is reduced the effective gain is increased, but the linear range is reduced.

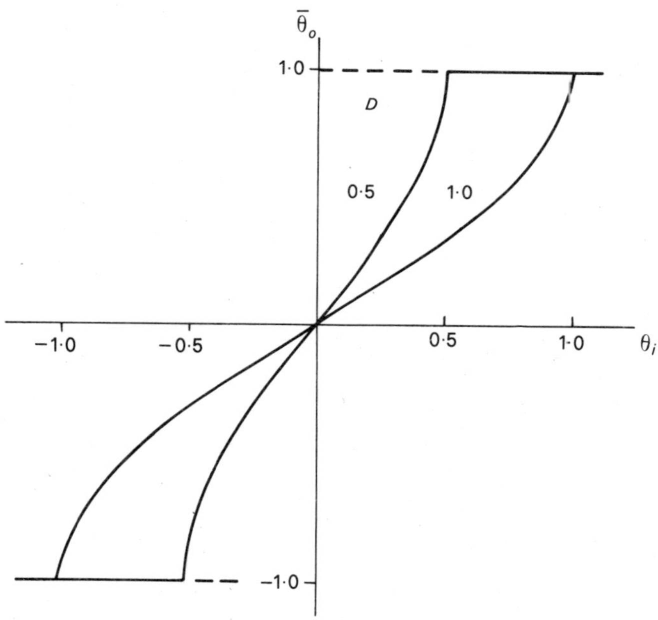

Fig. 17.11. The altered characteristic of a bistable using a sinusoidal modulator (after Taft and Nawaz [135])

This provides a means of varying system gain. Note that the device is saturated when the input signal magnitude exceeds the dither amplitude. The dither frequency should of course be high enough to avoid undesirable output oscillation. For a second order system, a dither frequency five or more times the system natural frequency gives satisfactory results.

Another method for obtaining pulse width modulation in practice uses fluidics [136]. Figure 17.12 shows that the dither is developed by a fluidic oscillator consisting of a bistable amplifier with positive feedback loops. The

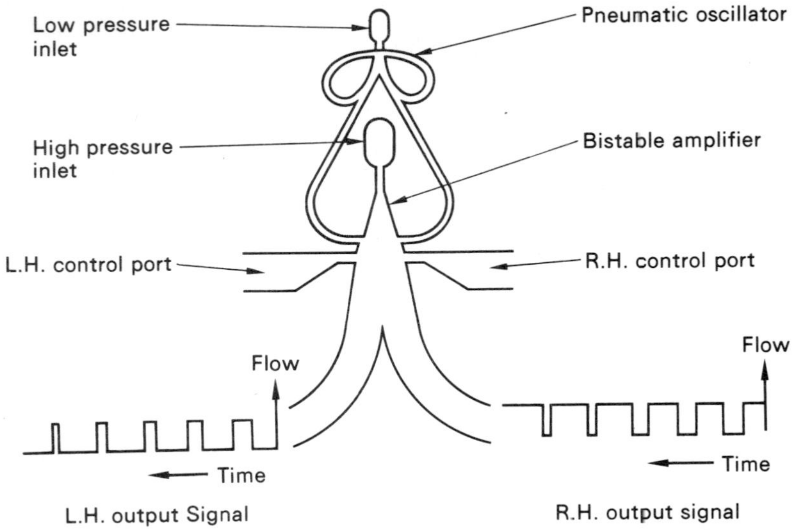

Fig. 17.12. Fluid logic circuitry to produce a pulse width modulated output proportional to a differential input

outputs of this device could be applied to the ends of the spool in a servo valve so providing a means of driving an actuator.

Besides the usual advantages of digital systems, pulse length modulated control of the servo valve is claimed to be superior to proportional control in several areas [134].

1. Valve gain is essentially constant for practically all inputs less than the saturation value.
2. There is little centre shift or gain change due to temperature effects.
3. Excellent resolution is obtained because of the dither effect of the carrier frequency.
4. Increased reliability, since the torque motor has a large force output and foreign bodies are thus unlikely to obstruct motion.
5. Pressure gain is much larger than proportional valves—hence system stiffness is improved and steady state errors due to actuator loading are reduced.

There is one significant disadvantage of this mode of operation. That is, the quiescent power drain when there is no input signal. Under these conditions, each side of the actuator is alternately pressurised and exhausted and this can lead, particularly in pneumatic systems, to substantial energy wastage. This can be overcome by using the technique of differential pulse length modulation which ensures that when there is no input signal, the control valve is closed. It has been shown [137] that this can be achieved by subtracting the carrier waveform from the modulated waveform, Fig. 17.13. A three-position valve, with closed centre position, can be used in this case. Reference [137] discusses a

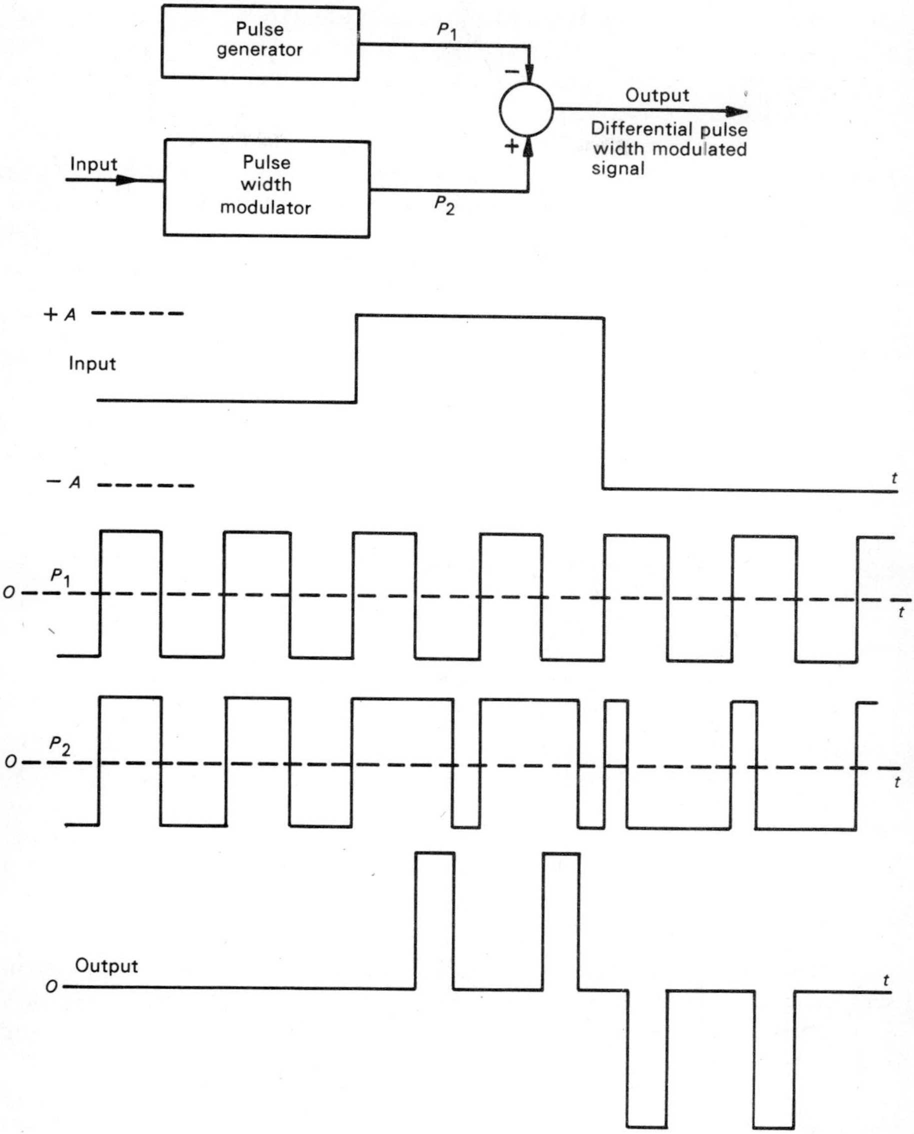

Fig. 17.13. Differential pulse length modulation

floating flapper disk valve which can be used in the implementation of pulse length modulation. It combines the simplicity of pure fluid elements with the shut off capability of the mechanical valve.

17.7. Response improvements using digital techniques

In analogue fluid power servos, stability compensation is achieved by using well established signal shaping techniques but, unfortunately, these cannot be used in digital systems unless digital to analogue conversion is used. Taft [135] has developed pure digital compensators which can be used in conjunction with the pulse width modulation technique. Compensation is achieved by adding or subtracting pulses to the bistable output every time it changes state. The so-called posicast controller is a particular form of this compensator in which a pulse is subtracted from the output of the bistable. Figure 17.14 shows how a posicast

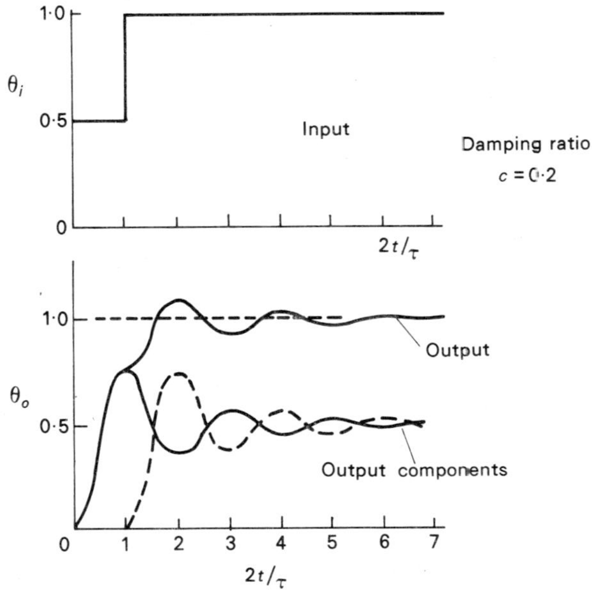

Fig. 17.14. Posicast control of a second order system with damping ratio $c = 0{\cdot}2$ and period T

controller [138] effectively improves the damping of a second order system with damping ratio $c = 0{\cdot}2$. The pulse duration is one-half of the system's periodic time. This method of compensation has been implemented by using fluidic elements and has operated satisfactorily.

One of the principal advantages of the hydraulic servo is its fast speed of response and in view of this it is clearly important that more effort be applied to design based on the theory of the time optimal controller. When moving a load from one position to another, on-off control can give faster response than proportional control. For a second order system, the valve is opened fully one way and the load moves with maximum acceleration. After an interval of time the valve is switched to its opposite extreme and the load decelerates at maximum rate. The switching point is chosen so that the load stops at the desired position. The number of switching points increases with the order of the system and, if valve dynamics and oil compressibility are all taken into account, time optimal control of a hydraulic servo would require a very complicated controller and the valve would need to switch position several times whilst driving the load from A to B in the shortest possible time. This is not feasible in practice. However, sub optimal controllers can be designed [139] using a reduced mathematical model and these, although being slower than the optimal, can still be appreciably faster than conventional controllers. For example, the system order can be reduced to two if valve dynamics and oil compressibility are ignored and a few possible controllers based on this model are shown in Fig. 17.15. The simplest

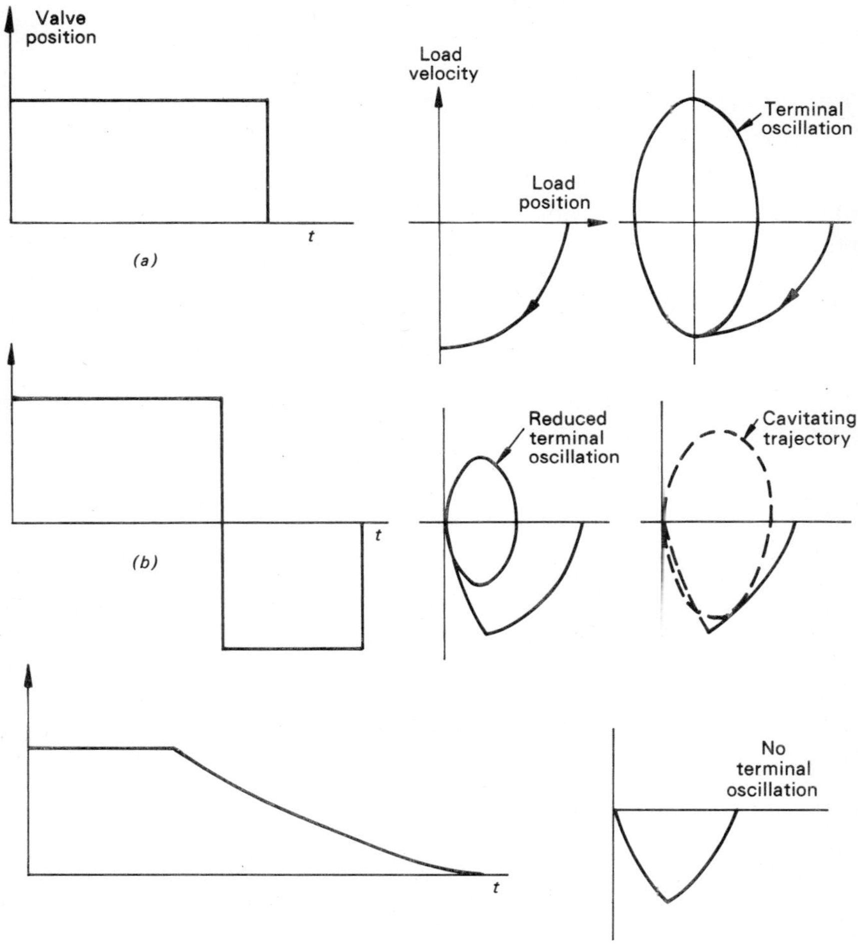

Fig. 17.15. Sub optimal control of a hydraulic actuator

controller closes the valve when the output has reached the desired position and if, as is assumed, the fluid is incompressible the output will remain fixed. In practice, of course, the hydraulic spring will cause the output to oscillate about the desired position. The magnitude of this terminal oscillation can be reduced if the load velocity is arranged to be zero when the valve is closed. This is the second controller illustrated and it can be seen that it is necessary to switch the valve twice so that the load reaches the desired position with zero velocity. However, although the terminal oscillation has been reduced in amplitude, it still exists because on valve closure there is still a pressure differential across the actuator piston. If cavitation occurs, the amplitude of the oscillation may still be large and if this is intolerable then a controlled valve closure must be used, so arranged that the load reaches the desired position both with zero velocity and zero load pressure. Although this is possible [139], the advantages of on-off control are lost.

17.8. Some phenomena associated with on-off controlled hydraulic seros

Discontinuous movements of the servo valve spool can lead to the occurrence of pressure peaks in the cylinder. In addition, cavitation is very likely to occur under these conditions and it is therefore of interest to examine the step response of the valve cylinder combination [140, 141]. Such discontinuous valve movements are encountered in on-off servos and in pulsed servos where the valve spool driving signal takes the form of a sequence of rectangular pulses generated by a sampler.

The valve actuator system is shown in Fig. 17.16. The system equations are repeated for the reader's sake.

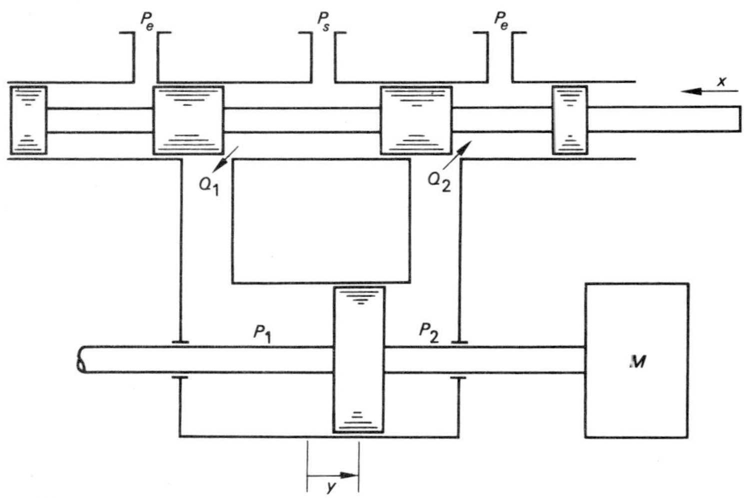

Fig. 17.16. The hydraulic valve and actuator

For spool displacement $x_v > 0$

$$\text{Inflow } Q_1 = Kx_v[\text{sgn}(P_s - P_1)]\sqrt{(|P_s - P_1|)} \tag{17.2}$$

$$\text{Outflow } Q_2 = Kx_v[\text{sgn}(P_2 - P_e)]\sqrt{(|P_2 - P_e|)} \tag{17.3}$$

For $x_v < 0$

$$Q_1 = -K\,|\,x\,|\,[\mathrm{sgn}(P_1 - P_e)]\sqrt{(\,|\,P_1 - P_e\,|\,)} \tag{17.4}$$
$$Q_2 = -K\,|\,x_v|\,[\mathrm{sgn}(P_s - P_2)]\sqrt{(\,|\,P_s - P_2\,|\,)} \tag{17.5}$$

Cavitation is at its worst when the load is predominately inertia and so for ease of analysis only this form of loading is considered. The load equation is then

$$(P_1 - P_2)A = M\,\mathrm{d}^2y/\mathrm{d}t^2 \tag{17.6}$$

Fluid compressibility effects will also be ignored at this stage. If there are no cavities, the continuity equations are then

$$Q_1 = Q_2 = Au \tag{17.7}$$

If a cavity exists then $Q = Au$ where Q is the flow rate into or out of the full cylinder side.

The response to a step input in spool position from $x_v = 0$ to $x_v = x_r$ will be considered first. The flow equations become

$$Q_1 = Kx_r\sqrt{(P_s - P_1)} = Au \tag{17.8}$$
$$Q_2 = Kx_r\sqrt{(P_2 - P_e)} = Au \tag{17.9}$$

These equations are made nondimensional using the method described in chapter 6. They reduce to

$$\text{velocity } \bar{u} = \sqrt{(1 - \mathrm{d}u/\mathrm{d}\bar{t})} \tag{17.10}$$

with load pressure

$$p_l = \mathrm{d}\bar{u}/\mathrm{d}\bar{t} \tag{17.11}$$

Equation 17.10 is of Riccati form and has solution

$$\bar{u} = \tanh\{\bar{t} + \tanh^{-1}\bar{u}(0)\} \tag{17.12}$$

where $u(o)$ is the initial velocity. Figure 17.17 shows the response for $u(o) = 0$ and for $p_e = 0$. When the spool step movement is initiated, the pressure P_1 jumps immediately to supply pressure, and P_2 falls to tank pressure. Ultimately both pressures settle at $\frac{1}{2}P_s$ and the load pressure is then zero in the new steady state. No cavities form as a result of this spool movement.

What effect does compressibility have? This has been determined from a full-scale digital computer simulation of the system and the results are shown in Fig. 17.18. The inertia factor α is an inverse measure of the degree of compressibility:

$$\alpha = \frac{Kx_r}{A}\sqrt{\left(\frac{2NM}{VP_s}\right)}$$

where

N = oil bulk modulus
V = total volume of oil in actuator

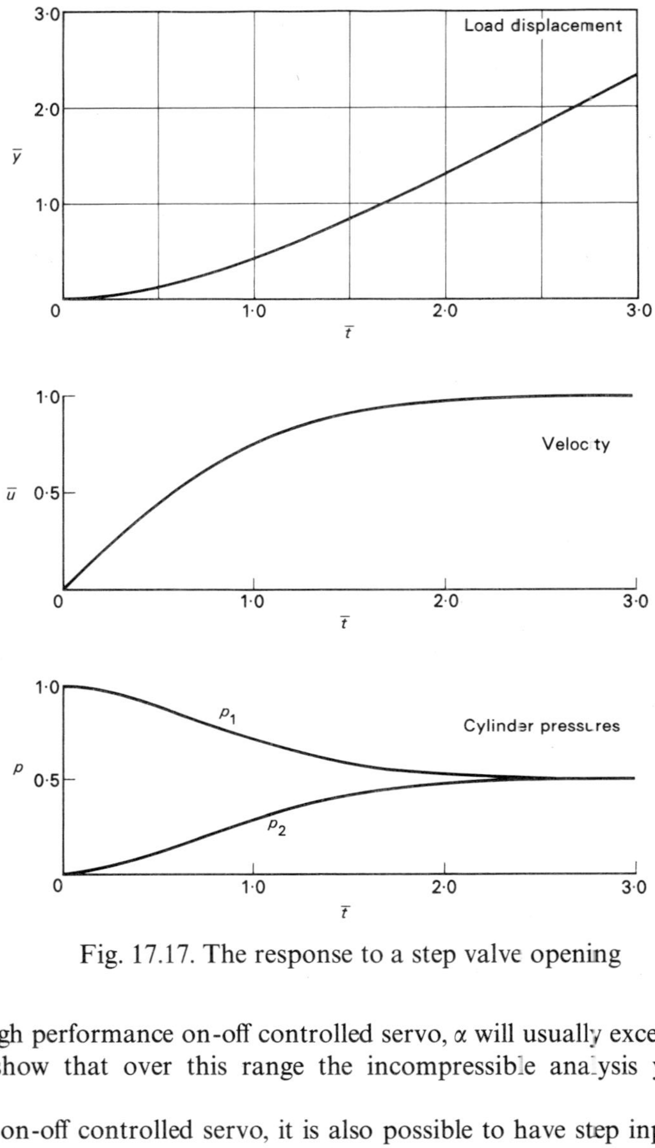

Fig. 17.17. The response to a step valve opening

For a high performance on-off controlled servo, α will usually exceed 2 and the results show that over this range the incompressible analysis yields useful results.

In an on-off controlled servo, it is also possible to have step inputs in valve position from one extreme to the other. Under these conditions cavities can form. Assume that the load is moving with steady velocity $+ u(0)$ corresponding to spool displacement $+ x_r$. There is no driving pressure required under these conditions so that $P_1 = P_2 = \frac{1}{2}P_s$. The spool is then moved instantaneously to $- x_r$ a step input of magnitude $- 2x_r$, Fig. 17.19. Since the velocity is initially positive and the fluid is assumed incompressible, then P_2 must exceed P_s for $t = O^+$. Q_2 and u are then positive, oil flowing from cylinder to supply. Side 1 is connected to tank and pressure P_1 falls instantaneously to oil vapour pressure (assumed zero) in an attempt to make Q_1 fill side 1 of the cylinder. A cavity

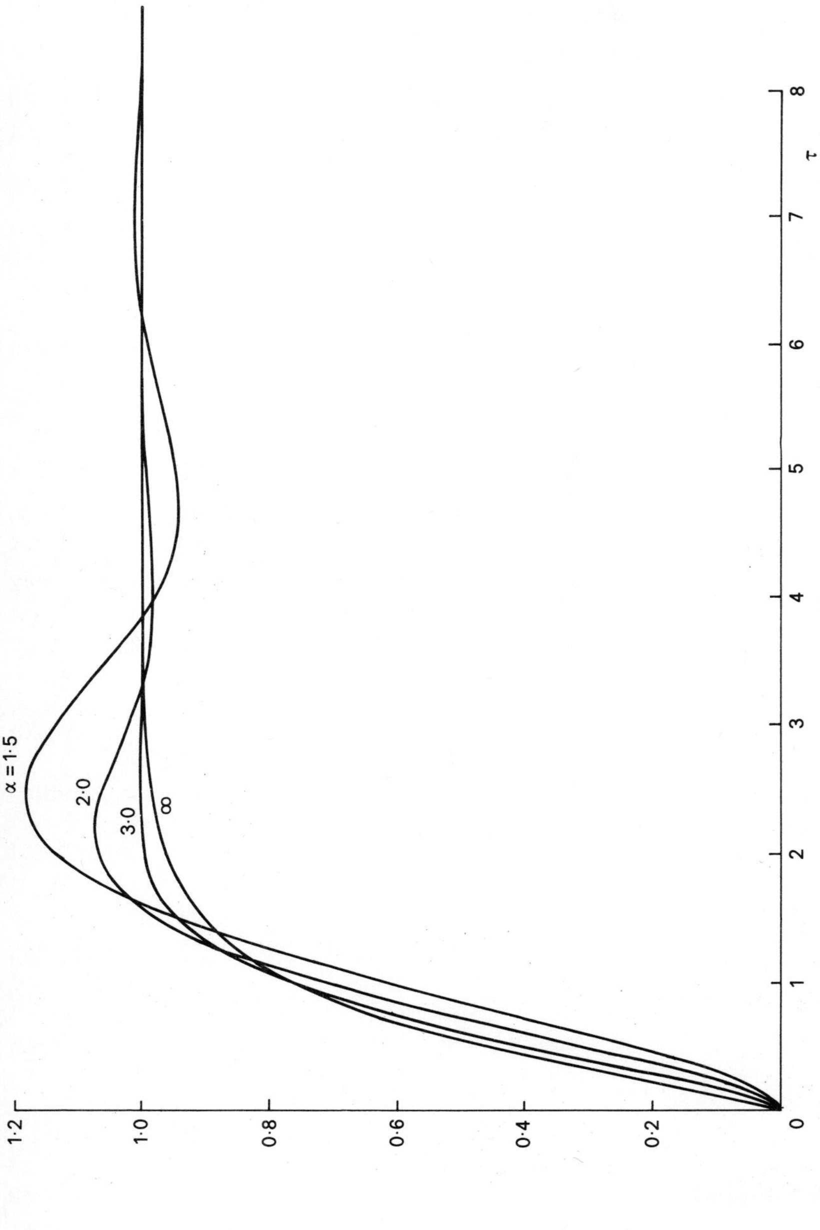

Fig. 17.18. Effect of oil compressibility on step response

forms on side 1. The equations become

$$Q_1 = Kx_r\sqrt{P_2} \tag{17.13}$$
$$Q_2 = Kx_r\sqrt{(P_2 - P_s)} = Au \tag{17.14}$$
$$P_2A = -M\,\mathrm{d}u/\mathrm{d}t \tag{17.15}$$

These equations reduce to

$$\bar{u} = \sqrt{\{-2(1 + \mathrm{d}\bar{u}/\mathrm{d}\bar{t})\}} \tag{17.16}$$

or, in Riccati form

$$\mathrm{d}\bar{u}/\mathrm{d}\bar{t} = -(1 + \bar{u}^2/2) \tag{17.17}$$

The solution is

$$0 < \bar{t} < \bar{t}_s$$

$$\bar{u} = \frac{\sqrt{2}(a \cos\sqrt{2}\bar{t} - b \sin\sqrt{2}\bar{t})}{1 + a \sin\sqrt{2}\bar{t} + b \cos\sqrt{2}\bar{t}} \tag{17.18}$$

$$a = \frac{2\sqrt{2}\bar{u}(0)}{2 + \bar{u}^2(0)}; \quad b = \frac{2 - u^2(0)}{2 + u^2(0)}$$

$$\bar{y} = \log\left[\frac{1 + a \sin\sqrt{2}\bar{t} + b \cos\sqrt{2}\bar{t}}{1 + b}\right] + y(0) \tag{17.19}$$

Note that at the instant of switching, pressure P_2 jumps to 1·5 times supply pressure. The load stops at $\bar{t}_s$ when pressure P_2 falls to supply pressure. At this point the load displacement is maximum positive. For $\bar{t} > \bar{t}_s$, the load velocity is negative, P_2 being less than P_s. Equation (17.17) then becomes

$$\mathrm{d}\bar{u}/\mathrm{d}\bar{t} = \bar{u}^2/2 - 1 \tag{17.20}$$

which describes the motion until the cavity on side 1 is filled by piston displacement and backflow from tank. The solution of equation (17.20) is $\bar{t}_s < \bar{t} < \bar{t}_c$

$$\bar{u} = \sqrt{2} - \frac{2\sqrt{2}}{1 + \mathrm{e}^{-\sqrt{2}(\bar{t} - \bar{t}_s)}} \tag{17.21}$$

$$y = \bar{y}(\bar{t}_s) + \sqrt{2}(\bar{t}_s - \bar{t}) - 2 \log\left[\tfrac{1}{2}(1 + \mathrm{e}^{-\sqrt{2}(\bar{t} - \bar{t}_s)})\right] \tag{17.22}$$

The cavity size H is

$$H = \int_o^t Au\,\mathrm{d}t - \int_o^t Kx_r\sqrt{P_e}\,\mathrm{d}t \tag{17.23}$$

or

$$h = H/Ay_r = \bar{y} - \sqrt{2}p_e\bar{t} \tag{17.24}$$

The cavity is filled at $\bar{t} = \bar{t}_c = \bar{y}/2p_e$. This is plotted in Fig. 17.20 which shows that the cavitation duration is strongly affected by back flow from tank. However, for the purpose of the present analysis, it is sufficient to assume that tank pressure is negligible compared to supply pressure, i.e. $p_e = 0$. In this case,

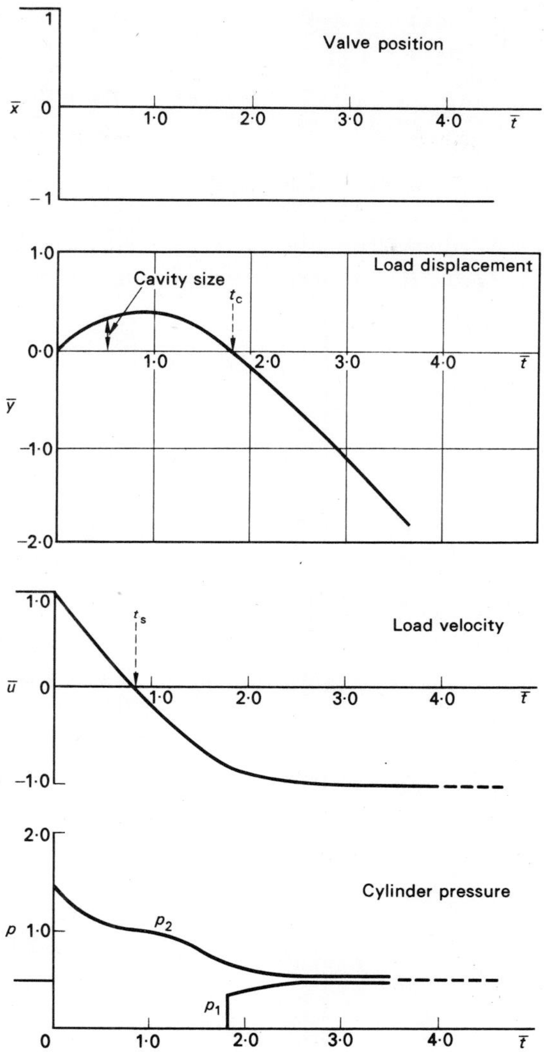

Fig. 17.19. Response to a step in valve position from $+x_r$ to $-x_r$

$\bar{t}_c = 1{\cdot}8$. For $\bar{t} > \bar{t}_c$, pressure P_1 can rise and the system equation is

$$\bar{u} = -\sqrt{(1 + \mathrm{d}\bar{u}/\mathrm{d}\bar{t})} \tag{17.25}$$

The solution of equation (17.25) is

$$\bar{t}_c < \bar{t} < \infty$$

$$\bar{u} = 1 + \left\{ \frac{2}{Be^{2(\bar{t}_c - \bar{t})} - 1} \right\} \tag{17.26}$$

$$B = \frac{\bar{u}(\bar{t}_c) + 1}{\bar{u}(\bar{t}_c) - 1}$$

$$\bar{y} = \bar{y}(0) - (\bar{t} - \bar{t}_c) - \log \frac{e^{2\bar{t}_c} - Be^{2(\bar{t}_c - \bar{t})}}{e^{2\bar{t}_c} - B} \tag{17.27}$$

Figure 17.19 shows that the new steady state is reached at approximately $\bar{t} = 3{\cdot}0$. If the valve were now switched back to its original position, $+ x_r$, the response could also be deduced from Fig. 17.19. The velocity would be the negative of that in Fig. 17.19 and the cylinder pressures would be interchanged.

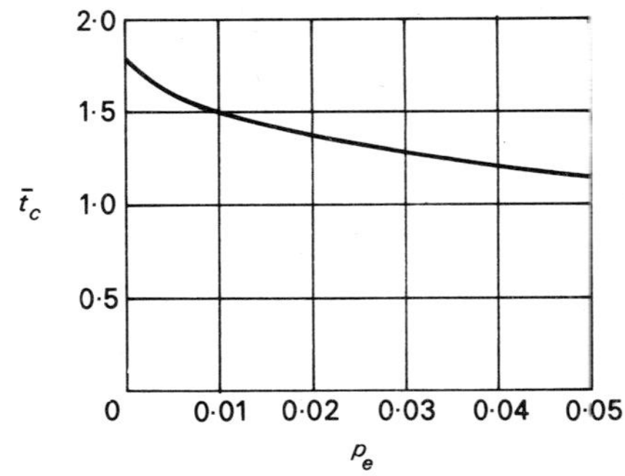

Fig. 17.20. Effect of tank pressure on cavity filling time

Appendix I

Units and the SI system

Newton's second law as applied to a body of constant mass is

$$\text{Force} \propto \text{Mass} \times \text{Acceleration}$$

This law cannot be used in calculations until the proportionality sign is replaced by something more specific. In the expression, four dynamical entities are involved; force, mass, length and time. If three of these are known the fourth can be derived. Standards of length and time have been established and it remains to decide whether force or mass is to be the derived quantity. Several systems have evolved

1. 1 derived unit of force (the poundal, pdl) = 1 pound mass (lb) × 1 ft s^{-2}; force has the units lb ft s^{-2}.
2. 1 pound force (lbf) = 1 derived unit of mass (the slug) × 1 ft s^{-2}; mass here has the units lbf s^2 ft^{-1}.
3. 1 pound force (lbf) = a derived numerical constant × 1 pound mass × 1 ft s^{-2}.

System 3, in particular, has led to much confusion in the past, mainly because the units of force and of mass were at one time both called the pound. It is not recommended here. Another system, the Système International d'Unités, (SI), is presently being introduced in Europe. It gives

4. 1 derived unit of force (the newton, N) = 1 kilogram mass (kg) × 1 ms^{-2}.

This is a coherent system in that the result of any manipulation (assuming a dimensionally correct formula) is in the correct units for the new quantity. For example, if a force of 2 newtons acts over a distance of 4 metres, then 2N × 4m = 8Nm = 8 joules of work are done.

In adition, the SI system has the advantage of being a decimal one. Multipliers and their prefixes are listed below.

Multiplier	Prefix	Symbol	Multiplier	Prefix	Symbol
10^{12}	tera	T	10^{-2}	centi	c
10^{9}	giga	G	10^{-3}	milli	m
10^{6}	mega	M	10^{-6}	micro	μ
10^{3}	kilo	k	10^{-9}	nano	n
10^{2}	hecto	h	10^{-12}	pico	p
10^{1}	deca	da	10^{-15}	femto	f
10^{-1}	deci	d	10^{-18}	atto	a

The units of those quantities of particular interest to the fluid power engineer are listed below.

Variable	System 1	System 2	System 4
Temperature	°R = °F + 460	°R = °F + 460	°K = °C + 273
Mass	lb	slug = 32·2 lb	kg
Length	ft	ft	m
Force	pdl	lbf	N
Pressure	pdl ft^{-2}	lbf ft^{-2}	Nm^{-2} (pascal, P)
Energy	pdl ft	lbf ft	Nm (joule, J)
Power	pdl ft s^{-1}	lbf ft s^{-1}	Nms^{-1} (watt, W)
Vol. flow rate	ft^3 s^{-1}	ft^3 s^{-1}	m^3 s^{-1}
Mass flow rate	lb s^{-1}	slug s^{-1}	kg s^{-1}
Spring rate	pdl ft^{-1}	lbf ft^{-1}	Nm^{-1}
Torque	pdl ft	lbf ft	Nm
Dynamic viscosity	pdl s ft^{-2}	lbf s ft^{-2}	Nsm^{-2} (poiseuille, P)
Kinematic viscosity	ft^2 s^{-1}	ft^2 s^{-1}	m^2 s^{-1}
Bulk modulus	pdl ft^{-2}	lbf ft^{-2}	Nm^{-2}
Gas constant R	pdl ft lb^{-1} R^{-1}	lbf ft $slug^{-1}$ R^{-1}	J kg^{-1} k^{-1}
Frequency	c.p.s.	c.p.s.	Hz = c.p.s.

Some useful conversion factors are

Length	1 in	= 25·4 mm
Length	1 ft	= 0·305 m
Area	1 in^2	= 0·000645 m^2
Vol. flow rate	1 ft^3 s^{-1}	= 0·0283 m^3 s^{-1}
Force	1 lbf	= 32·2 pdl = 4·45 N
Pressure	1 lbf in^{-2}	= 6894·8 Nm^{-2}
Energy	1 lbf ft	= 1·356 J
Power	1 hp	= 550 lbf ft s^{-1} = 0·746 kW
Mass	1 lb	= 0·4536 kg

Although the SI system is now being introduced it will take many years for the existing systems 1 and 2 to die out and unfortunately the present generation of students will need to be familiar with all systems. The following examples should be of help.

Example 1

Water flows through a circular orifice, of area 0·01 in^2, under a pressure drop of 100 lbf in^{-2}. If the orifice flow coefficient is 0·8, find (1) the velocity at the vena contracta, (2) the volumetric flow rate, (3) the mass flow rate and (4) the Reynolds number. The density of water is 62·4 lb ft^{-3}. The kinematic viscosity is 1·5 cSt.

The equations for the orifice are

$$\text{velocity } v = \sqrt{(2p/\rho)}$$
$$\text{vol. flow rate } Q = C_q av$$
$$\text{mass flow rate } W = \rho Q$$

Reynolds number $Re = vd/\nu$

System 1: $\rho = 62{\cdot}4$ lb ft^{-3}

$P = 100 \times 32{\cdot}2 \times 144 = 4{\cdot}64 \times 10^5$ pdl ft^{-2}

(1) $v = \sqrt{\{(9{\cdot}28 \times 10^5)/62{\cdot}4\}} = 122$ ft s^{-1}

(2) $Q = 0{\cdot}8\ av = (0{\cdot}01/144) \times 122 = 0{\cdot}00678$ ft^3 s^{-1}

(3) $W = \rho Q = 62{\cdot}4 \times 0{\cdot}00678 = 0{\cdot}423$ lb s^{-1}

(4) kinematic viscosity $\nu = 1{\cdot}5 \times 10^{-2}$ cm^2 s^{-1}

$= (1{\cdot}5 \times 10^{-2})/(30{\cdot}5)^2 = 1{\cdot}61 \times 10^{-5}$ ft^2 s^{-1}

diameter d $= \sqrt{(4a/\pi)} = 0{\cdot}113$ in $= 0{\cdot}0094$ ft

$Re = vd/\nu = 122 \times 0{\cdot}0094/1{\cdot}61 \times 10^{-5} = 7140$

System 2: $\rho = 62{\cdot}4/32{\cdot}2 = 1{\cdot}94$ slug ft^{-3}

$P = 100 \times 144 = 1{\cdot}44 \times 10^4$ lbf ft^{-2}

(1) $v = \sqrt{\{(2{\cdot}88 \times 10^4)/1{\cdot}94\}} = 122$ ft s^{-1}

(2) $Q = 0{\cdot}8 \times (0{\cdot}01/144) \times 122 = 0{\cdot}00678$ ft^3 s^{-1}

(3) $W = \rho\, Q = 0{\cdot}01315$ slug s^{-1}

(4) $\nu = 1{\cdot}61 \times 10^{-5}$ ft^2 s^{-1}

$d = 0{\cdot}0094$ ft

$Re = 7140$

System 4: $\rho = 1000$ kg m^{-3} (for all practical purposes)

$P = 100 \times 6894{\cdot}8$ N m^{-2} $= 689{\cdot}48$ kN m^{-2}

(1) $v = 2 \times 689{\cdot}48 \times 10^3/1000 = 37{\cdot}2$ ms^{-1}

(2) $a = 0{\cdot}01 \times 0{\cdot}000645\ m^2 = 6{\cdot}45 \times 10^{-6}$ m^2

$Q = 0{\cdot}8 \times 6{\cdot}45 \times 10^{-6} \times 372 = 0{\cdot}000192$ m^3 s^{-1}

(3) $W = 1000 \times 0{\cdot}000192 = 0{\cdot}192$ kg s^{-1}

(4) $\nu = 1{\cdot}5 \times 10^{-2}$ cm^2 s^{-1} $= 1{\cdot}5 \times 10^{-6}$ m^2 s^{-1}

$d = 0{\cdot}113 \times 0{\cdot}0254 = 0{\cdot}00287$ m

$Re = 37{\cdot}2 \times 0{\cdot}00287/1{\cdot}5 \times 10^{-6} = 7140$

This example was concerned with incompressible flow. When carrying out calculations concerning compressible flows every bit as much care must be taken when deciding on the units. Consider for example the equation of state

$$P = \rho RT$$

If system 1 is used, the units of the gas constant R are (pdl ft lb^{-1} R^{-1}). If system 2 is used, the units are (lbf ft slug^{-1} R^{-1}) and if the SI system is used the units are J kg^{-1} K^{-1}. The gas content for any gas can be found from

$$R = \mathscr{R}/M_w$$

where $\mathscr{R}$ is the universal gas constant

M_w is the molecular weight

The value of is given below

System 1	System 2	System 4
$\mathscr{R} = 49{\cdot}7 \times 10^3$	$49{\cdot}7 \times 10^3$	$8{\cdot}315 \times 10^3$

Example 2

Find the density of nitrogen at pressure 14·7 lbf in^{-2} and temperature 20°C. $M_w = 28{\cdot}02$.

System 1: $R = 49{\cdot}7 \times 10^3/28{\cdot}02 = 1780$ pdl ft lb^{-1} R^{-1}

$T = 460 + 68 = 528°\ R$

$P = 14{\cdot}7 \times 32{\cdot}2 \times 144 = 68\,300$ pdl ft^{-2}

$\rho = P/RT = 68\,300/(1780 \times 528) = 0{\cdot}0729$ lb ft^{-3}

System 2: $R = 1780$ lbf ft slug^{-1} R^{-1}
$P = 14{\cdot}7 \times 144 = 2120$ lbf ft^{-2}
$\rho = P/RT = 2120/(1780 \times 528) = 0{\cdot}00226$ slug ft^{-3}
System 4: $R = 8{\cdot}315 \times 10^3/28{\cdot}02 = 297$ J kg^{-1} K^{-1}
$T = 273 + 20 = 293$°K
$P = 14{\cdot}7 \times 4{\cdot}45/0{\cdot}000645 = 1{\cdot}015 \times 10^5$ Nm^{-2}
$\rho = P/RT = 1{\cdot}015 \times 10^5/(297 \times 293) = 1{\cdot}17$ kg m^{-3}

The student will find a lot more information about units in [143] and [144].

Appendix II

Bibliography

1. Proc. Conf. on Oil Hydraulic Power Transmission and Control, Inst. Mech. Engrs, London (1961).
2. McNeil, I., 'Hydraulic power transmission', *The Evolution of Engineering*, Inst. Mech. Engrs, London.
3. Stowers, A., Pumps and water-raising machinery, ibid.
4. Blackburn, J. F., *et al.*, *Fluid Power*, M.I.T. Press (1960).
5. Damasco, F. and Spar, C., 'High temperature hydraulic fluids', Nat. Conf. Ind. Hydraulics, (Oct., 1961), **15**.
6. Martin, H. R., *An Introduction to Fluid Power Control*, D.A.T.A. Technical Pubs., Richmond, Surrey.
7. Martin, H. R., 'A non-mathematical look at servomechanisms', *Hydraulic Pneumatic Power* (August, 1967).
8. von Mises, R., 'The calculation of flow coefficients for nozzle and orifice', 1917 V.D.A. **61**, Nr. 21, 22, 23 May–July.
9. Johansen, F. C., 'Flow through pipe orifices at low Reynolds number', *Proc. Roy. Soc.* (1941) **126**, No. 801.
10. Wuest, W., 'Stromung durch Schlitz-und Lochblenden bei kleinen Reynolds-Zahlen', *Ingenieur Archiv* (1954), No. 22. pp. 357–367.
11. Viersma, T. J., 'Designing load compensated fast response hydraulic servos', *Control Engng*. (May, 1962).
12. Lewis, E. and Stern, H., *Design of Hydraulic Control Systems*', McGraw-Hill (1962).
13. MacLellan, G. D. S., Mitchell, A. E. and Turnbull, D. E., 'Flow characteristics of piston type control valves', Proc. Symposium Recent Mech. Eng. Developments in Auto. Control, Inst. Mech. Eng. (London) (Jan. 1960).
14. Beck, A. and McCloy, D., 'Some cavitation effects in spool valve orifices', *Proc. Inst. Mech. Engrs,* London, (1967–68) **182**, Pt. I, No. 8 (1967–68).
15. McCloy, D., 'Discharge characteristics of servo-valve orifices', Paper 6, Proc. 1968 Fluid Power International Conf., Olympia.
16. McCloy, D., 'Cavitation and aeration—the effect on valves and systems', *Hydraulic Pneumatic Power*, (Jan.–Feb. 1966).
17. Stenning, A. H., 'An experimental study of two-dimensional gas flow through valve type orifices', A.S.M.E. paper no. 54–4–45, (1955).

18. Foster, K. and Kulkarni, M. M., 'The free oscillations of an inertia loaded hydraulic spool valve', University of Birmingham, Mech. Eng. Dept. Research Rep. 50 (May, 1966).
19. Rouse. H., 'Cavitation in the mixing zone of a submerged jet', *La Houille Blanche*, (Jan.–Feb., 1953) **8**, 1, pp. 9–18.
20. Turnbull, D. E., *Some Characteristics of Hydraulic Piston-type Control Valves in Relation to their Use as Components in a Feedback Position Control System*, Ph.D. Thesis, Cambridge University (1956).
21. Alpay, S. A. and Fleming, G. K., 'Stabilisation of the flow in spool valves', A.S.M.E. Conf., Philadelphia (1967).
22. Glaettli, H. H., Muller, H. R. and Zinag, R. H., 'Remarks on the limitations of pure fluid elements', Proc. fluid amplification symposium, (1964) **1**, p. 217, Washington.
23. McCloy, D. and Beck, A., 'Flow hysteresis in spool valves', First Fluid Power symposium, Cranfield (Jan., 1969).
24. McCloy, D. and Beck, A., 'Cavitation effects in poppet valves', A.S.M.E. paper no. 67–FE–27. Fluids Division meeting in Chicago (May, 1967).
25. Stone, J. A., 'Discharge coefficients and steady state flow forces for poppet type valves', *Trans. A.S.M.E.* (1960) **82**, Series D. No. 1.
26. Lichtarowicz, A. and Markland, E., 'Calculation of potential flow with separation in a right angled elbow with unequal branches', *J. Fluid Mech.* (1963) **17**, Pt. 4. pp. 596–606.
27. Feng, T. Y., 'Static and dynamic control characteristics of flapper nozzle valves', *Trans. A.S.M.E.* (1959) **81**, Series D.
28. Lichtarowicz, A., Duggins, R. K. and Markland, E., 'Discharge coefficients for incompressible non cavitating flow through long orifices', *J. Mech. Engng. Sci., London* (Jan., 1965). Vol. 7. No. 2.
29. Shearer, J. L., 'Resistance characteristics of control valve orifices' *Proc. Inst. Med. Engrs.,* London (Jan. 1960).
30. Perry, J. A., 'Critical flow through sharp edged orifices', *Trans. A.S.M.E.* (Oct., 1949).
31. Tsai, D. H. and Cassidy, E. C., 'Dynamic analysis of a simple pneumatic pressure reducer', *J. Basic. Engng.* (June, 1961) **83**, 253.
32. Norwood, R. E., *A Flapper Valve Study* M.S. Thesis Mech., Eng., Dept. M.I.T. (1959).
33. Anderson, B. W., *The Analysis and Design of Pneumatic Systems*, Wiley (1967).
34. Alpay, S. A., 'Design and performance characteristics of a hydraulic jet pipe valve', 9th J.A.C.C. University of Michigan (June, 1968).
35. Reid, K. N., 'Optimum design parameters of a pneumatic jet pipe valve', Proc. 1st. I.F.A.C. Congress, Moscow (1960).
36. Winslow, W. M., 'Method and means for translating electrical impulses into mechanical force', *U.S. Patent* 2417850 (March, 1947).
37. Standrud, H. T., 'Electric-field valves inside cylinder control vibrator', *Hydraulics and Pneumatics* (Sept., 1966).
38. Lee, S. Y. and Blackburn, J. F., 'Contributions to hydraulic control—axial forces on control valve pistons', *Trans. A.S.M.E.* (Aug., 1952) **74**.
39. Clark, R. N., 'Compensation of steady state flow forces in spool type hydraulic valves', *Trans. A.S.M.E.* (Nov., 1957).

40. Carrington, J. E., 'Developments in electrohydraulic valves', *Hydraulic Pneumatic Power* (Mar., 1966).
41. Lee, S. Y. and Blackburn J. F., 'Contributions to hydraulic control—transient flow forces and valve instability', *Trans. A.S.M.E.* (Aug., 1952) **74**.
42. Schrenk, E., *Disc Valves, Flow Patterns, Resistance and Loading* (1925). Abbreviated translation by British Hydromechanics Research Association Publication No. T.547.
43. McCloy, D. and McGuigan, R. H., 'Some static and dynamic characteristics of poppet valves', *Inst. Mech. Engrs, London* (April, 1965).
44. Green, W. L., 'The poppet valve-flow force compensation', Paper 2. Proc. 1970 Fluid Power International Conf., Olympia.
45. Pawlak, R. J., *A Continued Study of Pneumatic Flapper Valves,* M.Sc. thesis M.I.T. (1959).
46. Shearer, J. L., Murphy, A. J. and Richardson, H. H., *Introduction to System Dynamics*, Addison Wesley (1967).
47. Carrington, J. and Martin, H. R., 'Performance of an Ultrasonic Flowmeter under Pulsating Flow', Proc. Symp. on Pulsating Flow Measurement. Inst. Measurement and Control (April, 1970).
48. Gridley, J. H., *Principles of Electrical Transmission Lines in Power and Communications*, Pergamon Press (1967).
49. Harpur, N. F., 'Some design considerations of hydraulic servos of Jack type', Conf. on Hydraulic Servomechanisms, *Proc. Inst. Mech. Engrs, London* (1953), p. 41.
50. Nightingale, J. M., 'Hydraulic servo fundamentals', *Machine Design* (1956).
51. McCloy, D., 'Pressure peaks in the hydraulic actuator', *Trans. A.S.M.E.* (Dec., 1965); *J. Basic Engng.*
52. McCloy, D., 'Cavitation in hydraulic systems', *Tech. J. Fluid Power Inst.* (May, 1967) **7**, No. 2. Milwaukee School of Engineering.
53. Cunningham, W. J., 'The hydraulic control valve; an analysis of its performance', Yale University, Dept. Elec. Engineering, Tech. Rep. (June, 1950).
54. McCloy, D., 'Graphical analysis of the step response of hydraulic servomechanisms', *Proc. Inst. Mech. Engrs, London* (1967–68) **182**, Pt. 1.
55. Nikiforvk, P. N., Wilson, J. N. and R. M. Lepp "Transient Response of a time optimised hydraulic servomechanism operating under cavitation conditions' *Proc. Inst. Mech. Engrs.* (1970–71) **185**.
56. Royle, J. K., 'Inherent Non-Linear Effects in Hydraulic control systems with inertia loading', *Proc. Inst. Mech. Engrs, London* (1959), **173**, p. 257.
57. Wang, P. K. C. and Ma, J. T. S., 'Cavitation in valve controlled hydraulic actuators', *J. Appl. Math.* **29**; *Trans. A.S.M.E.* (1962) **84**, p. 435.
58. McCloy, D. and Martin, H. R., 'Some effects of cavitation and flow forces in the electrohydraulic servomechanisms', *Proc. Inst. Mech. Engrs, London* (1963–64) **178**.
59. McCloy, D., *Some Effects of Oil Compressibility and Valve Lap on the Performance and Stability of the Hydraulic Servomechanism*, Ph.D. Thesis, Queen's University, Belfast (1965).
60. Martin, H. R., *Some Factors Influencing the Performance of an Electrohydraulic Servomotor*, M.Sc. Thesis, Queen's University, Belfast (1962).

61. Williams, H., 'Effect of oil momentum forces on the performance of electrohydraulic servomechanisms', Proc. Conf. Recent Mechanical Engineering Developments in Automatic Control (1960), 31 (Inst. Mech. Engrs, London).
62. Eynon, G. T., 'Developments in high-performance electromechanical servomechanisms at the Royal Aircraft Establishment, Farnborough', Proc. Conf. Recent Mechanical Engineering Developments in Automatic Control (1960), 93 (Inst. Mech. Engrs, London).
63. Carrington, J. E. and Martin, H. R., 'Some uses of electrohydraulic servosystems in connection with aircraft cockpit simulations', Proc. of Conf. on Oil and Hydraulic Power, Transmission and Control, Inst. Mech. Engrs, London, (Nov., 1961) p. 110.
64. Guillon, M., *Hydraulic Servo Systems*, Butterworths (1961).
65. Martin, H. R. and Lichtarowicz, A., 'Theoretical investigations into the prevention of cavitation in hydraulic actuators', *Proc. Inst. Mech. Engrs, London* (1966–67) **181**, Pt. 1, No. 18.
66. Carrington, J. E. and Martin, H. R., 'Threshold problems in electro-hydraulic servomotors', *Proc. Inst. Mech. Engrs, London* (1966) **180**, Pt. 1, No. 37.
67. Martin, H. R and Lichtarowicz, A., 'Small amplitude response characteristics in hydraulic servomechanisms', Fourth Congress of I.F.A.C., Warsaw (June, 1969).
68. Royle, J. K., 'Hydraulic damping techniques at low velocity', Conf. Oil Hyd. Power Transmission and Control (1961), Inst. Mech. Engrs, London.
69. 'Dynamic Pressure feedback', *Aircraft Engng* (June, 1960).
70. Shearer, J. L., 'Study of pneumatic processes in the continuous control of motion with compressed air', Parts I and II *Trans. A.S.M.E.* (1956), **78**, 233.
71. Shearer, J. L., 'Non-linear analogue study of a high pressure servomechanism', *Trans. A.S.M.E.* (1957), **79**, 465.
72. Reethof, G., 'Analysis and design of a servomotor operating on high pressure compressed gas', *Trans. A.S.M.E.* (May, 1957).
73. Burrows, C. R. and Webb, C. R., 'Simulation of an ON-OFF pneumatic servomechanism', *Proc. Inst. Mech. Engrs, London* (1967–68), **182**, Pt. 1.
74. Mann, J. R., 'Hot gas bang-bang servo', WADD Tech. Report (1961).
75. Davison, E. and McCloy, D., 'Recent developments in gas servos', *Hydraulic Pneumatic Power* (Sept., 1965).
76. Halstenberg, R. V., 'Some linear and non-linear aspects of hot gas servo design', National Control Conf., Dallas (Nov., 1959).
77. Dummer, G. W., *Variable Resistors and Potentiometers*', Pitman (1963).
78. Scroggie, M. G., *Radio and Electric Laboratory Handbook*', Iliffe (1961).
79. Martin, H. R., *Introduction to Feedback Systems*' McGraw-Hill (1968).
80. Garner, K. C., *Introduction to Control System Performance Measurements*, Pergamon Press (1968).
81. Davies, W. D. T. 'Random signal testing for evaluating system dynamics', *Wireless World* (Aug., 1966).
82. Dudgeon, D. R. and McLean, D., 'Control system analysers' (Control Survey, 23), *Control* (April, 1962).

83. Martin, H. R., 'Instrumentation relating to mechanical engineering', D.A.T.A. Tech. Pubs. Richmond, Surrey.
84. Bode, H. W., *Network Analysis Feedback Amplifier Design*, D. Van Nostrand (1945).
85. Lorenz, M. H. and Stringer, J. D., 'Oil hydraulic spool valve operating times', *Proc. Inst. Mech. Engrs, London* (1966–67), **181**, Pt. 1.
86. Ma, C. Y., 'The analysis and design of hydraulic pressure reducing valves', *Trans. A.S.M.E.* (1966) paper 66–WA/MD–4.
87. Merritt, H. E., *Hydraulic Control Systems*, Wiley (1967).
88. Foster, K., 'Dynamic analysis of a two-stage relief valve', Proc. of Conf. on Oil and Hydraulic Power, Transmission and control, Inst. Mech. Engrs, London (1961).
89. Takenaka, T. and Urata, E., 'Static and dynamic characteristics of oil hydraulic control valves' Fluid Power International Conf. (Sept., 1968).
90. 'Pneumatics—dynamic behaviour of double acting air cylinders', P.E.R.A. rep. no., 135 (Feb., 1965).
91. Healey, A. J. and Stringer, J. D., 'Dynamic characteristics of an oil hydraulic constant speed drive', *Proc. Inst. Mech. Engrs, London* (1968–69) **183**, Pt. 1.
92. Stewart, H. L. and Pritchett, S. L., *Fluid Power*, Howard W. Sams and Co. Inc. Indianapolis (1968).
93. Brown, F. T., foreword to *Advances in Fluidics* (1967) American Society of Mechanical Engineers, Fluidics Symposium, Chicago (May, 1967).
94. Kirshner, J. M., *Fluid Amplifiers*, McGraw-Hill (1966).
95. Parker, G. A., 'Some aspects of fluidic sequencing and digital control systems for machine tools', 7th International M.T.D.R. Conference, Birmingham (Sept., 1966).
96. Auger, R. N., 'Turbulence amplifier design and application', Fluid Amplification symposium proc. D.O.F.L. (Oct., 1962).
97. Dexter, E. M., 'No moving parts—features of new valves', *S.A.E. Journal* (1961) 102.
98. Bjornsen, B. G., 'The impact modulator', *Proc. 2nd Harry Diamond Lab. Symposium* (May, 1964), *II*, p. 5.
99. Shinn, J. N., 'No moving parts needed', *S.A.E. Journal* (Aug., 1964).
100. Kompass, E. J., 'Perspective on fluid amplifiers—capabilities and applications', *Control Engng* (Feb., 1964).
101. Gray, W. E. and Stern, H., 'Fluid amplifiers—capabilities and applications', *Control Engng* (Feb., 1964).
102. Hoernes, G. E. and Heilweil, M. R., *Introduction to Boolean Algebra and Logic Design*, McGraw-Hill (1964).
103. Proceedings of the Cranfield Fluidics Conferences. B.H.R.A. 1965, 1967, 1968, 1970.
104. Proceedings of the International Federation of Auto Control Conferences on Fluidics.
105. Proceedings on the Fluid Amplification Symposia Harry Diamond Labs. 1963 on.
106. *Fluidic Applications*, B.H.R.A. (1968).
107. *Fluidics*, Fluid amplifier associates (1966).

108. Moylan, M. J., *Fluid Logic in Simple Terms*, Machinery Publishing Co., Ltd. (1968).
109. McCloy, D., *An Introduction to Fluidics*, Draughtsmens' and Allied Technicians' Association (1969).
110. Parker, G. A., 'Digital fluid position encoders', Paper E2, Proc. 1st Cranfield Fluidics Conf., Cranfield (1965).
111. Reeves, D., Inglis, M. E and Airey, L., 'The fluid oscillator as a temperature sensor', Paper D1, Proc. Ist Cranfield Fluidics Conf., Cranfield (1965).
112. Halbach, C. R., Otsap, B. A. and Thomas R. A., 'A pressure insensitive fluidic temperature sensor', *Advances in Fluidics Proc* (1967), Fluidics Symposium A.S.M.E., pp. 298–312.
113. Pavlin, C. and Kadosch, M., 'Mechanical characteristics of a pure fluid respirator with curved walls', Paper F3, Proc. 1st Cranfield Fluidics Conf., Cranfield (1965).
114. Monge, M., 'Handling fluidics for practical purposes', Paper E5, Proc. 1st Cranfield Fluidics Conf., Cranfield (1965).
115. Cooper, B. J., 'Applications for fluidics', Paper 4, Proc. Fluid Power Inter. Conf. (1968).
116. 'Gas servos for engine actuation', Fluid amp. application studies (Dec., 1964). General Electric Co., New York. N 65–12443–CR–137.
117. Rose, R. K., 'Fluidic power control systems', Paper L3, Proc. 4th Cranfield Fluidics Conf., Coventry (1970).
118. 'Fluidics provide safety circuits for detonator crimping', *Instrum. Control Engng* (Feb., 1967).
119. *Pneumatic Handbook*, Trade and Technical Press (England). 2nd ed.
120. *Hydraulic Handbook*. Trade and Technical Press (England), 4th ed.
121. *A Course in Applied Pneumatics*, Martonair (1966).
122. McCloy, D., 'Low cost automation and fluid power', *Hydraulic Pneumatic Power* (Feb., 1970).
123. Parker, G. A., 'Fluidic digital to analogue converters', Paper K1, 3rd Cranfield Fluidics Conf., Turin (1968).
124. Wells, J. C., 'Digital techniques in fluid power engineering', *Hydraulics and Pneumatics* (Sept., 1966), pp. 108–112.
125. Himmler, C. R., 'Comparative performance of servo valves with fluidic and electrical inputs', Proc. 1968 Fluid Power International Conf., Olympia, London (Sept., 1968).
126. Toshimasa Kaiwa and Seivemon Inaba, 'Electrohydraulic pulse motor', *Control Engng* (Jan., 1962), pp. 73–75.
127. Ott, M., 'Digital control equipment for impulse controlled linear and rotary motors', *Hydraulic Pneumatic Power* (Nov., 1967), pp. 635–640.
128. Griffin, W. S., 'Development of high speed flueric logic circuitry for a novel pneumatic stepping motor', Advances in Fluidics, Proc. A.S.M.E. (1967), pp. 402–414.
129. Miles, K. E., 'A new type incremental piston', *Fluid Power Int.* (April, 1970), pp. 33–34.
130. Delmege, A. H. and Tremblay, M. J., 'Hydraulic digital actuator converts binary inputs to lineal motion', *Control Engng* (Feb., 1965), pp. 69–70.
131. Bidlack, J. D., 'Digital servo actuators work with compressible fluids', *Space and Aeronautics* (Aug., 1963), pp. 118–121.

132. Raider, J. W. and Scott, G. W., 'A digital hydraulic actuator', Paper 25a. Proc. 3rd Congress Inter. Fed. of Auto Control (June, 1966).
133. Baeck, H. A., 'Hydraulic on-off servo: simple, rugged positioner', *Control Engng* (Dec., 1967), pp. 79–82.
134. Murtaugh, S. A., 'An introduction to the time modulated acceleration switching electrohydraulic servomechanism', *A.S.M.E. J. Basic Engng* (June, 1959), pp. 263–271.
135. Taft, C. K. and Nawaz, S., 'Fluidic pneumatic control system compensator', Paper E2, 4th Cranfield Fluidics Conf., Coventry (1970).
136. 'Basic applied research in fluid power control', Reports Nos. 8998–5, 6. Dept. Mech. Eng. Project Lab., M.I.T. Cambridge, Massachusetts (Feb., 1964).
137. 'Basic applied research in fluid power control' Report 8998–7. Dept. Mech. Eng. Project Lab., M.I.T. Cambridge, Massachusetts (June, 1964).
138. Gorbatenko, G. G., 'Posicast control by delayed gain', *Control Engng* (Feb., 1965) pp. 74–77.
139. Armstrong, P. J. and McCloy, D., 'Sub optimal solutions for time optimal control of a hydraulic servomechanism', Paper 6B2 (1970). J.A.C.C. Atlanta, Georgia.
140. McCloy, D., 'Cavity formation in valve controlled hydraulic cylinders', *Proc. Inst. Mech. Engrs, London* (1969–70), **184**, Pt. 1.
141. McCloy, D., 'Cavitation effects in on-off controlled hydraulic servos,' Jour. of Dynamic systems, measurement and control. A.S.M.F., March 1972.
142. Martin, H. A. and McCloy, D., 'Some aspects of the response of bistable hydraulic servos', 2nd International Fluid Power Symposium. B.H.R.A. (Jan., 1971).
143. Morrison, J. L. M. and Crossland, B., *An Introduction to the Mechanics of Machines*, Longmans (1964). SI Edition (1970).
144. Blackman, D. R., *SI Units in Engineering*, Macmillan (1969).
145. Leroy, W. W., and R. L. Leslie, 'Hydraulic Fluid Properties and System Performance', 18th National Conference Industrial Hydraulics, U.S.A., 1962, Vol. 16.
146. Hayward, A. T. J., 'Aeration in Hydraulic Systems; its Assessment and Control', Conf. Oil Hydraulic Power Transmission and Control, Inst. Mech. Engrs., London, Nov., 1961.
147. Smith, L. H., R. L. Peeler, L. H. Bernd, 'Hydraulic Fluid Bulk Modulus; its Effect on System Performance and Techniques for Physical Measurement'. 16th National Conf. on Industrial Hydraulics, 1960, p. 179, Vol. 14.
148. Schanzlin, E. H., 'Higher speeds and Pressures for the Hydraulic Pump', 12th National Conf. on Industrial Hydraulics, 1956, Vol. 10.
149. Pigott, J., 'Cavitation in Reciprocating and Rotary Pumps', 6th National Conf. on Industrial Hydraulics, 1950, Vol. 4.
150. Harvey, E., W. D. Newton, E. McElroy, A. Whiteley, 'On Cavity Formation in Water', *Journal of Applied Physics*, 18, 1947.
151. Knapp, R. T., J. W. Daily, F. G. Hammitt, *Cavitation*, McGraw-Hill Book Co., 1970.
152. Thiruvengadam, A., 'A Unified Theory of Cavitation Damage', *Jour. of Basic Eng.*, A.S.M.E., Sept. 1963.

153. Plesset, M. S., A. T. Ellis, 'On the Mechanism of Cavitation Damage', *Trans. A.S.M.E.*, 1955, Vol. 77, p. 1055.
154. Hobbs, J. M., 'Practical Aspects of Cavitation', *Phil. Trans. Royal Soc.* (London) A. 260, pp. 267–275, 1966.
155. Beatty, E. K., and Markland, E., 'Feasibility study of laminar jet deflection fluidic elements', Paper H1, Third Cranfield Fluidics Conf., May 1968.
156. McCloy, D., and I. J. Stevenson, 'Some experiments with oil hydraulic vortex valves', Paper D6, Fifth Cranfield Fluidics Conf., June 1972.
157. McCloy, D., 'Oil hydraulic fluid logic', Paper 6, Proc. 1972 Fluid Power International Conf., Olympia.
158. Hobbs, J. M., and D. McCloy, 'Cavitation erosion in oil hydraulic equipment', Metals and Materials, Jan. 1972.
159. Brown, F. T., 'The transient response of fluid lines' *A.S.M.E. Journal of Basic Engineering,* Trans. A.S.M.E. Dec., 1962 pp 547–553.

Index

D. McCloy is Reader in Mechanical Engineering at the Queen's University, Belfast, and is Assistant Dean of the Faculty of Applied Science and Technology. In 1971 he was Visiting Professor in the Sir George Williams University, Montreal, Canada. He has had industrial experience in research with Short Brothers and Harland Ltd., and with the Lockheed-Georgia Aircraft Co. He is Director of the Northern Ireland Low Cost Automation Centre and maintains his links with industry as an Industrial Consultant.

H. R. Martin is Associate Professor in Mechanical Engineering at the University of Waterloo, Canada. He is a course co-ordinator for industrial part-time M.A.Sc. programmes for industry. He has had much industrial experience and is a consultant to the Ontario Government on industrial low cost automation techniques. As Joint Director he was a Founder of the Northern Ireland Low Cost Automation Centre.